ENGINEERING PROPERTIES OF SOILS AND THEIR MEASUREMENT

Also Available from McGraw-Hill

Schaum's Outline Series in Civil Engineering

Most outlines include basic theory, definitions, and hundreds of solved problems and supplementary problems with answers.

Titles on the Current List Include:

Advanced Structural Analysis
Basic Equations of Engineering
Descriptive Geometry
Dynamic Structural Analysis
Engineering Mechanics, 4th edition
Fluid Dynamics
Fluid Mechanics & Hydraulics
Introduction to Engineering Calculations
Introductory Surveying
Mathematical Handbook of Formulas & Tables
Mechanical Vibrations
Reinforced Concrete Design, 2d edition
Space Structural Analysis
State Space & Linear Systems
Statics and Strength of Materials
Strength of Materials, 2d edition
Structural Analysis
Structural Steel Design, LFRD Method
Theoretical Mechanics

Schaum's Solved Problems Books

Each title in this series is a complete and expert source of solved problems containing thousands of problems with worked out solutions.

Related Titles on the Current List Include:

3000 Solved Problems in Calculus
2500 Solved Problems in Differential Equations
2500 Solved Problems in Fluid Mechanics & Hydraulics
3000 Solved Problems in Linear Algebra
2000 Solved Problems in Numerical Analysis
800 Solved Problems in Vector Mechanics for Engineers: Dynamics
700 Solved Problems in Vector Mechanics for Engineers: Statics

Available at your College Bookstore. A complete list of Schaum titles may be obtained by writing to: Schaum Division
McGraw-Hill, Inc.
Princeton Road, S-1
Hightstown, NJ 08520

ENGINEERING PROPERTIES OF SOILS AND THEIR MEASUREMENT

FOURTH EDITION

JOSEPH E. BOWLES

Consulting Engineer / Software Consultant
Engineering Computer Software

Boston, Massachusetts Burr Ridge, Illinios Dubuque, Iowa
Madison, Wisconsin New York, New York San Francisco, California St. Louis, Missouri

Irwin/McGraw-Hill

A Division of The McGraw-Hill Companies

ENGINEERING PROPERTIES OF SOILS AND THEIR MEASUREMENT

ISBN 978-0-07-006778-3
MHID 0-07-006778-3

16 17 18 19 20 QPD QPD 09 08

P/N 006778-3

This book was set in Century Expanded by Publication Services.
The editors were B. J. Clark, Kiran Kimbell, and John M. Morriss;
the production supervisor was Leroy A. Young.
The cover was designed by Joan Greenfield.
Project supervision was done by Publication Services.
Quebecor Printing/Dubuque was printer and binder.

Library of Congress Cataloging-in-Publication Data

Bowles, Joseph E.
Engineering properties of soils and their measurement / Joseph E. Bowles.—4th ed.
p. cm.
ISBN 0-07-911266-8 (set)
1. Soil mechanics—Laboratory manuals. 2. Soils—Testing—Laboratory manuals. I. Title.
TA710.5.B67 1992
620.1'91—dc20 91-38035

CONTENTS

PREFACE

To the Instructor:

This manual has been substantially rewritten for a more logical flow of text topics within a given test. Some rewriting was also required so that the covered laboratory tests adhere more closely to the ASTM and AASHTO standards for testing of soils. The labeling of the tests as "Experiments" has been deleted and they are referred to in the text as "Tests" except for Project 8 concerning soil classification, which is termed a "project."

Please note that the data sheets and other notation have been revised to comply with correct SI units of *mass* determinations rather than "weights." The term "weigh" is consistently used for "determining the mass" but there is a clear distinction between unit density ρ and the computed unit weight γ.

Major test revisions include:

1. Introducing with a revised data sheet the wax alternative to mercury for the Shrinkage Limit (Test 4).
2. Making a correct combination of the hydrometer test (Test 6) with the mechanical sieve analysis test (Test 5).
3. Including the equipment compression into the consolidation test (Test 13) together with both alternatives for computing void ratio or strain and for computing c_v.
4. Expanding the use of a membrane correction in triaxial testing (Test 15 and Test 16).
5. Rewriting all of the test "procedures" to improve on both details and clarity.

Although the cost is increased somewhat, the publisher was encouraged to retain the spiral binding of the previous editions. While cost reduction is certainly a worthy consideration, the greater convenience of the spiral binding should also be of concern. A spiral binding allows the user to fold the manual back so it lies flat on the work table in a minimum of space, making it easier to read the procedure sequence while doing the test. Since the number of tests given in this manual—together with the additional data sheets—makes it useful for at least two courses, this binding keeps both the text and remaining data sheets intact for later use.

Data sheets have been revised both to replace Wt. with Mass and to number them for that test. The use of, for example, 13*a*, 15*a* and so on informs the user that there is more than one data sheet for that test and that "*a*" is the first.

The general format of providing some discussion of the test, sources of error, and limitations, and—where appropriate—select references, makes the manual of some reference shelf value to the user. The discussion of equipment modifications and alternative approaches and having the student comment on test improvements are of value in encouraging at least minimal interest in further study. An abbreviated "cookbook" approach is of little value to a user except for getting through a course.

The diskette provided replaces the Fortran computer program listings given in the third edition. Those programs have been edited for screen prompts, and two programs for soil classification (by ASTM and AASHTO methods) are added. All programs have been compiled for immediate execution either with or without a math co-processor chip present. The diskette provides the user with programs for Tests 5, 6, 13, 14, 15, and 16 and for Project 8. The same program is used for Tests 14, 15, and 16 because of the similarity in data reduction. There is a README.LAB text file on the disk for the user to print which gives details on using these several programs. There are also sample data sets for all but the two soil classification programs. The computer programs include the major text revisions for the hydrometer analysis, consolidation test, and triaxial membrane correction.

If the student is allowed to use the computer programs, you probably should require him or her to show at least one line of data reduction done by hand. Since both the laboratory test and its data reduction are learning experiences, the student may not gain much experience in data reduction and any resulting problems by just inputting the data into a computer program.

If your laboratory uses electronic data acquisition equipment the basic test details outlined in this manual are valid. The major difference is in what is recorded by the student (load dial or load and deflection dial versus displacement) in the appropriate columns of the data sheet. At the present time there is probably more manual than electronic test equipment. The changeover in commercial laboratories is slow because

a. Equipment amenable to electronic replacement of load rings and dial gages with load cells and LVDTs is durable.
b. There is no savings in personnel but a substantial increase in equipment costs.
c. Dial gages that one can visually see move inspire more confidence than digital readouts.

To the Student or Other User

This laboratory manual is a simplified digest of the principal details of the most common laboratory soil tests you will encounter in geotechnical practice. You should make an attempt to look at both the ASTM Standards vol. 4.08 and the AASHTO Materials Standards.

Included is a diskette with six programs written in Fortran and compiled so you can directly use them on any DOS-based PC that has PC- or MS-DOS version 2.0 or higher. If the system has a co-processor chip the programs will use it. If it does not have the co-processor chip the programs use a floating point emulator method. There is a README.LAB file giving you some information on using each of the programs. It is in ASCII format so you can read it at the screen but you should use a print utility and print a copy for easier reference.

If you have access to a Graph (or CAD) program, you may save the required output to a disk file for later plotting. You should only use these programs in the course if allowed by the instructor. If you do use these programs in data reduction you should *always* make one or two typical computations to show that you understand how to do them by hand (and how the program works).

The tests in this manual follow standardized procedures sufficiently that the experience you will have gained following its use will be of considerable value if you go into geotechnical work upon graduation.

General Comments for All Users

This manual is widely used in the U.S. and is in several foreign translations. The teaching experience of the users (Professor, Instructor, or Teaching Assistant), can vary widely. For this reason, I will make a few comments which may be of value to some of the users.

1. Most soil laboratories have a series of work tables or benches where the students do the assigned soil projects (or tests). These may or may not be equipped as work stations with certain basic test equipment. If they are, you should consider numbering the equipment, where appropriate, and making a list of the equipment at each station. Doing so ensures that students performing tests will not grab equipment from an adjacent station where there are other students also trying to do the tests.

   At the end of the period the person in charge should routinely check that the equipment has been cleaned and returned to the storage drawers (or wherever). This allows one to make specific comments to the group using that station rather than to make general comments which irritate the responsible student(s). Having the equipment clean is extremely important if there is a following lab section, since some equipment is used in nearly every lab period. Furthermore, metal that is contaminated with wet soil can rust overnight and equipment with corrosion pits or rust is difficult to work with.

2. If your laboratory uses work stations it is generally efficient to divide the laboratory section into groups which are assigned work stations. Some of the tests are better done by each person but several are suggested as group projects and some may have to be done as "laboratory" tests.

   Group (or lab) projects are usually necessary where major equipment is required and the laboratory only has one or two pieces. For a laboratory project it is usual for some students to volunteer for select parts of the data acquisition and the others watch. When more than one test is done in the laboratory project (for example, not many laboratories have over one triaxial test apparatus and you usually will want to run two and preferably three tests) other students should be assigned tasks on subsequent tests so there is maximum participation.

   Please note that my indication of the project as individual or group is a guide. You may run your projects as you wish. Based on personal experience, however, if the tests are not run about as suggested it is very easy for the laboratory session to extend well beyond the assigned time (usually three hr).

   Most of the projects also require additional outside work on the part of the student. Typically to return the following day for water content determinations or to return at later hours to take test readings.

3. Most of the soil tests adhere closely to the ASTM or AASHTO test standards. If distilled water is not easily obtained, it is suggested that you use ordinary tap water with almost no loss in data accuracy. The compaction test complies with the AASHTO test method, which allows re-use of soil. It is suggested you use this method since there are the problems both of obtaining large soil quantities for testing by several groups and its later disposal—particularly if your university is in an urban location.

4. *You should post a set of safety regulations* concerning specifics of the equipment in your laboratory in a prominent location. If you have not posted regulations and a student gets seriously hurt there may be severe legal repercussions.

5. You should also both hand out and post a set of laboratory rules concerning due dates for reports, general clean-up of the work area and consequences of not doing so. You should also include rules concerning equipment breakage due to excessive carelessness (or horseplay) and the like. If you want deviations from the suggested report format you should also list them.

6. You should indicate at the beginning of the course whether the students may use the computer programs on the enclosed diskette.

7. There may be some question about some of the equipment shortcuts. Soil laboratories at many universities operate on very limited budgets so that some of the suggestions may be of considerable value in operating the laboratory efficiently.

8. I have made some suggestions for advance preparations needed for upcoming tests. If there is not some advance planning the laboratory period can easily extend beyond the allotted time. Many students simply cannot stay because of course or work conflicts. These suggestions are not necessary for the experienced instructor but may be of substantial aid for those with less experience or for the teaching assistants who may be used for laboratory preparation.

9. There may be some question about the inclusion of Tests 21 (Volumetric-Gravimetric Relationships) and 22 (Unit Density of a Cohesive Soil). These are not intended as student projects, rather they were included so that the instructor could schedule, about the second class period in the laboratory, with Test 21 as a demonstration. It is most instructive for the class to see a dry soil visually (the bell jar full of small gravel) then see it "saturated" by pouring in water until the pores are visually full (except for possibly a few air bubbles which indicate limitations on "saturation"). They can use their pocket calculators to compute both dry and saturated densities, to check the approximate volume of voids as the volume of water poured into the bell jar, and to back-compute a specific gravity, G_s. In addition to the instructional value of Test 21, the

laboratory meeting gives you a chance to introduce them to the laboratory and some of the equipment they will be using.

I have found Test 22 (Unit Density of Cohesive Soils) also to be useful if there is time to run it. It also gives the student a visual feel for what unit density and unit weight are. With only a small amount of equipment required this can be done fairly easily in the classroom.

McGraw-Hill and I would like to thank the following reviewers for their many helpful comments and suggestions: Thomas L. Brandon, Virginia Polytechnic Institute; Tuncer B. Edil, Geotechnical Consultant; Ralph J. Hodek, Michigan Technological University; Norman L. Jones, Brigham Young University; Derek Morris, Texas A & M University; Dawit Nagussey, Syracuse University; and Manooch Zoghi, University of Dayton.

Joseph E. Bowles

ENGINEERING PROPERTIES OF SOILS AND THEIR MEASUREMENT

INTRODUCTION

This section provides information of a general reference nature. It contains a section on soil mechanics definitions as well as volumetric and gravimetric relationships. Other sections give suggestions on laboratory procedures and report preparation. These are intended as suggestions but may be requirements if so stated by the instructor. Reports prepared by geotechnical firms for clients generally follow the report format suggested here. Comments on graph construction have both student and commercial application.

I-1. Soil Mechanics Definitions

This section of the Introduction references select soil mechanics definitions and equations. The equations given are sufficient to enable the user to derive any others that may be needed.

Also given in this section is an example which uses some of these equations and is in a format that is quite similar to the Water Content Determination test outlined as the first laboratory exercise (Test 1). Since a number of the laboratory tests require the determination of the water content you should reference this page with a tab.

SI/METRIC UNITS USED IN THIS TEXT

No foot-pound-second (fps) laboratory units are used in this text. Commonly, soils laboratory equipment will measure in the standard SI unit of *mass*, the kilogram (kg), or in the smaller mass unit of grams (g). Weight as used by engineers is a force unit (but is hardly ever stated as such). We have the following considerations:

$$F = \text{mass} \cdot \text{acceleration} = (W/g)a$$

and if we substitute the standard gravitational acceleration g for the acceleration a in the above equation we have $F = W$ (where W is the body force or weight produced by the gravitational effect of the earth on the mass of the body under consideration). The standard gravitational acceleration (which varies slightly with location) is based on elevation of sea level at 45°N latitude and is very nearly 9.807 m/s^2 (or 32.17 ft/s^2).

The standard SI force unit is the newton (N), which is equal to 100,000 dynes. The newton is also the force that will accelerate a mass of 1 kg at a rate of 1 m/s^2. One gram force = 980.7 dynes (also the acceleration of gravity in cm/s^2), and thus we have:

$$F = 1 \text{ g (mass)} \times 980.7 \text{ cm/s}^2 \qquad \text{(units of force = dynes)}$$

If we divide by 980.7 dynes/g, it is clear that grams mass and grams force are interchangeable (also kilograms) as long as we deal with the standard gravitational constant. This has caused considerable confusion in the past, but it may become less of a problem in the future since the newton (force) uses an acceleration of 1 m/s^2 (which is not the acceleration of gravity).

For soils work we will use the following units (with standard abbreviations shown). Note that the intermediate units used will be consistent with the laboratory equipment, and the final data will be reported in SI units.

Quantity	Nonstandard Intermediate Unit	SI Unit
Length	centimeter (cm)	millimeter (mm) or meter (m)
Volume	cubic centimeter (cm^3, cu cm or cc—also milliliter (mL))	cu meter (m^3)
Mass	gram (g)	kilogram (kg)
Weight	g or kg (force)	newton (N) or kilonewton (kN)
Unit density ρ	g/cm^3; sometimes tonnes/m^3 (1 tonne = 1000 kg)	(kg/m^3)
Unit weight[1] γ	None	kN/m^3 (or k/ft^3)
Pressure	kg/cm^2	N/m^2 [pascal (Pa) in soil work use kilopascal (kPa)]
Energy		Newton · meter = joule (J)
Moment		N · m (but is not a joule)

[1] k = abbreviation for kilo in SI and kips (or 1000 Pounds) in fps.

Using the above definitions and the conversion factors on the inside of the front cover of this text the *weight* of 1 m^3 of water at 4°C is computed as follows:

$$1 \text{ cu ft of water} = 62.43 \text{ pcf} \qquad 1 \text{ lb} = 4.448 \text{ N}$$

$$1 \text{ cu m} = \frac{1}{0.3048^3} = 35.3147 \text{ ft}^3$$

Thus,

$$1 \text{ m}^3 \text{ of water} = 62.43(4.448)(35.3147) = 9806.5 \text{ N/m}^3$$

$$\text{Mass of } 1 \text{ m}^3 \text{ of water} = 1{,}000{,}000 \text{ g/m}^3$$
$$= 1 \text{ Mg/m}^3, \text{ as given in table on inside of cover}$$

FUNDAMENTAL DEFINITIONS

Referring to Fig. I-1*a*, we have a volume of soil removed from a field location. Further, it is assumed that the soil was removed in the form of a cube with lateral dimensions of 1 cm (volume = 1 cm^3). Actually it would be difficult to do this in practice, but for illustrating the volume-weight relationships that follow, we shall assume that we could remove this soil volume as a perfect cube. This cube is made up of a soil *skeleton* with water and air in the interstices, or *pores*, between contact points of the soil particles (Fig. I-1). It should be evident that, depending on the location of the cube of soil in the field (in situ) and on climatic factors, the quantity of water and air can vary from all the pores full of water and no air present to all the pores full of air and no water. Depending also on the instantaneous temperature, the water could be present as ice or as an ice-water mixture.

A *block diagram* as shown in Fig. I-1*d* should always be used when deriving any volumetric-gravimetric relationships. This provides an easy means to identify both what is known and the relationship between known and desired quantities. Make the block diagram a two-dimensional model since the third dimension is the area. In most relationships some volume (= height x area) is in both numerator and denominator, allowing the area to cancel. This allows height ratios to be used instead of actual volumes.

For the purposes of better visualizing the make-up of the cube of soil, let us first drain the cube of all the water present and place it in a container. Next we heat the soil skeleton until it melts and flows together to form a solid, nonskeletal (nonporous) mass to occupy the volume V_s in a container with a total volume of 1 cm^3, as in Fig. I-1*c*. Notice that the soil skeleton occupied a volume of 1 cm^3. The actual volume of soil *solids*, V_s, is less than 1 cm^3. Now we pour the previously collected volume of water V_w, into the container on top of the

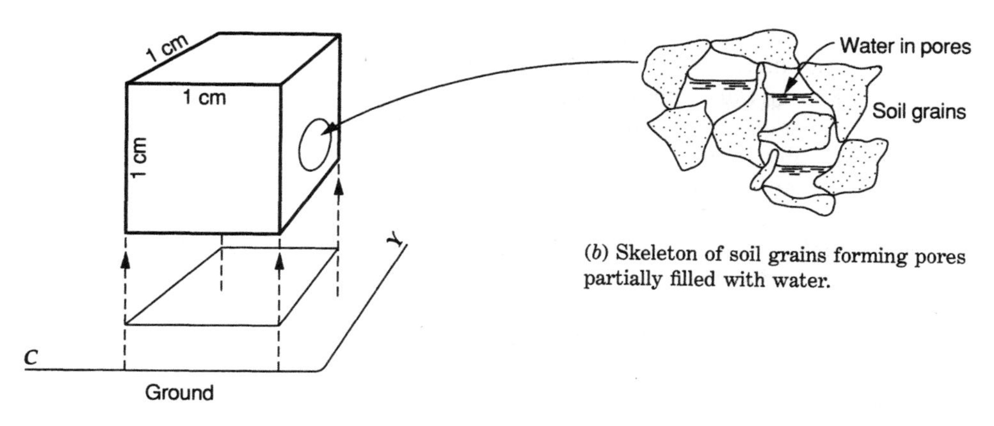

(*a*) Cube of soil removed from ground.

(*b*) Skeleton of soil grains forming pores partially filled with water.

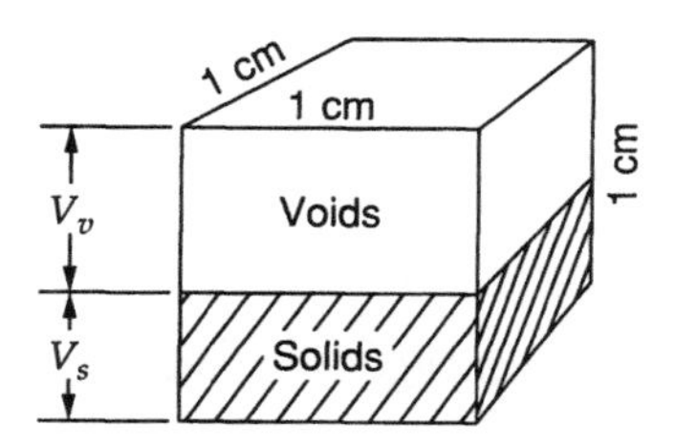

(*c*) Soil solids formed into a non-porous volume of less than 1 cm^3.

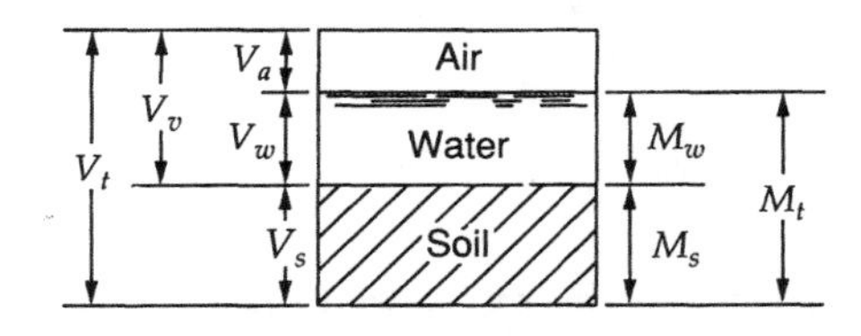

(*d*) Two-dimensional block diagram showing volumetric and mass relationships for the original soil volume.

Figure I-1
Volumetric-gravimetric relationships.

solids volume (Fig. I-1*d*). If the pores of the soil skeleton had been full of water, it should be evident that the volume of soil solids plus the volume of pore water would fill the 1 cm^3 container. Since in this illustrative case they were not full of water, the remaining volume required to fill the cube must be the air volume, V_a, present in the original soil skeleton.

The assumptions listed below will be made for convenience in the following material:

1. Soil has mass (and weight).
2. Air has no mass (for the volumes we shall be using).
3. Water has mass. Generally we shall take this as 1g/cm^3 (Mg/m^3), although this value is correct only at a temperature of 4°C. At temperatures of 18 to 25°C, the mass density ranges from 0.9986 to 0.9971 g/cm^3 (see Table 6-1).

The following standard symbols will be used in the definitions to follow:

e = void ratio
G = specific gravity of any substance
G_s = specific gravity of the soil solids
G_w = specific gravity of water
n = porosity
w = water content
S = degree of saturation
V_a = volume of air present in a soil mass.
V_s = volume of soil solids in a soil mass
V_t = total volume of a soil mass $= V_s + V_w + V_a$
V_w = volume of water in a soil mass
V_v = volume of voids in a soil mass $= V_w + V_a$

M_s = mass of dry soil solids
M_t = total mass of soil including any water present
M_w = mass of water present in the soil
ρ = mass density of a material (g/cm³ or Mg/m³)
ρ_d = dry unit density of soil
ρ_{sat} = saturated unit density of soil
ρ_w = unit density of water (approximately 1 g/cm³)
γ_i = symbol for unit weight of a soil; $i = d$ for "dry"; $= sat$ for "saturated"; $= wet$ for "wet"

Referring to Fig. I-1*d*, we define the *void ratio e* as

$$e = \frac{V_v}{V_s} \tag{I-1}$$

The void ratio is usually expressed as a decimal.

The *porosity n* is defined as

$$n = \frac{V_v}{V_t} \tag{I-2}$$

Porosity may be expressed either as a decimal or as a percentage; decimal usage is preferred.

By substituting Eq. (I-1) into Eq. (I-2), we can obtain the void ratio in terms of the porosity as

$$e = \frac{n}{1 - n} \tag{I-3}$$

The *degree of saturation S* is defined as

$$S = \frac{V_w}{V_v} \times 100 \quad \text{(percent)} \tag{I-4}$$

Notice that the range of S is from 0 to 100 percent; if

$S = 0\%$, the soil is *dry*

and if

$S = 100\%$, the soil is *saturated*

The *mass density* ρ of a soil (or any other material such as iron, copper, coal, etc.) is

$$\rho_i = \frac{M_i}{V_t} = \frac{M_s + M_w}{V_t} \tag{I-5}$$

For the special case of soil, M_i can range from a *dry* to a *saturated* state. If the soil is saturated, change the i subscript in Eq. (I-5) to *sat*; if the soil is dry $M_w = 0$ ($S = 0$) and Eq. (I-5) reduces to

$$\rho_d = \frac{M_s}{V_t} \tag{I-6}$$

If $0 < S < 100$ percent, M_t is still computed as $M_t = M_s + M_w$ and Eq. (I-5) gives the *wet unit density*, ρ_{wet}, of the soil. Generally one must look at the context of a given problem to see which unit density is being described or is to be determined, since, as the preceding material indicates, more than one unit density value may be computed for a given soil depending on the mass state being used in the numerator of Eq. (I-5).

UNIT WEIGHT

The *unit weight* γ of a material is a derived force unit. When ρ is expressed in g/cm^3 the unit weight is obtained by manipulation of the local gravitational constant to obtain on the planet earth to sufficient precision

$$\gamma = \rho \times 9.807 \quad kN/m^3$$

or in fps units

$$\gamma = \rho \times 62.4 \quad lbs/ft^3$$

From this it is clear that the use of mass in g/cm^3 is of considerable value in simplifying unit weight computations in geotechnical work.

From the definition of specific gravity G found in physics textbooks, the volume of any material is

$$V = \frac{M}{G \times \rho_w} \tag{I-7}$$

Using this relationship and Fig. 1-1*d* the volume of soil solids V_s is

$$V_s = \frac{M_s}{G_s \times \rho_w} \tag{I-8}$$

The method of determining the specific gravity, G_s, of a soil is given as one of the laboratory projects (Test No. 7).

The volume of water, V_w (using g and cm units), is

$$V_w = \frac{M_w}{G_w \times \rho_w} = \frac{M_w}{1 \cdot 1} = M_w \tag{I-9}$$

since both G_w and ρ_w are approximately 1.0 for water at the usual laboratory and/or field temperatures of 10 to 20°C. In Eqs. (I-7) through (I-9) the desired volume can also be obtained by using W_i instead of M_i in the numerator and using γ_w instead of ρ_w in the denominator.

Water content w is defined and the laboratory procedure for determining it is given in the Water Content Determination Test (Test 1). For convenience, the equation defining water content is repeated here:

$$w = \frac{M_w}{M_s} \times 100 \quad \text{(percent)} \tag{I-10}$$

With appropriate assumptions and substitution of equations previously derived (and using a block diagram) various other relationships can be derived as in the following table.

Given quantities	To find	Derived expression	Limitations
ρ_d, G_s	e	$e = \frac{G_s}{\gamma_d} - 1$	γ_d in g/cm³
G_s, w	n, e	$n = \frac{wG_s}{1 + wG_s}$	$S = 100\%$
		also $Se = wG_s$	any S
		$e = \left(\frac{w}{S}\right)G_s$	
G_s, e	γ_d, γ_{sat}, n	$\gamma_d = \frac{G_s\gamma_w}{1+e}$	
		$\gamma_{sat} = \frac{(G_s + e)\gamma_w}{1+e}$	
		$n = \frac{e}{1+e}$	

Other relationships can be derived, but they will be left as an exercise for the student.

Example

For this example you should refer to *Water Content Determination* (Test 1) and Data Sheet 1 (found at the end of the text part of the manual where data sheets for all of the tests are located).

Given:

a. Mass of container + wet sand = 248.5 g
b. Mass of container + dry sand = 231.2
c. Mass of container = 63.7
d. Mass of dry sand = 167.5 g (b − c)
e. Mass of water = 17.3 g (a − b)
f. Volume of container = 100 cm³ (= V_t)

Find: (1) w, percent; (2) dry unit weight γ_d; (3) wet unit weight γ_{wet}; and (4) void ratio e if $G_s = 2.68$.

SOLUTION **(We will use equations previously given.)**

1. The water content is (see Eq. (I-10))

$$w = M_w/M_s \times 100 = 17.3/167.5 \times 100 = 10.3\%$$

2. For dry unit density refer to Eq. (I-6)

$$\rho_d = M_s/V_t = 167.5/100 = 1.675\ \text{g/cm}^3$$

Unit weight γ is usually reported in kN/m³ (or k/ft³)

$$\gamma_d = 1.675 \times 9.807 = 16.43\ \text{kN/m}^3$$

or

$$\gamma_d = 1.675 \times 62.4 = 104.5\ \text{k/ft}^3$$

Typically one reports γ to 0.01 kN/m³ or 0.1 k/ft³.

3. The wet unit density is (using Eq. (I-5)

$$\rho_{wet} = M_{wet}/V_t = (248.5 - 63.7)/100$$
$$= 184.8/100 = 1.848 \text{ g/cm}^3;$$

and

$$\gamma_{wet} = 1.848 \times 9.807 = 18.12 \text{ kN/m}^3$$

You should verify that $\gamma_{wet} = (1 + w)\gamma_d = 18.12 \text{ kN/m}^3$

4. Now compute the void ratio e using Eq. (I-1).

$$e = V_v/V_s$$

Volume of solids

$$V_s = M_s/(G_s\rho_w) \qquad \text{(Eq. (I-8))}$$
$$= 167.5/(2.68 \times 1.0) = 62.5 \text{ cm}^3$$

The volume of voids is:

$$V_v = V_t - V_s$$
$$= 100 - 62.5 = 37.5 \text{ cm}^3$$

giving void ratio

$$e = 37.5/62.5 = 0.60$$

Void ratio is usually reported and/or used in decimal form.

Note again that using the mass density of water ρ in the non-SI unit of 1 g/cm^3 and M_s in grams for these computations greatly reduces the computational effort (and potential for errors).

I-2. Laboratory Procedures

Laboratory equipment is expensive; this can be more properly appreciated by looking at a current copy of a supplier's price list. The costs of various items of equipment may be indicated at various times throughout the term. Equipment may be damaged by careless handling, and damaged equipment can produce serious test errors. Some laboratories may require the student to pay for any equipment damaged through extreme carelessness.

Scales are especially susceptible to damage. For this reason, they are not to be moved to other laboratory locations except under the instructor's supervision. The triple-beam balances used primarily for water-content determinations should be locked at the end of each laboratory period. The balances should be checked for zero reading before using, and any series of weighings should *always* be taken on the same balance to avoid zero errors between two balances. Small quantities are used in certain tests, and in such tests weighing errors from differences in balance zero corrections can be very important. If your laboratory is equipped with electronic balances you should follow instructions in their use and care given by your instructor. Periodically check the balances for dust or soil particles clinging to critical parts that can affect mass determinations.

The drying ovens are preset to a temperature of 105 to 110°C, and the thermostats should not be manipulated without instructor authorization as it takes considerable time to stabilize the oven temperature so that the thermostats can be set properly. It is essential in soil mechanics work that this particular value of oven temperature be maintained. If you use a microwave oven for any soil drying carefully follow your instructor's directions for use.

Many ovens have a thermometer that protrudes into the oven cavity. Exercise care that you do not push a sample into the oven which breaks off this bottom piece. These are specially made for ovens and cannot be replaced with regular laboratory thermometers.

Each group is responsible for equipment and cleanup of its work area. Never leave a piece of equipment with soil (wet or dry) on it. Depending on the equipment it can rust overnight and if it is not used again for several months may corrode so badly it will have to be replaced.

If there are multiple laboratory sections the equipment should be ready for use by the next section that will be using the laboratory. Valuable time is lost if a piece of equipment must first be washed and dried before it can be used.

LABORATORY SAFETY

Potential for accident always exists when doing any of the tests in this manual. In the Shrinkage Limit Test one may use mercury, which is classified as a hazardous material. Various equipment that is used can be broken or used carelessly to produce cuts and/or abrasions. If you get cut (accidentally or otherwise) immediately check with the lab supervisor to get first aid.

Vibratory equipment may produce high noise levels that require using ear plugs.

Do not put hot materials removed from the oven where others can get burned by accidental contact. If a hot plate has been used be sure you locate (or identify) it so someone does not touch it until it has cooled to a safe level.

In any case the testing personnel should adhere to all safety requirements in accordance with individual laboratory policies. Be sure to read any posted safety rules before you use a piece of laboratory equipment.

NEVER HORSEPLAY IN THE LABORATORY. In addition to someone possibly getting hurt, equipment may be broken or damaged so that the assigned lab project cannot be done. An entire class may be deprived of a learning experience from a trivial "I didn't mean to do that" event.

I-3. Laboratory Reports

A laboratory report is required for all projects unless otherwise stated by the instructor. This report will be in a folder (one report to a folder) with the following information shown on the cover:

1. Name of student
2. Title of project
3. Course number and laboratory section
4. Laboratory group number
5. Date of project and date of report submission

The report should be produced on a word processor or typewriter; if neither is available use neat hand lettering with a ballpoint pen. Include the following material in the given sequence:

1. *Flyleaf*—showing title of project, name of student, course number, date of work, date of submission of report, and the names of the other group members.

2. *Introduction*—a brief (one-half page or less) summary of the objectives of the work. In a commercial report this is a summary of the contractual agreement as understood by the geotechnical consultant.
3. *Discussion of the work*—including any special techniques used or changes from laboratory manual instructions. Any equipment limitations or possible sources of error should be discussed. If your results are not very good, give *your* analysis of the probable cause. If you use any equations other than basic definitions, include both the equation and its derivation. The "discussion" should generally take less than 1 1/2 pages. In a commercial setting you would indicate any deviations from the ASTM (or other standards) and the reasons.
4. *Conclusions*—a brief summary *and* tabulation of the project results. Any improvements in work procedures that can (or you think could) be made should be listed under this heading. Do not say you got a lot (or nothing) out of the experiment; this is not considered a "conclusion." This part of the report should be about one-half to one page long, depending on the amount of information to be presented.

   In a commercial report this would be those items in the first sentence together with a summary of recommendations for use by the client later in the project design. Appropriate reference would be made to any graphs produced by the testing.
5. Show any graphs next, and follow these with your collected data, which will generally be on one of the data sheets furnished in the back of this manual. If you use one of the computer programs, or your own computer program, to reduce your data, be sure to include the worksheet that was used for data collection. Do not recopy your data; get in the habit of taking neat (or at least reasonably neat) laboratory notes. The report is to contain original data; if it is recopied, list the group member(s) who took the original data.

   In a commercial report you would include here any graphs and significant test results from any tests not requiring graphs.
6. *Sample calculations*—It is not necessary to show every calculation, but at least one typical calculation must be shown for each step required in the data reduction. For example, take the second and tenth lines of data that are to be reduced and show every step required to produce the data in the columns of the data sheet. Be sure to show the derivation of any equations you used in data reduction. This is also to be done if one of the computer programs (or your own) was used to reduce the data.

   In a commercial report this section would probably be omitted, with an Appendix provided which included any boring logs or other field data on which the laboratory work is based.
7. Where a computer program is used, the output sheet can be included in your report in lieu of using one of the data sheets from this manual. This is also applicable in the commercial situation.

A soil report should follow good technical report writing form, including the citing of any references used. Do not use the first person (I, me, we, our, etc.) in writing a technical report. Do not write such statements as, "I found that . . ." or, "My group found . . . ;" instead use "It was found that. . . ." Use correct spelling—when in doubt, consult the dictionary. Try to use good sentence construction, and do not change from past tense to present tense in the same sentence or paragraph. Many of these problems can be avoided by:

1. Writing the report and then reading it over a day or so later. A neatly marked-through word or two will be better received by the instructor than a report that is poorly written. It will also indicate that you went over the report before submitting it.
2. Not writing in such a hurry that you tend to drop letters off the ends of words or spell a word as it sounds.
3. Thinking about what you are going to write and being brief. A few well-put-together sentences are far more impressive than a long, poorly written report that says nothing.

4. Obtaining (perhaps from the library) a text on technical report writing and consulting it for writing tips.

A primary purpose of the report is to give the instructor an indication of what you learned from the project. Other major benefits are obtained from practice in report writing and presenting engineering data. While you are graded primarily on the project and the presentation of the data (and not report writing), a poorly written report will generally result in a lower grade than a well-written one.

In a commercial environment the report is the project information record that documents to the client the scope of work performed, both for design use and for payment authorization. Great care is required in its preparation since the contents may be used later in any project peer review or as evidence if legal action is brought against the Geotechnical Engineer.

Student users may not yet fully appreciate the formality involved in the reports; however, be aware that report writing is an essential part of the work of most engineers. Further, if you are not involved in soil mechanics work until several years after graduation you will have some refresher material ready at hand—old laboratory reports—which, if well written, will be easy to follow.

DRAWING GRAPHS

Summarizing laboratory (and much other) work is greatly facilitated by presenting the results in the form of some kind of graph. A number of the tests in this manual rely on a graph for presentation of essential data, and a number of sheets of graph paper are provided for that purpose at the end of the data sheet sets. A graph should be legible, neat, and easy to understand. Graph sheets, both regular grid and semi-logarithmic with the necessary number of cycles, are furnished in the data sheet section. If you use other graph paper it should be of about the same size as that furnished here (205 × 280 mm). It is undesirable to paste several pieces of semilog paper together to obtain the required number of cycles because of the page folds that result when the report is bound. To encourage the use of graph paper in cm/mm divisions (commonly used for graphing) the sheets furnished in this manual are in those units.

When drawing the graph axes, always place them on the graph sheet so that the *left* and *lower* margins are both at least 2 cm wide. All lettering on the axes *should be within the ruled portion of the sheet*, and the line rulings should be used as lettering guidelines. Always use as large a scale as possible, but one that is easy to plot and/or read. For example, scales of either 5 (or 10) divisions/cm plot well in multiples of 1 and 5 units (sometimes 4), but scales of 3, 6, or 7 units/division are awkward for both plotting and reading.

Show a title block (see Fig. I-2) on all graphs that includes:

1. Title of project (*Example:* Dry density vs. water content)
2. Type of soil (*Example:* Brown silty clay, $w_L = 42.1\%$; $w_P = 21.3\%$)
3. Date of work (*Example:* January 12, 1995)
4. Scale (if appropriate). A graph reduced on a copying machine may be difficult to use without a scale (which is also reduced).
5. Name of person preparing the graph

Locate the title block in a region of the graph where it does not interfere with the data being presented—but not in the center. The upper right-hand or lower right-hand corners are most commonly used. When the results of several tests are shown as curves on a single graph, a legend should be used to identify the data plotted from the different tests, as shown in Fig. I-2. Several curves on a sheet generally make a better presentation if shown using lines of different colors or types (solid, dashed, dash-dots, etc.).

Always show your plotted points somewhat as illustrated in Fig. I-2 and never extend your curve(s) through the points. This allows the points to be identified so that the reader

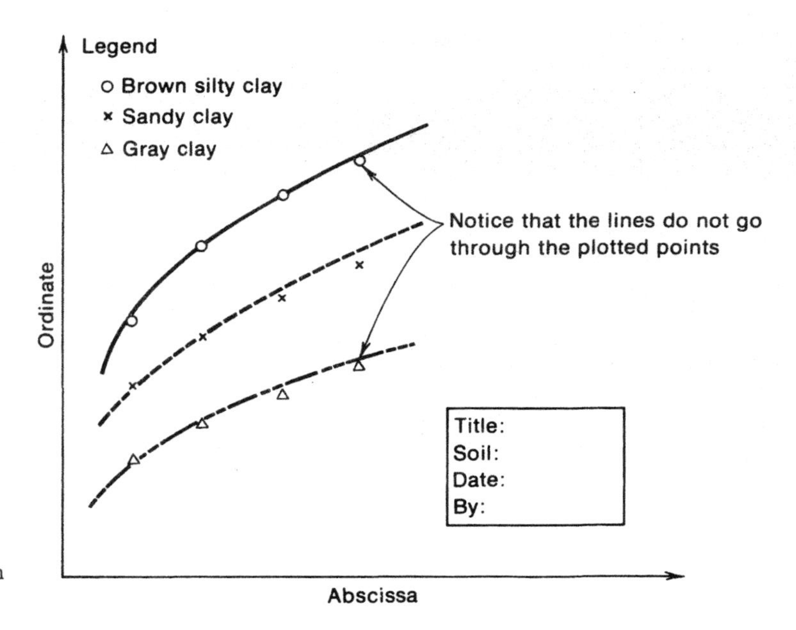

Figure I-2
Method of displaying test data on a graph when more than one test is to be shown.

can verify your interpretation of the plotted data. Always draw smooth curves using a French curve (never free-hand) unless the instructor tells you to use straight lines. All graphs should be oriented to read with the binding to the left side or at the top of the displayed page.

Before plotting a graph, give thought to what information the graph is to present. Is it qualitative (relative) or quantitative (numerical) or both? Obviously, quantitative information requires a better scale than qualitative information. How have others (in textbooks, etc.) presented the same type of data; i.e., does the ordinate or abscissa best control the plot? As an example, a curve that is asymptotic generally (but not always) displays better with the asymptote horizontal rather than vertical. Note that:

1. The most accurate numerical results are obtained from plots that use the largest possible scale.
2. You can reduce the appearance of data scatter by reducing the scale of the plot. Use this type plot when the curve is to represent the "best average" of values.

If a graph is used to compute material constants (such as the stress-strain modulus), display the values and how they were obtained. Also show any computed value(s). This allows others to verify your work quickly.

DATA SHEETS

There is a data sheet numbered to match the test in the back of your manual which can be detached to summarize data collection and reduction. If there is no data sheet then none is required, and you submit the project data using whatever format you think is appropriate. The data sheet, or the first sheet of a set (numbered with *a*, *b*, . . .), contains spaces for general project information that should always be filled in.

The "Description of soil" is somewhat approximate. Here you should note the color; and if it is sand or gravel, include this fact. If it is sticky when wet or has hard lumps call it a clay or sandy clay, gravelly clay, and similar. You could not formally classify the soil until you complete Project 8, and for the purposes of the data collection the visual color and grain size description are adequate.

There are enough data sheets to complete two tests and in some cases, such as water content, four or five. The general format of all of the data sheets is suitable for use in commercial laboratories.

I-3. Laboratory Practice

The procedures outlined in the following tests are reasonably standard. For example, distilled water is often specified for use in the ASTM[1] and AASHTO[2] versions of the tests. This can be omitted and ordinary tap water used for the tests (*at the discretion of the instructor*), yielding in most cases a negligible error.

Methods of soil preparation and length of testing may be modified from the ASTM Standards in the interest of saving student time. A copy of the current ASTM Standards, section 4, volume 4.08 (all the soil tests that have been standardized by ASTM to date are in volume 4.08) should be available in the soils laboratory or library for the interested user. AASHTO also has standard procedures for performing soil tests for transportation facilities. These are given in the two-volume publication *AASHTO Materials*, 15th ed. (published 1990).

The soil tests outlined in the following pages of this manual list the current ASTM and AASHTO Standards designation (if there is one) for reference and reader convenience. For example, the ASTM Standards designation for the test in Water Content Determination is D 2216-XX, where D 2216 identifies the test and XX is the year of adoption or readoption of the test standard.

Note that although nearly every soil test that is done in a commercial laboratory is "standardized," many laboratories in the interest of time use procedures essentially as outlined in this text. This is done for two reasons:

1. Time is saved (and costs to the client are reduced).
2. The procedures outlined here (based on the author's experience in both university and commercial laboratories) do not produce results that differ enough from the standard tests to justify the additional costs.

Soil is highly variable, and this variability cannot be controlled to any great extent by engineers, who must work with the material as they find it. Laboratory tests are necessary to identify and classify the soil so that design decisions can be made. Classification and/or identification enables the geotechnical engineer to utilize experience (and reported experience of others with similar soils) to make the necessary design recommendations. Other tests enable necessary follow-up quality control to be made to ensure that the design recommendations are followed by the construction personnel.

It is obviously impossible to test the entire soil mass of a project, so it becomes necessary to perform a few tests on small quantities of soil and extrapolate the results to the entire soil mass. For the tests to be valid for the soil mass, they must be on samples that are *representative* of the soil mass. Obtaining truly *representative* samples requires great care in sampling operations. Bags of samples from the field are collected by the field personnel to be representative.

Likewise the smaller samples for individual tests should be representative of the bag of soil; thus picking out a few lumps from the top of the bag or other container of field soil is very probably not going to yield a representative sample. Keep constantly in mind the fact that you are or will be, as a geotechnical engineer, estimating with tests on a few grams of material the soil performance of huge masses of a soil that may vary widely in particle composition. This variation can occur within very small distances both horizontally and vertically.

I-4. Computation Precision

Electronic pocket calculators and/or personal computers (or PCs) are routinely used for engineering computations: both tend to introduce a fictitious precision to computed results. You should be aware of the data limitations when making computations.

[1] American Society for Testing and Materials, 1916 Race Street, Philadelphia, PA 19103

[2] American Association of State Highway and Transportation Officials, 444 N. Capitol Street, N.W., Washington, DC, 20001.

The only computations in this manual that justify using two decimal places (to the nearest 0.01) in the final answer are the specific gravity test and—if done in SI units—the compaction test and the field density test. The other tests listed in this manual as possible laboratory exercises can hardly justify more than one decimal (to the nearest 0.1). Therefore, even when averaging two or more test results, round off and report the answer to the nearest 0.1 (or to 0.01 for the three tests cited).

ENGINEERING UNITS AND DATA COLLECTION

When using the data sheets (from the data sheet section) for your laboratory work, *always insert the units* of the dial gauges and/or load rings (if used) and whether a load cell (for loads) or a linear voltage displacement transducer (LVDT) is used for displacement measurements, as well as any other information that may be needed later when reducing the data to complete the test. *Never omit these items*, as you may later forget or, more importantly, if you need to refer to your report several weeks or months later, you will immediately see the units used and what you did to obtain the test results shown.

This will be especially critical in the next few years as soil laboratories replace equipment with SI/metric or electronic data acquisition devices. The data sheets are structured so that if you attend to necessary details they can be readily used for almost any type of test equipment. Of course the data sheet is not needed if the test has been completely automated so that the necessary information is tabulated and graphs directly generated during the test or immediately after completion. Few laboratories, either university or commercial, at present are this completely automated.

1: WATER-CONTENT DETERMINATION

References

ASTM D 2216-90 (*ASTM Standards* vol. 4.08)
AASHTO T 265 (*AASHTO Materials:* Part II: Tests)

Equipment

Moisture cups (tin or aluminum—also called moisture cans or boxes)
Oven[1] with accurate temperature control (a forced-draft type, with a fan for circulation, is preferred)

General Discussion

Water-content determination is a routine laboratory test to determine the amount of water present in a quantity of soil *in terms of its dry mass.* As a definition,

$$w = \frac{M_w}{M_s} \times 100 \qquad \text{(percent)} \tag{1-1}$$

where M_w is the mass of water present in the soil mass and M_s is the mass of soil solids. One could define the water content as the ratio of the mass of water present to the total mass (i.e., mass of water plus soil); however, this would give a quantity in the denominator of the fraction that would depend on the amount of water present:

$$w' = \frac{M_w}{M_s + M_w} = \frac{M_w}{M_t} \tag{1-2}$$

Equation (1-2) is not used in geotechnical work because with the denominator including a variable water content w' is not an independent quantity.

The soil is to be dried in an oven at a drying temperature of 110 ± 5°C to a *constant mass;* that is, as long as water is present to evaporate, the mass will continue to become smaller with successive weighings. Since it is usually not practical to make several weighings to determine if a constant mass state has been obtained, what is usually done is to assume that after oven-drying the soil for a period of 12 to 18 h (often over night), the sample will be in a constant mass state; and we will use that value for the mass of "dry soil + cup" (see Data Sheet 1).

Experience indicates that this method of sample drying is sufficient for the small (10 to 200 g) samples used in routine laboratory work. It is usually adequate for much larger samples, 1000 g or more, but large samples should be stirred occasionally so uniform drying is obtained.

It is usual to remove the water-content samples from the oven and weigh them immediately (use tongs or heat-resistant gloves, as they are quite hot). If it is not feasible to

[1]ASTM D 4643 allows use of a microwave oven for water content determination. This procedure allows a rapid drying of the soil but in order to use it correctly one must make several weighings to assure that the drying has been done to a constant weight. For student laboratory (and often commercial) work it is usually more convenient simply to oven-dry overnight (12 to 16 h) in the standard oven and assume the sample is "dry." If you do use a microwave oven be sure to follow instructions carefully so that no injuries result.

weigh the dry samples immediately,[2] put the lid on the cup as soon as it is cool enough to handle and/or place the can of dry soil in a desiccator so that the soil does not adsorb water from the laboratory atmosphere.

To provide reasonably reliable water-content determinations, the following minimum *representative* wet-soil sample masses are recommended:

Sieve retaining more than about 10 percent of sample	Recommended mass of moist specimen, g minimum	maximum*
No. 10 (or smaller)	20	20
No. 4	20	100
20 mm (3/4-in)	250	2,500
38 mm (1 1/2-in)	1,000	10,000
75 mm (3-in)	5,000	50,000

* ASTM recommendations: If the minimum mass is used report w to 1%; for maximum mass report w to nearest 0.1%.

The oven temperature of 110°C is too hot for certain organic (peaty) soils, soils containing appreciable amounts of gypsum or some other minerals, certain clays, and some tropical soils. These soils contain loosely bound water of hydration, or molecular water, that can be lost. As a result there may be changes in the soil characteristics—notably the Atterberg Limits of Test No. 3 as well as alterations in the grain size and/or reduction in the specific gravity. ASTM suggests an oven temperature of 60°C for drying these soils.

The data sheet for this test (Data Sheet 1) uses a format convenient for making the necessary computations for water content and includes provision for up to 10 tests on the sheet. As you proceed through the manual you will note that this test is part of a number of other tests—some of the data sheets for those tests have a similar format for computing water content. Certain other tests in the manual require that you use one of the data sheets of this test to supplement the data sheets for that test. Extra sheets are provided for that purpose.

SAFETY

This test may be hazardous if the soil being dried contains certain chemicals. For student laboratories reliable soil sources should be used to avoid creating a hazardous environment.

Procedure

This may be done by a group or an individual as required.

1. Weigh[3] a clean, dry tin or aluminum cup *and lid* for small samples (under 100 g). Weigh an evaporating dish or something similar for drying large samples. Be sure to record the cup and lid (or dish) number on your data sheet together with the mass from weighing. Carefully note that when a metal moisture can is used, "Mass of cup" always includes the lid. The metal moisture cups are available in several sizes, with the 50-mm diameter by 30-mm height and the 64-mm diameter by 40-mm height being most commonly used. Refer to Data Sheet No. 1 for typical data organization.

[2]ASTM suggests allowing the sample to cool to room temperature or until the cup of soil is sufficiently cool that convection currents over the cup do not affect the weighing. The author's experience is convection currents are of no consequence and you can weigh the samples as soon as you have access to the scales.

[3]Throughout this test the term "weight" is used to describe the process of determining the mass of an item (or material). When using SI units, *mass* is a quantity measured using scales and *weight* is a quantity computed using mass and conversion factors.

Small soil samples (usually in the range of 20 to 40 g) are often used in this test. It is essential that weighing errors be minimized by

a. Always using the same set of balances (or scales).
b. Always checking the first time you use the balances that they read zero before a mass is put on them. If they do not read zero, report this to your instructor so they can be adjusted before they are used.

2. Place a *representative* sample of wet soil in the cup and determine the mass of wet soil + cup. If the mass is determined immediately, it is usually not necessary to place the lid on the cup. If there is a delay of 1 to 5 min or more, put the lid on the cup so soil does not dry; and place the cup under a damp paper towel to maintain the humidity in its vicinity.
3. After weighing the wet sample + cup, *remove the lid*—it is usual practice to place it under the bottom of the cup—and place the sample in the oven.
4. When the sample has dried to a *constant mass*, obtain the mass of the cup + dry soil. Be sure to use the same scales for all weighings. Routinely check that the mass here is less than in Step 3 above and that the water loss from evaporation seems reasonable for the mass of wet soil used.

   If the mass appears to be in error, first reweigh the cup + dry soil, and, if the new value appears satisfactory, continue. If it is not, next check if you are using the same balances and/or if they are in adjustment; also be sure the lid is included if a metal moisture cup is being used. If these items all check and there is too much mass, you will have to discard this sample because it is impossible for a "dry" sample to have a greater mass than a "wet" one—an earlier weighing error has been made. There may still be a weighing error, but at this point you have no means to identify it. Later, if there are additional tests on the same soil sample, you may discover a water content value that differs significantly from the rest and decide to discard it.

   Be sure to empty the cup of soil and wash it so that is is clean and ready for use when again needed.
5. Compute the water content w. The difference between the mass of wet soil + cup and the mass of dry soil + cup is the mass of water M_w that was present in the sample.

   The difference between the mass of dry soil + cup and the mass of cup is the mass of dry soil M_s. You directly compute the water content by inserting known mass quantities into Eq. (1-1).

   You should also refer to the Example given in the **Introduction** at this point if you have any uncertainty about making the computation.

   Always report water content to the nearest 0.1 percent but in computations w is used as a decimal quantity.

2: FIELD COLLECTION OF A SOIL SAMPLE

References

ASTM D 420
AASHTO T 86
Read: "Water-Content Determination," Test No. 1

Objectives

This project is for the purposes of:

1. Introducing the student to one simple method of obtaining disturbed soil samples.
2. Enabling the student to obtain an indication of the variation of natural soil moisture with depth.
3. Collecting information to plot a simple soil profile.
4. Obtaining a bag of soil for future laboratory testing.[1]

Equipment

50- or 76-mm hand auger (or small gasoline-powered continuous-flight auger)
Extension rods for the auger to obtain borings to a depth of at least 6 m
2 sample bags per group
12 clean, dry moisture cans per group (obtain empty mass before going to the auguring site)
Wrenches to add on auger extension rods
30-m tape to locate borings

Procedure

This is suggested as a group project.

Note: So the students have an idea of particle sizes that are used in a soil description, display a range of sieves before going to the field.

1. Each group should auger a hole at least 6 m deep (unless prevented by underground conditions). Record all data on the enclosed Data Sheet No. 2, which is a composite of several used commercially.
2. Take two moisture-content samples at each 1 m of boring depth or where visual soil stratum or moisture changes occur. Place the lids on the moisture cans immediately on obtaining the samples to minimize moisture loss.

   Upon return to the laboratory, weigh the cans of moist soil, *remove the lids* and place the samples in the oven for drying. Return to the laboratory the following day and weigh the dried soil and compute the water content. Use water content data sheets furnished in the back of this manual. Average the two values of water content as the reported water content at each elevation. Fill in as much of Data Sheet No. 2 as possible.

[1]This project may be deleted at the discretion of the instructor. If it is omitted, it is suggested that a quantity of somewhat cohesive soil, enough to provide at least 15 kg of air-dry soil per group, be obtained from a local construction site during good weather and *prior to classes. This soil should be all (−) No. 4 sieve size.* Sieve this soil while it is still damp and easy to work through a 6-mm ($1/4$-in) wire screen to break the lumps. Discard the gravel retained on the screen device. Store this "soil" for later use in the several lab tests. A different soil (color or texture) is recommended for each group—or at least for each lab section.

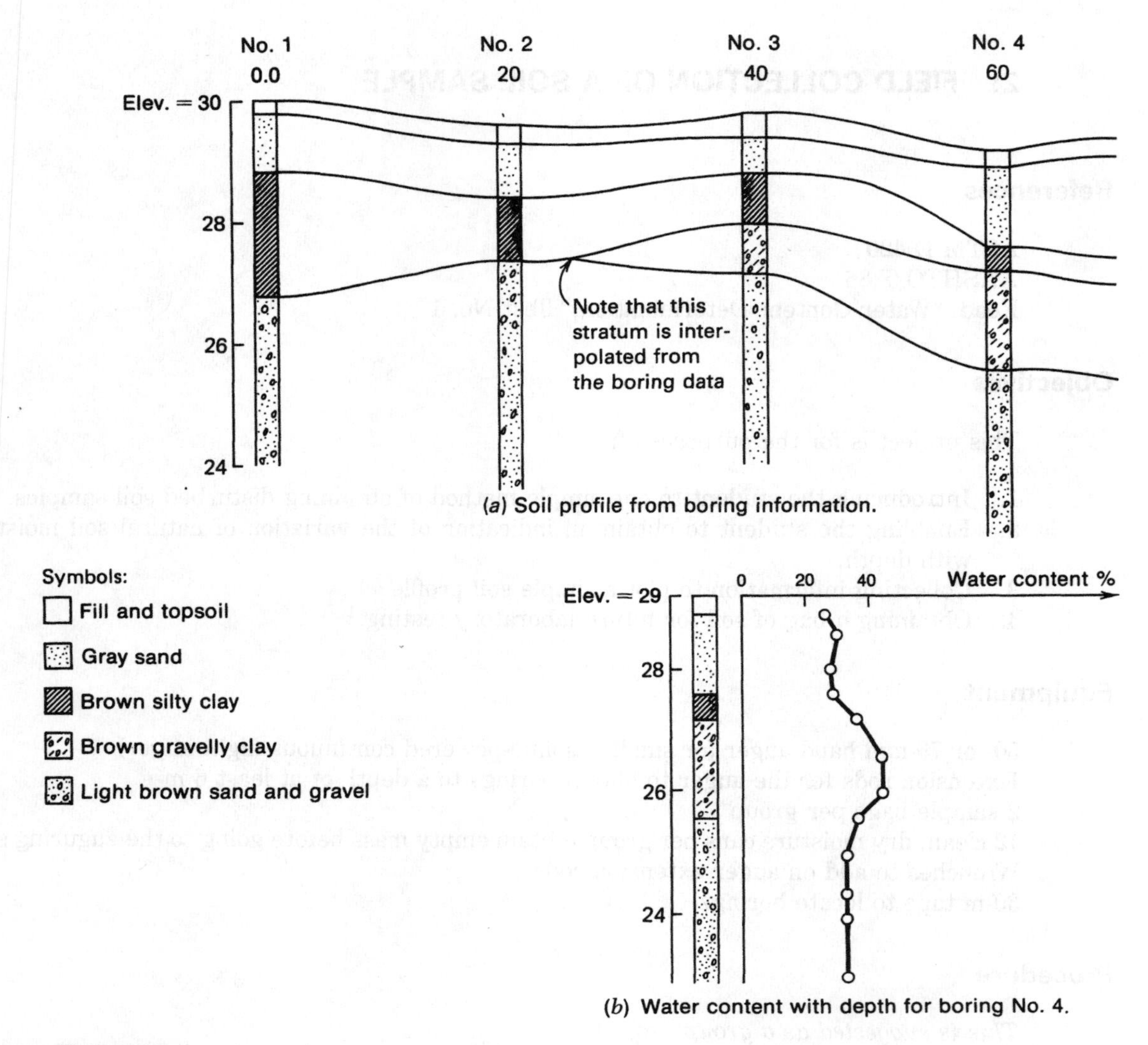

Figure 2-1
Typical boring-log presentation. Note the plotting of the ground-surface profile. Some persons (as you will do in this project) use the ground surface as the plotting reference for each boring. Exact details may vary somewhat from laboratory to laboratory. All dimensions are shown in meters.

3. Collect two bags of clayey soil per group (or enough material to yield at least 15 kg of air-dry soil) from the boring. All of this soil should pass the No. 4 sieve. This will be taken back to the laboratory, labeled, and stored for later soil testing, except that 5 kg will be immediately placed in a large pan to air-dry for use in the next laboratory period.
4. For this laboratory report, each group will, on the day following the field work, place on the blackboard in the soil laboratory the profile and location of their boring. Each student will use these data to plot a simple soil profile using a scale of 2 cm = 1 m vertically and 2 cm = 15 m horizontally on a sheet of 205 × 280 mm graph paper from the back of this manual. Use a visual description of the soil in each stratum (e.g., gray, sandy clay; black topsoil; gravelly sand; etc.). Each student will also present on a second sheet of 205 × 280 mm graph paper a profile of the individual boring with a plot of water content with depth as shown in Fig. 2-1. A legend similar to that shown in the figure should be used.

In the brief report to be submitted for this project (including all data sheets and your graphic plots), can you draw any conclusions on the variation of water content with depth based on any recent rainfall, lack of rainfall, temperature or other environmental factors?

3: LIQUID AND PLASTIC LIMITS OF A SOIL

References

ASTM D 4318 (Liquid Limit, Plastic Limit, and Plasticity Index of Soils)
AASHTO T 89 and T 90
Armstrong, J. C., and T. M. Petry (1986), "Significance of Specimen Preparation upon Soil Plasticity," *Geotechnical Testing Journal, ASTM*, Vol. 9, No. 3, September, pp. 147–153). This reference contains both historical and recent references and should be a starting point for any user-desired research on this topic.

Objective

To introduce the reader to the test procedures for determining the liquid and plastic limits of a soil. These are sometimes called the "Atterberg Limits."

Equipment[1]

Liquid-limit device with grooving tool (Fig. 3-1) or fall cone equipment (not shown)
Moisture cans
Plastic-limit plate (optional)
Soil-mixing equipment (porcelain dish, spatula, plastic squeeze bottle to add controlled amount of water)
Balance sensitive to 0.01 g for mass determinations
Sieve, pan, and lid (US No. 40, BS No. 36, AFNOR No. 27, or DIN No. 400; see Table 5-1)

General Discussion

The liquid and plastic limits are two of five "limits" proposed (ca. 1911) by A. Atterberg, a Swedish agricultural scientist. These limits are:

1. *Cohesion limit*—that moisture content at which soil crumbs just stick together.
2. *Sticky limit*—that moisture content at which soil just sticks to a metal surface such as a spatula blade. This would have some significance to the agricultural engineer, since it is related to soil sticking to the moldboard of a plow or disc in cultivating soil.

Figure 3-1
Usual liquid- and plastic-limit test equipment. Shown are the following: No. 40 sieve and pan; liquid-limit machine with both the ASTM (formerly "Casagrande") (left) and wedge grooving tools; plastic squeeze bottle for adding controlled amounts of mixing water; porcelain dish and spatula for mixing soil; glass plate for plastic-limit test with optional 3-mm welding rod for visual comparison of soil threads.

[1]Use stock laboratory soil (or soil from Project No. 2) that was seived through a wire screen and stored in a large container such as a 20 gal garbage can. There should be enough soil for a lab section to perform Tests 3, 6, 7, 9, and 14 and have enough for completion of the classification of this soil in Project No. 8.

3. *Shrinkage limit*—that moisture content below which no further soil volume reduction (or shrinkage) occurs. The method of determining this moisture content is presented as Soil Test No. 4.
4. *Plastic limit*—moisture content below which the soil is nonplastic.
5. *Liquid limit*—moisture content below which the soil behaves as a plastic material. At this moisture content, the soil is on the verge of becoming a viscous fluid.

The liquid and plastic limits are used internationally for soil identification and classification and for strength correlations. The shrinkage limit is useful in certain geographical areas where soils undergo large volume changes when going through wet and dry cycles.

The potential for volume change can often be detected from the liquid- and plastic-limit tests. The liquid limit is sometimes used to estimate settlement in consolidation problems (Soil Test No. 13), and both limits may be useful in predicting maximum density in compaction studies (Soil Test No. 9). The two methods of soil classification presented in Project No. 8 incorporate the use of the liquid and plastic limits (see Figs. 8-1 and 8-2).

The cohesion and sticky limits are not generally used in geotechnical engineering work.

The relative locations of the shrinkage limit w_S plastic limit w_P and liquid limit w_L are shown on the water-content scale in Fig. 3-2.

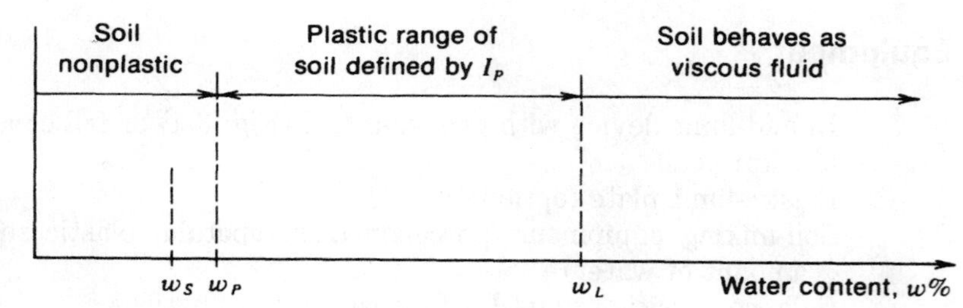

Figure 3-2
Relative locations on a water-content scale of shrinkage, plastic, and liquid limits. Note that the w_S location may vary to the right of w_P for some soils.

In order to place definite, reproducible values on these limits, it was proposed that the *liquid limit be arbitrarily defined* as that water content at which a pat of soil placed in a brass cup (configured as shown in Fig. 3-3), cut with a standard groove, and then dropped from a height of 10 mm will undergo a groove closure of 12.7 mm when the cup of soil is dropped 25 times at the rate of 120 drops/minute. Several variables affect the liquid-limit test, or the number of blows required to close the standard groove 12.7 mm, including:

1. Size of soil pat in brass cup (thickness and quantity)
2. Rate of blows (crank rate should be 120 rpm)—there is a natural operator tendency to reduce the rate as the groove begins to close, causing an increase in N.
3. Length of time soil is in cup before beginning test blow count; and cleanliness of cup before soil is added for the test
4. Laboratory humidity and speed of performing the test
5. Degree of care with which the soil is mixed for the test. Careful mixing may take from 10 to 30 min or more.
6. Type of material used for liquid-limit device base (commonly Micarta or hard rubber)
7. Accuracy of height-of-fall calibration (should be exactly 1 cm)
8. Type of grooving tool (see two types in Fig. 3-3)
9. Condition of liquid-limit device (worn pins, loose connections, etc.)

These variables can all be controlled by the technician doing the tests. Since there are so many, there is a considerable incentive for using the fall cone procedure, described later, for the liquid-limit determination.

To reduce the test variables somewhat, test equipment has been standardized. For example, the Casagrande grooving tool is now required by ASTM. The wedge-shaped

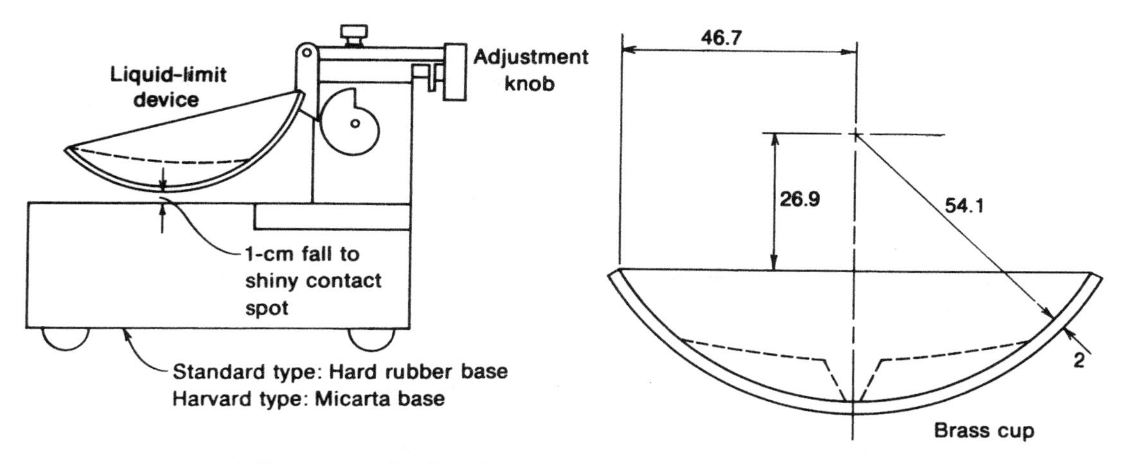

(*a*) Construction details and dimensions of the liquid-limit device.

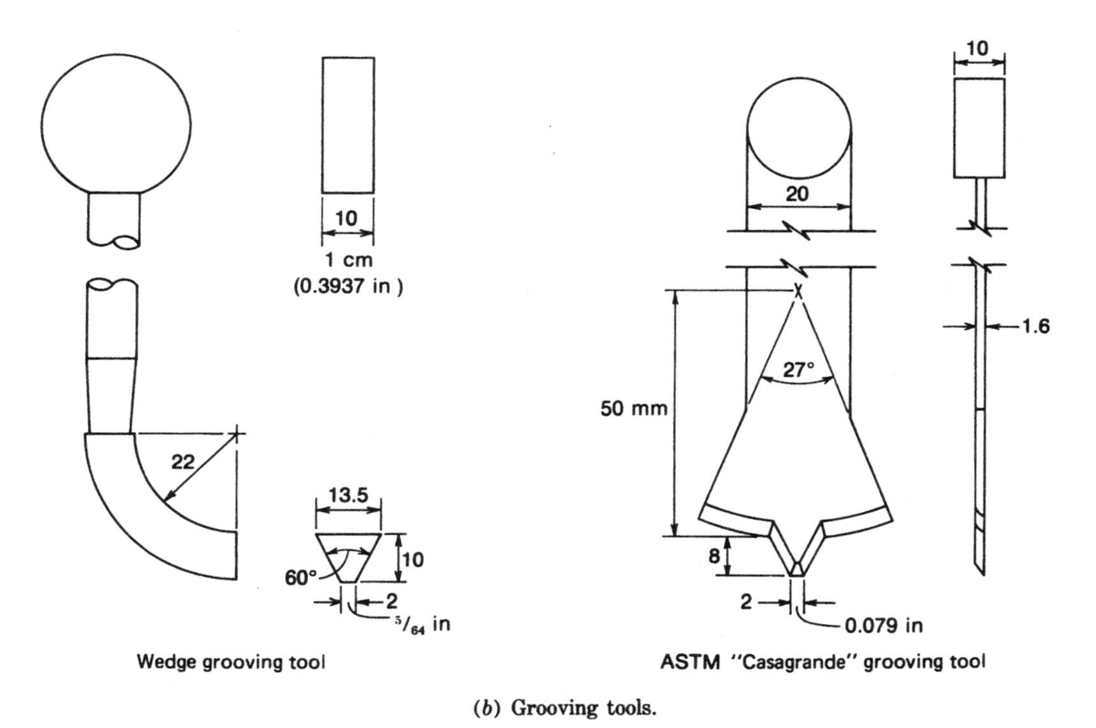

(*b*) Grooving tools.

Figure 3-3
Liquid-limit test equipment. All dimensions are in millimeters unless otherwise indicated.

grooving tool, however, is required by AASHTO and may be preferred in cases where the sample is difficult to groove without tearing the soil pat (sandy or silty with low plasticity). In this case it may be necessary to first use a spatula to trim the groove to the approximate size and shape produced by the grooving tool then finish the groove with one of the tools.

The blow rate is specified at 120 blows/minute. Some operator experience is required to obtain this rate consistently; too fast reduces N and too slow increases N (but not in any quantifiable manner). Motorized liquid-limit machines are available that can accurately reproduce this blow rate.

The ASTM standard requires the use of distilled water in mixing the soil for testing. In most cases the author has found that ordinary tap water is satisfactory.

The liquid and plastic limits of a soil are determined on the soil fraction passing the No. 40 (or equivalent) sieve. In order to obtain this soil fraction, one must choose one of two methods:

1. After minimal-to-no drying, wash the soil through the No. 40 sieve, collect the residue, and use it for testing. This procedure is required by ASTM Method A.
2. Air-dry (never oven-dry) the soil, pulverize it and then sieve it through the No. 40 sieve. This is ASTM Method B and was the only procedure allowed until this standard was revised for the 1984 edition of Vol. 4.08.

Both of these methods have advantages and potential errors. The author has found that air-drying the soil can reduce the liquid limit from 2 to 6 percent unless the soil is remixed with water and allowed to stand (or *cure*) for 24 to 48. If curing is done there is usually no discernible difference between the liquid limits by Method A or B.

Sieving air-dry soil through the No. 40 sieve may not fully recover clay and silt particles that cling as dust to the sides of larger particles. Also, the pulverizing process may not reduce clay lumps that remain on the No. 40 sieve in the form of larger soil particles. Since the clay mineral fraction has a significant influence on the liquid limit, it is essential that all of it be recovered.

In an attempt to recover more of the (−) No. 40 sieve fraction, ASTM Standard D 4318 strongly recommends Method A, which requires washing the soil through the No. 40 sieve. There is one major disadvantage with this procedure: The soil particles are substantially segregated, and a very careful mixing and long re-mix time are required to redistribute the clay minerals through the test sample. Adequate mixing has been achieved when the soil has a distinctive creamy, uniform appearance.

The liquid limit is defined as that water content which produces a standard groove closure of 12.7 mm (1/2 in) for 25 blows (or cup impacts on the standard base) in the liquid-limit machine. It is a coincidence if one can mix the soil to this water content in a reasonable number of trials (also if you retried the same soil from which 25 blows produced the groove closure, you might very well obtain 23 to 27 blows). This variance requires some alternative means to obtain the liquid limit.

It has been found experimentally that plotting blow count N on a semilogarithmic scale versus water content produces a nearly linear plot over one log cycle. Thus, one can obtain the water contents for several reasonably well spaced points from about N = 15 to 35, plot these points, and "best" fit a straight line (the flow curve). Next enter the abscissa (the log scale) and at N = 25 make a projection to the flow curve and then to the ordinate to obtain the water content corresponding to the liquid limit w_L. This procedure is illustrated in Fig. 3-6, where a value of w_L is obtained.

In practice, the total time for the testing using either method will usually be about the same. The principal difference is the way you prepare the soil sample. The preparation steps are as follows:

METHOD A (method favored by ASTM)

a. First wash the soil into a large dish (and then decant as much excess water as practical).
b. Let the sample air-dry[2] down to the first test point (about 30 to 35 blows or fall cone penetration of 10 to 15 mm).
c. Carefully mix the test sample and start testing.

[2]You may speed this process using a hand held hair dryer. Use care that you do not get the soil over about 40–45°C and keep the dryer moving while continually remixing the soil.

Figure 3-4
Steps in the liquid-limit test.

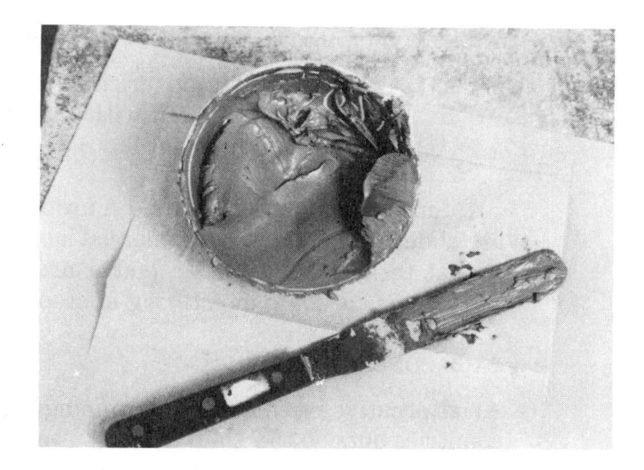

(*a*) Carefully blended soil. Note uniform, creamy texture.

(*b*) Soil placed in liquid-limit cup for test. Note that cup is not nearly full and that only front portion is used.

(*c*) Part of soil is grooved with ASTM grooving tool. Depth is such that tool has only barely trimmed soil pat at deepest part when held perpendicular to cup.

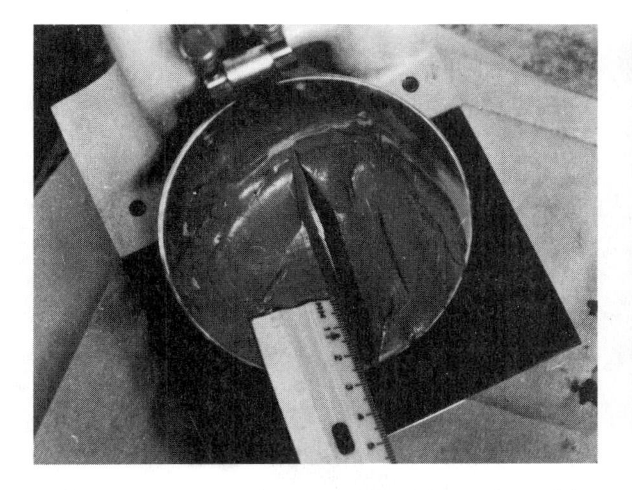

(*d*) Soil groove closure of 12.7 mm (visual, but scale used for numerical comparison is photograph) while rotating crank at 120 rpm and counting drops of cup onto base of liquid limit machine.

(*e*) Water-content sample recovered from location of groove closure. Quantity taken is about 35 to 45 g.

METHOD B (method best suited for student laboratories)

a. Air-dry the soil, pulverize it, and sieve it through the No. 40 sieve.
b. Carefully mix the soil with sufficient water to obtain your first test point as in part b of Method A (about 30 to 35 blows).

Since times $a + b$ of Method A are about equal to time a of Method B, the test economics are about the same for a commercial laboratory. The major difference in student laboratories is that tasks a and b of Method A are required in advance of the laboratory period, and there must be some monitoring of step b so the soil does not become too dry.

FALL CONE

An alternative method for determining the liquid limit that has been standardized in several countries outside the United States uses a fall cone (or cone penetrometer). This test uses a standard cone with a 30° apex angle and a total mass of 80 g. A dish (55 mm diam by 40 mm deep) is filled with soil in the consistency range of the Atterberg limit test and struck flush. The cone point is carefully brought into contact with the soil. The cone is then released so the 80-g mass produces some penetration for 5 s.

The liquid limit is defined as that water content at which 20 mm of cone penetration occurs in the 5 s test time. Since having the soil at that water content would be a coincidence, a semilog plot is made of penetration D (instead of blow count N) versus water content, and a linear curve is drawn. The curve is entered on the log scale at D = 20 mm, and the corresponding water content is obtained. This is illustrated with the dashed curve on Fig. 3-6 labeled "Fall Cone".

Some authorities are of the opinion that the fall cone liquid limit is more reproducible (or repeatable) than that obtained using the liquid-limit machine. The principal item requiring attention is that the point of the cone not become blunted from excess use, in which case a new cone must be obtained.

ACTIVITY

In geotechnical work the term *activity* is used to indicate:

1. The percentage of clay in the (−) No. 40 sieve fraction of soil as used for the Atterberg limits, and
2. The potential swell and shrinkage (or volume change) of a soil, with larger values indicating an increasing potential.

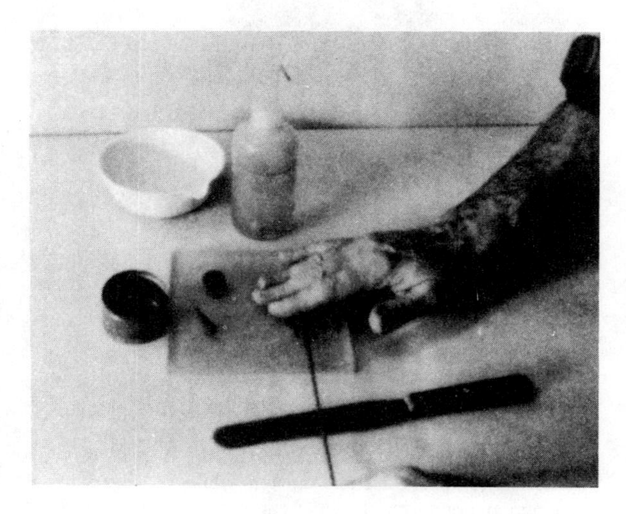

Figure 3-5
Plastic-limit test. Soil is set aside when blow count is around 50 blows. Use 3-mm diameter rod for visual comparison with thread diameter. Soil on side of plate is for additional trials.

Activity of a clayey soil is defined as:

$$A = \frac{\text{Plasticity Index } I_p}{PC - C_0} \tag{3-1}$$

where $I_P =$ plasticity index ($w_L - w_P$) shown in Fig. 3-2.

$PC =$ percent clay (the soil particles $\leq$ 0.002 mm).

$C_0 =$ constant ranging from 0 to 10; most persons use 0 [see the Seed et al. (1962) reference for the Shrinkage Limit Test No. 4].

The smaller the value of A in Eq. (3-1), the smaller the potential volume change of the soil. The approximate range of A is from 0.3 to about 5.5 and depends on the type of clay minerals present; for example, predominately kaolinite clay minerals has a lower value of A than either illite or montmorillonite minerals.

Research has found that the liquid limit w_L increases as the average size of the particles in the minus No. 40 sieve fraction decreases. This implies that the liquid limit increases with an increase in PC in Eq. (3-1).

FLOW INDEX

It has been established that both w_L and w_P of a soil are dependent on the percentage of clay (PC) in the (−) No. 40 sieve fraction, so the following linear equation can be written for the liquid limit:

$$W_L = K(PC - C_1) \tag{3-2}$$

where K and C_1 are constants to be determined. Solving Eqs. (3-1) and (3-2) for PC, redefining $A/K = M$ and $K(C_0 - C_1) = C_2$ we can obtain an expression for I_P as

$$I_P = M(w_L - C_2) \tag{3-3}$$

Casagrande's classification diagram (in Fig. 8-2) uses $M = 0.73$ and $C_2 = 20$.

The equation of the linear semilogarithmic plot of N vs. water content is an equation of the general form

$$w = -F_i \log N \pm C_3 \tag{3-4}$$

where $F_i =$ flow index, which is negative since the log N versus w_L line slopes down from left to right

$C_3 =$ constant to be determined

$N =$ blow count at corresponding water content w

and by analogy for the fall cone:

$$w = F_i' \log D \pm C_3 \tag{3-5}$$

The F-value here is positive since the fall cone plot of D versus $\log w$ slopes upward from left to right.

In practice one obtains the flow index from the appropriate plot of test data as

$$F_i = -\frac{\Delta w}{\log N_2/N_1} \tag{3-6a}$$

and for the Fall Cone

$$F_i = \frac{\Delta w}{\log D_2/D_1} \tag{3-6b}$$

with Δw normally taken over one log cycle (extend the plot if necessary) so that the denominator in the above equations is log 100/10 = 1.0 as a computation convenience.

ONE-POINT LIQUID LIMIT METHOD

Based on regression analysis of 767 soils, the U.S. Army Corps of Engineers Waterways Experiment Station, Vicksburg, Mississippi, in Technical Memorandum No. 3-286, June 1949, concluded that the liquid limit could be estimated from a single-point test using the following equation:

$$w_L = w_N \left(\frac{N}{25}\right)^{\tan\beta} \tag{3-7}$$

where w_N = water content at the blow count N of the single test

β = slope of the semilog plot of w vs. $\log N$

For all of these soils a best approximation for the exponent was $\tan\beta = 0.121$, resulting in

$$w_L = w_N \left(\frac{N}{25}\right)^{0.121} \tag{3-8}$$

The value of $\tan\beta$ is not 0.121 for all soils; however, negligible error is obtained using this single value if N is between 20 and 30 (i.e., the closer to 25, the smaller the error).

PLASTIC LIMIT

The plastic limit has been arbitrarily defined as that water content at which a soil thread just crumbles when it is rolled down to a diameter of 3 mm. The *original* requirement was a thread diameter of 1/8 in. Note that 1/8 in implies some rounding but an exact soft conversion to SI is 3.2 mm (as given in the ASTM D 4318 standard). At best the diameter is "eyed" to size by the operator. Since 3 mm is a rather small diameter, Fig. 3-5 illustrates using a 1/8 in diameter welding rod for making a visual thread comparison.

This test is somewhat more operator-dependent than the liquid-limit test, since what constitutes crumbling and a visual detection of a 3-mm diameter are subject to some interpretation (thus 3 mm is adequate instead of the 3.2 mm given by ASTM). With some practice, it appears that the plastic-limit values can be reproduced within 1 to 3 percent by different laboratory technicians on the same soil.

Procedure

This is suggested as an individual project.

A. LIQUID LIMIT DETERMINATION (Refer to Fig.3-4)

1. Pulverize a sufficient quantity of air-dry soil (from the 5 kg put out for drying the previous week) to obtain a 250 ± 10 g of representative (−) No. 40 sieve material. Discard the (+) No. 40 sieve residue (do not put back in stock container). Be sure to break down all soil lumps to elemental particles.

Alternatively: Prior to this laboratory period, wash sufficient soil on the No. 40 sieve to produce about 250 g (dry weight) of representative soil. Decant the excess water and air-dry the soil.[3]

Monitor the drying process and stir the soil sample occasionally so that the surface does not crust over. When the soil is dry enough that it can be molded (pinch a groove) with the fingers, cover it with a damp towel or place it in an airtight container until the laboratory period.

Mix the prepared sample with a small amount of water in a porcelain (also called an evaporating) dish of about 250 ml capacity and set aside while you weigh your moisture cans and adjust the liquid limit machine (or inspect the fall cone device—especially the point). You will need at least five moisture cans and probably should weigh at least six.

2. Next, check the height of fall of the liquid-limit machine and also check for excessive play in the hinge assembly. Adjust the height of fall to exactly 10 mm, using the 10 mm block on the end of your grooving tool. Measure the fall from the worn spot on the base of the cup to the worn space on the impact base. The fall is set accurately when there is no discernible movement or click when the 10 mm block is in place and the crank is rotated. Height of fall is very critical, and as little as 1 mm can affect the water content by several percent. In fact, fall height is one of the largest sources of error in this test.
3. Now return to your soil sample in the dish that you earlier mixed with a small amount of water and set aside. Add more water in small amounts and thoroughly mix to a uniform color and consistency (no lumps). A major source of error is poor mixing. When the consistency is uniform throughout and the surface appearance is creamy, a proper blending of soil and water has generally been achieved (this may take 15 to 20 minutes).

   When the soil consistency (stickiness) is where you estimate a trial test would produce about 50 blows to close the groove 12.7 mm (you may make an actual test but do not take a water content sample) set aside about 20 to 30 g of soil on the plastic limit plate. This will be used later for the plastic limit test.

   Now add additional water and continue the careful mixing. When you have the consistency to between 30 and 35 blows (D = 10 to 15 mm) as estimated by inspection (or trial tests), you are ready to make your first test of record.
4. Remove the *clean* brass cup from the liquid-limit machine, and place a small amount of soil to the correct depth of the grooving tool and centered with respect to the hinge (refer to Fig. 3-4*b* and *c*). Smooth the surface of the pat, and carefully cut a groove using one of your grooving tools. If you use the ASTM tool, be sure the tool is perpendicular to the cup at all points.

   The tool shoulders will also trim the sample to the correct depth. Be very careful (and use Fig. 3-4*b* as a guide) not to put too much soil into the brass cup (you cannot overfill the fall cone cup since you use a straight edge to strike it flush). If the brass cup contains too much soil the extra mass will reduce the blow count.

   If you trim more than about 2 mm of soil at the deepest point, remove the soil and start over. If the soil does not touch the tool shoulders at any point, it is not deep enough. Remove the soil, clean the cup and make a new trial. If you use the wedge tool, observe the depth, and if the soil depth is below the tool at the deepest point or above by about 2 mm, start over.

   Now quickly place the cup into the liquid-limit machine, securely fasten the hinge points, and perform the test by rotating the crank at approximately 120 rpm. Observe the groove, and when it has closed 12.7 mm (visual observation) record the number of blows.

[3]Cf. Footnote 2, page 24, on using a portable hair dryer.

Do not delay the test at this point because the soil will adhere to the brass cup and/or the surface will dry. *Do not slow the rate of crank rotation as the soil appears to develop the required closure for this will increase the N value—a source of error.*

If you use the fall cone, fill the sample cup just to overflowing and, using a steel straightedge, strike the soil flush with the cup rim. Quickly place the cup beneath the cone tip, bring the point just into contact with the soil, initialize timing, release the cone and record the depth of penetration D at the end of 5 s.

5. Take a moisture content sample of about 40 g from the closed part of the soil pat in the brass cup (or where the fall cone penetrated the soil in the dish) using a spatula. Place this soil into your pre-weighed moisture can, put the lid on, and temporarily set it to one side for later weighing. If the humidity in the laboratory is low (seems very dry), cover the can with a damp paper towel.

   Remove the remainder of soil from the test cup and add it to the remaining soil in your mixing dish. Wash and dry the test cup—never use a cup that is contaminated from a previous test trial.
6. Next add additional water in small amounts, using the squeeze bottle, to the soil in the mixing dish to reduce the consistency, and mix thoroughly, usually about 3 to 5 minutes. When you have a consistency to produce an N in the range of 25 to 30 blows (D in a range of 15 to 20 mm), repeat Steps 4 and 5.

   Be sure to wash and dry the brass cup or fall cone cup before each test.
7. Repeat Steps 4 to 6 for two additional tests for N between 20 and 25 and between 15 and 20 (D between 20 and 25 and between 25 and 30). If the blow counts or penetrations are in this range, you should have a reasonable spread of points around and close to the critical value for the most accurate determination of w_L.
8. Weigh the four (or five) moisture containers and record their "container + wet soil" masses on your Data Sheet 3 from back of this manual on the appropriate line (see Fig. 3.6). *Now remove the lids* and place the containers in the oven (set at 110°C) to dry overnight.

   It should be evident that adding water to the soil and doing the blow count or fall cone penetration sequence in this manner will give the operator some control over the tests:

   a. By adding water the technician should anticipate that N will decrease (or D will increase) If this does not happen something is being done incorrectly, not enough water was added, or there was too much time delay as well as insufficient added water for the next test point.
   b. Note that it is much easier obtain a well blended soil-water mixture for the next test point when you add water than when you add dry soil. Further, when you add dry soil there is little to suggest where in the testing range the next test point lies until you make the test.

   Of course, if due to inexperience you get your second point in the range of 10 to 15 blows save the test; but you will have to add soil (or hand dry) the mixture to get your intermediate points.

B. PLASTIC LIMIT DETERMINATION

On completion of the liquid limit test perform the plastic limit test as follows:

1. Break the 20- to 30-g "peanut" of soil set aside in Step 3 above into several smaller samples.
2. Roll one of these smaller samples between the fingers, on a glass plate or on a piece of paper lying on a smooth surface. Use sufficient pressure to form a thread of uniform diameter using about 80 to 90 rolling strokes per minute (forward and back = 1 stroke) (see Fig. 3-5). If the soil thread diameter becomes 3 mm (1/8 in) and is not crumbling, break this thread into several smaller pieces, re-form into a ball and repeat the process.

Continue this rolling and reballing until you roll the thread down to 3 mm and it is in a "just crumbling" state. Put this soil into one of your weighed cans and put the lid on. Store under a damp paper towel if necessary.

If the thread crumbles at a diameter > 3 mm, this is satisfactory to define w_P if the previous thread was rolled to 3 mm. Thread failure is said to occur when:

a. The thread simply falls apart, or
b. An outer tubular (hollow roll) layer splits from the ends inward, or
c. The thread forms barrel-shaped solid pieces 6 to 8 mm long (for heavy clays).

Do not produce a failure by allowing thread to reach 3 mm as a result of reducing the rate of rolling and/or hand pressure. As an exception, with soils approaching non-plastic, the initial ball may be shaped closer to 3 mm by hand prior to rolling.

3. Repeat this sequence several times, adding each test to the same moisture can.[4]
4. Weigh the covered can, *remove the lid* and place the can in the oven with your liquid limit cans. Be sure to record your can number (or other identification mark) and the "Mass of wet soil + can" in the appropriate space on your data sheet.

Computations

1. Return to the laboratory on the following day, weigh all the oven-dried samples. Record these data on your data sheet and compute the several water contents. Refer to the Example in the *Introduction* if you do not know how to do this.

   Plot the liquid-limit data on the small graph[5] contained on the data sheet from the data sheet section of this manual. Use as large a vertical scale for w as possible as in Fig. 3-6. Note that you can shift the origin of the ordinate to obtain a best scale.

   Draw the flow curve as a best fit using the several points.

   Obtain the liquid limit w_L by entering the abscissa at $N = 25$, project vertically to the flow curve and then horizontally to the water content axis. The latter intercept is the liquid limit. Use a scale that allows you to read w_L to 0.1%.

   Compute the flow index F_i and show this on your data sheet (as shown on Fig. 3-6). Also compute the plastic limit and the plasticity index I_P as

$$I_P = w_L - w_P \tag{3-9}$$

   Show this computation on your data sheet in the space provided.
2. If you used the liquid-limit machine, compute the liquid limit for each value of N and w_N using Eq. (3-8); tabulate in your Discussion, and compare with w_L from the semilog plot.
3. Compute the activity of the soil using the percent finer from the hydrometer analysis for grain size diameter = 0.002 mm if the Hydrometer Test (Test No. 5) is done in parallel with this test. If you use Eq. (3-1) for the activity A take $C_0 = 0$.
4. In your report Discussion, give six uses for the w_L and w_P data you have just obtained.

[4]ASTM (1984 and later) requires that this sequence be done to obtain two cups of plastic-limit threads. This effectively does the test 6 to 10 times. For student laboratories one cup that contains the results of 3 or 4 trials is adequate to illustrate the procedure. Prior to 1984 the standard procedure was to put each test into a separate can. The author suggested in the first edition of this manual (1970) combining tests in a single cup to minimize weighing errors since a single trial may only be 2 or 3 g of soil.

[5]This graph is adequate for the accuracy inherent in this test. You should make the vertical scale as large as possible so that you can estimate w_L to the nearest 0.1 percent. Do not use a large semi-log graph sheet in an attempt to increase the accuracy you think might be obtained in reading the w_L water content.

ATTERBERG LIMITS DETERMINATION Data Sheet 3

Project *Exp. No. 3* Job No. ~

Location of Project *Soil Laboratory* Boring No. ~ Sample No. ~

Description of Soil *Brown silty Clay*

Depth of Sample ~ Tested By *JEB* Date *5/4/19—*

Liquid Limit Determination

Can no.	27	28	31	34	
Mass of wet soil + can	48.61	55.53	51.71	50.51	
Mass of dry soil + can	41.19	46.05	42.98	41.54	
Mass of can	17.33	17.41	17.45	17.36	
Mass of dry soil	23.86	28.64	25.53	24.18	
Mass of moisture	7.42	9.48	8.73	8.97	
Water content, w%	31.1	33.1	34.2	37.1	
No. of blows N	34	27	22	17	
Penetration D, mm					

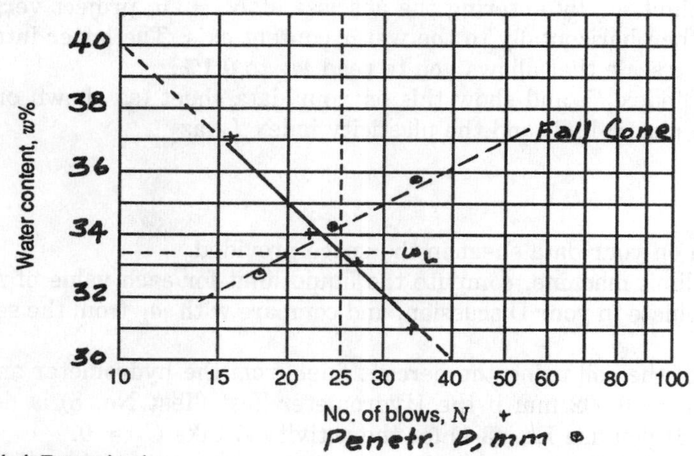

Flow index F_i = *−16.6*

Liquid limit = *33.4 %*

Plastic limit = *18.9 %*

Plasticity index I_P = *14.5 %*

$$F_i = \frac{\Delta w}{\log N_2/N_1} = \frac{30-40}{\log 40/10} = -16.6$$

Plastic Limit Determination

Can no.	35	37			
Mass of wet soil + can	29.26	30.03			
Mass of dry soil + can	27.90	28.53	Av. $w_P = \frac{18.7+19.1}{2} = 18.9$ %		
Mass of can	20.63	20.66			
Mass of dry soil	7.27	7.87			
Mass of moisture	1.36	1.50			
Water content, $w\% = w_P$	18.7	19.1			

Figure 3-6
A set of liquid- and plastic-limit test data.

4: SHRINKAGE LIMIT

References

ASTM D 427 (Mercury method)
ASTM D 4943 (Wax method)
AASHTO T 92
Chen, F. H. (1975), *Foundations on Expansive Soils*, Elsevier, Amsterdam, 280 pages.
Holtz, W. G., and H. J. Gibbs (1956), Engineering Properties of Expansive Clays, *Transactions, ASCE*, Vol. 121, pp. 641–677.
Seed, H. B., R. J. Woodward, Jr., and R. Lundgren (1962), Prediction of Swelling Potential of Compacted Clays, *J. Soil Mech. Found. Div., ASCE*, SM 3, June, pp. 53–87.
See also Proceedings of International Conferences on Expansive Soils, e.g., 1st (1965) and 2nd (1969) were held at Texas A & M University, the 4th (1980) in Denver, Colorado, and so forth.

Objective

To determine the moisture content below which no further volume change of a soil mass occurs and to obtain a quantitative indication of the amount of volume change that can occur in a cohesive soil.

Equipment

Petroleum jelly, silicone grease, or aerosol container of Teflon™ powder
Shrinkage dish of porcelain or monel metal (refer to Fig. 4-1)—monel metal is non-reactive with mercury and is preferred to reduce breakage
Balance sensitive to 0.1 g
Circular ground glass strike-off plate
Small funnel, glass or plastic

and if the mercury method is used

Glass or plastic plate with three monel metal prongs
Volume dish (called a "glass cup" or "crystallizing dish" in lab suppliers' catalogs)
Mercury metal—**this is a hazardous material**
Cardboard overflow container[1]

or if the wax method is used

Graduated beaker of about 200–400 mL capacity
Wax (paraffin may be suitable)
Hot plate and small metal pan or melting pot to melt paraffin
Glass graduated cylinder accurate to 1 mL
Small pipet to contain ≤ 10 mL
Thermometer to monitor water temperature

[1]A suitable container can be made using part of a file folder with the four edges bent up 10 to 20 mm and fastened with masking tape as shown in Fig. 4-1 and Fig. 4-3*b*.

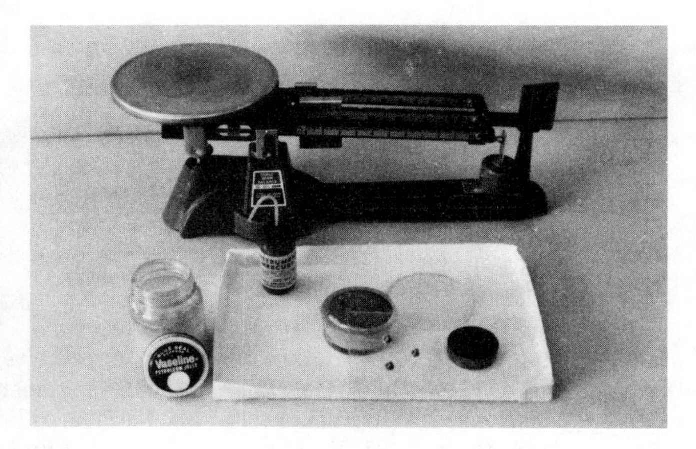

Figure 4-1
Shrinkage-limit equipment. Balance sensitive to 0.1 g (0.01 by estimation); monel metal shrinkage dish; glass volumetric dish (filled with mercury); circular ground glass plate to strike off mercury in glass volumetric dish; glass plate with three monel-metal prongs used to submerge dry soil cake into volumetric dish; container of extra mercury metal; petroleum jelly to lubricate shrinkage dish so soil cake does not stick during drying. Use a handmade shallow cardboard tray (as shown) to contain any mercury spill. Make tray of sufficient size that measurements can be easily made. Mercury is not used in wax method.

General Discussion

See Test 3: Liquid and Plastic Limits of a Soil.

Any soil that undergoes a volume change (expands or contracts) with change in water content may be troublesome

a. if used for highway or railroad fills—produces a bumpy road.
b. if a structural foundation is placed on it—produces uneven floors and/or structural cracks.
c. if used as a backfill behind a retaining wall—produces excessive thrust against the wall, which may cause it to fail.

Volume expansion and/or contraction occurs over a period of time and depends both on soil type (e.g., clay minerals) and changes in water content from the reference value. The reference water content is that at the time of construction. Most of the damage with time is from differential water contents which produce different amounts of volume change.

Soil shrinkage (or contraction) is produced by what is called *soil suction.* This is the cumulative effect of water surface tensions on the soil grains making up the mass. Suction is the phenomenon which produces a capillary rise of water in soil pores above the water table. A 1000 mm height of capillary water produces a gauge pressure of −9.807 kPa; in other words; the tension in the water at the top of the capillary column on those soil grains is 9.807 kPa, which pulls the grains together. Of course at the surface of the water table the gauge pressure is 0 (or atmospheric). Similarly when soil dries and the pores start to empty of water, surface tension (soil suction) develops which pulls the grains ever closer together to some limiting value called the shrinkage limit.

The liquid and plastic limits may be used to predict potential for volume change. However, to obtain a quantitative indication of how much volume change can occur and the amount of moisture necessary to initiate volume change a *shrinkage limit test* is sometimes performed. This test begins with a given volume of *fully saturated cohesive soil*, preferably (but this is not absolutely necessary) at a water content above the liquid limit. This volume of saturated soil is then carefully dried.

It is assumed during drying that down to a certain limiting value of water content, any loss of water is accompanied by a corresponding change in bulk volume (or void ratio). Below this limiting value of water content called the *shrinkage limit* w_S, no further change in volume occurs with loss of pore water. This is graphically illustrated in Fig. 4-2. For example, if $w_S = 8\%$ then a water content ranging from $w = 0$ (dry) to $w = 8\%$ produces no volume change; if $w = 9\%$ volume change occurs and if $w = 12\%$ even more volume change would be expected. From this it would appear that the preferred soil would have a large shrinkage limit w_S.

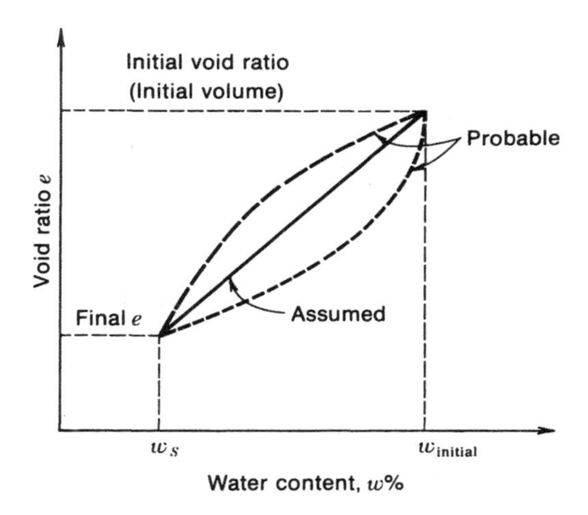

Figure 4-2
Qualitative plot of water content vs. void ratio. Dashed lines indicate possible nonlinear shrinkage paths compared with linear assumption used in derivation of equations.

The *shrinkage ratio* SR gives an indication of how much volume change may occur with changes in water content. The shrinkage ratio is defined by the following equation:

$$\text{SR} = \frac{\Delta V/V_f}{\Delta_w/M_s} = \frac{M_s}{\rho_w V_f} \tag{4-1}$$

since the volume change $\Delta V = \Delta w/\rho_w$. Symbols in Eq. (4-1) and the following equations are defined as follows:

V_f = dry volume of soil skeleton (pores + solids)
V_o = initial volume of wet soil at a water content w_o
ΔV = change in soil volume from V_o to V_f
w_o = initial water content of soil (S = 100% for test)
w_S = shrinkage limit water content
M_s = mass of dry soil skeleton of volume V_f
ρ_w = unit density of water in consistent units (generally 1 g/cm^3)
Δ_w = change in water content from initial w_o to final $w_f (w_o - w_f)$

The *volumetric shrinkage* S_v is defined as

$$S_v = (w_o - w_S)\text{SR} \tag{4-2}$$

Linear shrinkage can be computed from the volumetric shrinkage as in the following equation.

$$\text{LS} = \left[1 - \left(\frac{1.0}{S_v + 1.0}\right)^{1/3}\right] \times 100 \qquad \text{(percent)} \tag{4-3}$$

The volumetric shrinkage S_v is used as a decimal quantity in Eq. (4-3). This equation assumes that the reduction in soil volume is both linear and uniform in all three directions.

Linear shrinkage can be directly determined by a test; however, this test has not yet been standardized in the United States. The British standard BS 1377 uses a half-cylinder mold of diam = 12.5 mm × 140 mm length L_o. The wet sample is dried and the final length L_f is obtained; from this the linear shrinkage LS is computed as

$$\text{LS} = \frac{L_0 - L_f}{L_0} \tag{4-4}$$

Procedure

This is suggested as a group project.

NOTE: AVOID DIRECT SKIN CONTACT WITH MERCURY, AS IT IS A HAZARDOUS MATERIAL.

For both the mercury and wax method volume determination you must first find the volume of the shrinkage dish (monel metal or porcelain). Be sure to number or otherwise identify the dish, record this on your data sheet, then proceed as follows:

1. Completely grease the inside of the shrinkage dish very lightly with a coat of petroleum jelly, silicone grease or spray with Teflon™ powder (to keep the soil from sticking to it) and determine its mass as M_d. Next:

   By Mercury Method: Put the shrinkage dish into your cardboard overflow container (if you have one). Fill the dish to just overlowing with mercury from your plastic squeeze bottle supply source. Next take the ground glass plate and strike the mercury flush with the dish. Now determine the mass of dish + mercury as M_{dm} and record on your data sheet. Now empty the mercury back into its storage container using the small funnel. The volume of the dish is computed as

   $$V_d = (M_{dm} - M_d)/\rho_m = \Delta M/13.53 = V_o \qquad (\text{cm}^3)$$

   since the density of mercury is approximately 13.53 g/cm³.

   By Wax Method: Fill the greased shrinkage dish exactly full with room-temperature water and put in a weighed graduated cylinder using the small funnel. It may be necessary to use either a wetting agent or a glass plate to strike the dish level. Do this between 5 and 10 times for N_t. Observe after each trial that the increase in accumulated mass in the graduated cylinder is approximately equal. Read the final volume of water in cylinder as V_{fc}. Determine the mass of water in cylinder as

   $$M_{cw} = \text{Final mass} - \text{mass of cylinder}$$

   Determine the volume of shrinkage dish as follows:

   $$V_1 = V_{fc}/N_t$$
   $$V_2 = M_{cw}/(\rho_w N_t)$$

   Take the volume of the shrinkage dish as the average of the two dish volumes just computed, or

   $$V_d = (V_1 + V_2)/2$$

   If there is a large difference between V_1 and V_2 you should repeat the volume determination (but save this data in case one of the two values is correct—if so include it as either V_1 or V_2 in the above computation.

   Now for either method:

2. Again determine the mass of the greased (or regreased if using wax) shrinkage dish as M_d and record on Data Sheet 4 (lines c and h) along with its volume (and computations) which you have just found as V_d. Note this volume is also the initial volume of wet soil. If the mass differs from step 1 check for the reason. If necessary regrease or remove excess grease or other contamination.

3. Now take about 40 g of *representative* material passing the No. 40 sieve[2] used for the liquid- and plastic-limit tests, and carefully mix with distilled (or tap) water to make a creamy paste that can be placed in the shrinkage dish without any air voids. The soil consistency should be obtained by using a water content estimated to be somewhat above the liquid limit. Do not make a watery slurry.
4. Next put a layer of soil in the shrinkage dish, filling it approximately one-third full. Tap the dish gently on a firm base padded with two or three paper towels until the soil settles and no apparent air bubbles exist (Fig. 4-3). Then repeat with second and third layers, allowing the soil to flow over the top. Strike the dish off smooth using a medium-size spatula, and determine the mass of the coated dish + wet soil. Record this on your data sheet on line labeled (a).
5. Next air-dry the wet soil cake in the laboratory until the surface changes to a light color (say, 6 to 8 h); this step is taken to reduce the possibility of the soil cracking apart on drying. Then oven-dry the air-dried dish of soil at 105 to 110°C to constant weight (10+ h). Remove the dish of dried soil from the oven, and obtain the mass of the coated dish + dry soil. Record this on the line labeled (b) of your data sheet.

Figure 4-3
Shrinkage limit test.
(*a*) Filling the shrinkage dish by rapping against table to remove air bubbles. An already oven-dried soil cake is in shrinkage dish on the left.

(*b*) Preparing to determine volume of dried soil cake by using the three-pronged glass plate to submerge cake into glass volumetric cup filled with mercury. The soil cake floats on the surface until forced under. Use the shallow cardboard tray to catch mercury overflow. Use balance to obtain before and after masses of glass volumetric cup to find mass of mercury displaced by soil cake on submergence.

[2] See Table 5-1 for standard sieve sizes and designations in countries outside the United States.

6. Now compute the mass of water M_w in the soil cake and the mass of dry soil M_s remaining. The mass of water, line (e), is directly obtained as

$$M_w = \text{line (a)} - \text{line (b)}$$

The mass of the greased cup is in both lines (a) and (b) and cancels. The mass of dry soil [line (d)] is computed as

$$M_s = \text{line (b)} - \text{line (c)}$$

Typical computations are illustrated on Fig. 4-5 where you can verify that the mass of water and mass of dry soil are obtained in the manner just outlined. The data sheet is specifically organized so these computations can be easily made.
   The initial water content (line (f)) is

$$w_o = M_w/M_s \times 100 \qquad \text{(percent)}$$

7. *The next step is to find the volume of the dry soil cake V_f. The method depends on whether you use mercury or wax.* If the cake came apart during drying see "Special Note" at end of *Procedure.*

   Mercury Method:

   a. Put the volume cup onto the cardboard overflow box and fill it to overflowing with mercury. Take the plate with the three prongs and with the prongs down strike the shrinkage dish exactly flush. Be careful not to trap air between the plate and mercury surface. Put the excess mercury from the cardboard container back into the flask using the small funnel. Now determine the mass of the volume cup + mercury as M_{vcm} and record on your data sheet. If your mercury flask has a tube as illustrated in Fig. 4-1 siphon off about one-half the volume of the soil cake (Step 6, above) by compressing the flask, sticking the tube into the mercury and releasing so mercury is drawn back into the flask.
   b. Next remove the soil cake from the shrinkage dish and carefully place it on top of the mercury (it will float) in the volume dish (see Fig. 4-3*b*). Take the plate with the three prongs and use the prongs to submerge the soil cake. Tilt the cake slightly while submerging so no air is trapped beneath it. Push the cake down (Fig. 4-4) until the plate is flush with the top of the shrinkage dish and no air is visibly trapped. Use care not to squirt or splash the mercury so the overflow is contained in only in the cardboard box. Carefully remove the plate and soil cake. Put the soil cake to one side (in case you made some kind of error and have to redo part of the test).
   c. Now weigh the volume cup + remaining mercury as M_{vcms}. Compute the cake volume V_f as

$$V_f = (M_{vcm} - M_{vcms})/13.53 \qquad (\text{cm}^3)$$

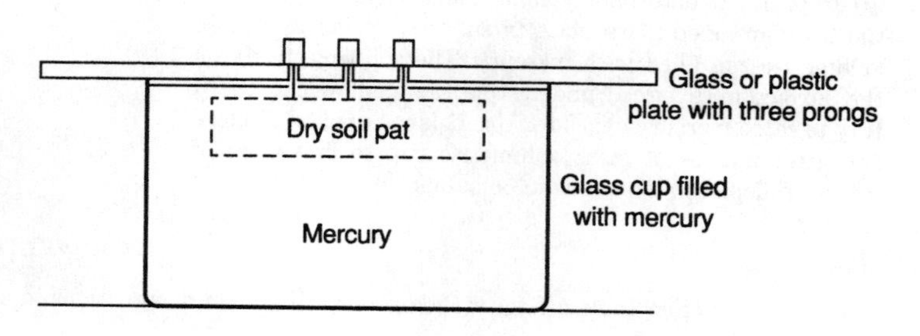

Figure 4-4
Dry soil cake submerged in mercury to find cake volume. Even when using the cardboard tray to catch the overflow, you should siphon some of the mercury from the volumetric cup after filling and weighing so that there is not a large quantity of loose mercury to recover.

Wax Method:

a. Take the soil cake from its shrinkage dish and securely attach threads to it so you can dip it into the melted wax. This produces a waterproof coat so you can then determine the volume using water displacement. Check the coat for air bubbles and if any are visible cut them and then re-dip the cake. Trim the excess thread as much as possible so a minimum of thread volume is included. Do not use wax at a temperature much above its melting point (50 to 55°C) as it will take too long for it to harden on your cake and may crack in the process—but more importantly if it is very hot, wax may spontaneously ignite.

   Alternatively take tongs and dip part of the cake into the wax, let it cool, rotate and again dip to coat the remainder of the cake. Check that there are no air bubbles—especially where the coats overlap.

b. Now obtain the mass of the soil cake + wax coat as M_{scw}.

c. Next fill the graduated beaker to one of the volume marks with water at room temperature and determine its mass. Carefully check that the water surface (not the meniscus) is at the volume mark. Call this mass M_{bw}.

d. Now pour off some of the water into your small graduated cylinder (about the estimated volume of the soil cake). Next carefully place the wax coated soil cake into the beaker and add back water from the small graduated cylinder to the volume mark. Use the pipet to draw off any excess water as necessary. Dry any excess water from the beaker and above the water level using a paper towel. Now weigh the beaker and obtain the mass of water + soil cake and call this M_{bws}.

e. Now directly compute the soil cake volume as follows:

$$\text{Volume of wax on soil cake } V_{wax} = (M_{scw} - M_s)/\rho_{wax}$$

where ρ_{wax} = density of wax[3] in g/cm^3.

M_s = cake mass, g (see "Computations" step 6 above)

Compute the cake volume using Archimedes' principle that a body displaces a mass that equals its volume when submerged in water (this is same procedure as used in the Specific Gravity Test No. 7). This method gives the total cake volume as

$$\text{Total soil cake volume } V_t = (M_{bw} + M_s - M_{bws})/\rho_w$$

but $\rho_w = 1$ so the total volume of soil cake + wax is simply the mass difference as shown. The cake volume is the total volume reduced by the volume of the wax coat, giving

$$\text{Volume of the soil cake } V_f = V_t - V_{wax}$$

Put as many of these computations on your data sheet in the spaces provided as possible.

Special Note. If the soil cake cracks apart on drying, the test should be redone. At the discretion of the instructor, however, one can estimate V_f by first weighing the entire amount of dry soil as M_t then determine the volume of the largest piece as outlined above. Next by proportion obtain

$$\frac{V_f}{V_{\text{piece}}} = \frac{\text{total dry mass}}{\text{mass of largest piece}} \tag{4-5}$$

[3]Obtain wax density from supplier. If a value is not available cut a block from the wax cake of about 5 × 5 cm and machine the sides to exact dimensions. Also check that thickness is constant. Compute wax density by computing wax block volume and weighing the block to determine the mass.

Computations

The shrinkage limit is defined as the water content below which no further volume change occurs. We will use this definition to compute the shrinkage limit w_S as follows:

1. The volume at which no further volume change occurs is the final volume of the test cake, V_f.
2. The initial volume of the test cake is the volume of the shrinkage dish V_d and we now have a value.
3. The difference in volumes $V_d - V_f$ represents water lost (a change in water content) as the volume of soil decreased to the shrinkage limit.
4. The initial water content w_o and the dry mass of soil solids M_s is also known.

With these data and using a block diagram one can derive the following equation to define the shrinkage limit:

$$w_S = w_o - \frac{V_d - V_f}{M_s} \times 100 \quad (\%) \tag{4-6}$$

You should show select computations as illustrated in Fig. 4-5. Note the adjustments in collection of data depending on whether you use the mercury or wax method for obtaining the several volumes necessary in these computations.

Compute the shrinkage ratio SR using Eq. (4-1).

The Report

In preparing the report for this test discuss the advantages and disadvantages of mercury versus wax. Your sample computation sheet should include the derivations of Eq. (4-1) and Eq. (4-6). It will be convenient to use a block diagram to make the derivations.

Use a transparent sheet and overlay Fig. 8-2. Extend the UL and A lines until they intersect at a point O below the chart. Now plot the w_L and I_P of this soil on your overlay as point X. Connect points 0 and X, and where the line crosses the w_L axis obtain an approximate w_S. Comment on how this value compares with the experimentally determined value.

Compute the linear shrinkage using Eq. (4-3) for this soil and using the volumetric shrinkage S_v obtained from Eq. (4-2). Next assume several additional values of S_v and solve Eq. (4-3) for each. Plot these results along with the test value and comment on the resulting curve.

If you used the wax method, why are you required to use 5 to 10 dishes of water accumulated into a graduated cylinder to obtain the shrinkage dish volume V_d? Also if you had to determine the density of the wax, comment on the probable accuracy and how (if practical) the accuracy can be improved.

Project Shrinkage Limit Job No. —

Location of Project Soil Laboratory Boring No. — Sample No. —

Description of Soil Dark gray silty clay Depth of Sample —

Tested By JEB Date of Testing 6/30/19

Mass of coated dish + wet soil = 37.91 g (a)

Mass of coated dish + dry soil = 31.85 g (b)

Mass of coated dish, M_d = 10.43 g (c) dish # = 4

Mass of dry soil, M_s = 21.42 g (d) (b – c)

Total mass of water, M_w = 6.06 g (e) (a – b)

Initial water content, w_o = 28.29 % $\left(\frac{6.06}{21.42} \times 100\right)$ (f) (e/d·100)

Data for volume of soil cake

Initial cake volume (wet) *Wax Method*

Mass coat dish + Hg, M_{dm} = 25.38 g (g) M_{dw}

Mass of coat dish, M_d = 10.43 g (h) M_d

Vol. of shrink dish V_o = (g – h)/ρ = 14.95/1 = 14.95 cm³

Dry cake volume (final)

Mass of vol. cup + Hg, M_{vcm} = 210.35 g (i) M_{bw}

Mass of vol. cup + Hg – soil, M_{vcms} = 220.15 g (j) M_{bws}

Mass of displaced Hg (i – j) = 21.42 g ← Line "d" (k) M_s

Vol. of soil cake V_f (with Hg = k/ρ) = — cm³

Vol. of soil cake V_f (w/wax = i + k – j) = 210.35 + 21.42 – 220.15 = 11.62 cm³

$w_s = w_o - \frac{(V_o - V_f)\rho_w}{M_s} \times 100$ = $28.29 - \frac{(14.95 - 11.62)(100)}{21.42} = 12.74\%$

$SR = \frac{M_s}{\rho_w V_f}$ = $\frac{21.42}{(1)(11.62)} = 1.84$

Figure 4-5
Typical shrinkage limit test data and computations using the wax method for V_0 and V_f.

5: PARTICLE-SIZE ANALYSIS—MECHANICAL METHOD

References

ASTM D 421 (Sample Preparation)
ASTM D 422 (Test Procedures)
AASHTO T 87 (Sample Preparation)
AASHTO T 88 (Test Procedures)
U.S. Army Corps of Engineers, *Laboratory Soils Testing*, EM 1110-2-1906, Appendix V: *Grain-size Analysis*

Objective

To introduce the reader to methods of making a mechanical particle-size analysis of a soil and presenting the resulting data.

Equipment

Set of sieves (see typical list in "General Discussion" below; see also Fig. 5-1 and Table 5-1)
Mortar and pestle or a mechanical soil pulverizer
Balance sensitive to 0.1 g
Supply of thoroughly air-dried (or oven-dried) soil[1]

General Discussion

Some type of particle-size analysis is universally used in the engineering classification of soils (see Project No. 8). It is also used in concrete and asphalt mix design for pavements.

Figure 5-1
Typical stack of sieves in a mechanical sieve shaker.

[1]It is suggested that a stock coarse-grained material, using a quantity of sand from the concrete laboratory and "contaminated" with some very fine sand and/or other fines, be used in this test. The "fines" should be such that not over 10% passes the No. 200 sieve so the student can compute C_U and C_C. This material will not require washing so the test is readily performed and, more importantly, understood. It is suggested to use the cohesive soil previously used in Test 3 in Test 6 (Hydrometer analysis), 7, 9 and 14. *If Tests 5 and 6 are done as outlined here the student obtains sufficient data to classify two soils for Project No. 8 with minimal additional effort.*

Table 5-1 Standard Sieve Sizes

All U.S. sieves are available in 200 mm and most in 305 mm diameter. The current U.S. designation uses 100 down to 6.3 mm as a size designation; from No. 4 (4.75 mm) to No. 18 (1.00 mm), the mesh opening is in millimeters; from No. 20 to No. 400, the mesh opening is in micrometers (1μ m = 0.001 mm). The No. 20 sieve is 850 μm or 0.850 mm. From the No. 4 sieve and smaller, the size is approximately $N_i = (N - 1)/\sqrt[4]{2}$ (*Example*: No. 5 = $4.75/\sqrt[4]{2} = 4.00$ mm).

U.S.[a]		BRITISH STANDARD[b]		FRENCH[c]		GERMAN DIN[d]	
Size or No.	Opening, mm or µm	No.	Opening, mm	No.	Opening, mm	Designation, µm	Opening, mm
	100.0 mm						
	75.0						
	63.0						
	50.0						
	45.0						
	37.5						
	31.5						
	25.0						25.0
							20.0
	19.0						
							18.0
	16.0						16.0
	12.5						12.5
							10.0
	9.5						
	8.0						8.0
	6.3						6.3
				38[e]	5.000		5.0
4[c]	4.75						
5	4.00			37	4.000		4.0
		5[e]	3.353				
6	3.35						
				36	3.150		3.150
7	2.80	6	2.812				
				35	2.500		2.500
		7	2.411				
8	2.36						
10	2.00	8	2.057	34	2.000		2.000
12	1.70	10	1.676				
				33	1.600		1.600
14	1.40	12	1.405				
				32	1.250		1.250
		14	1.204				
16	1.18						
18	1.00	16	1.003	31	1.000		1.000

Particle-size is one of the suitability criteria of soils for road, airfield, levee, dam, and other embankment construction. Information obtained from particle-size analysis can be used to predict soil–water movement, although permeability tests (Tests No. 11 and 12) are more generally used. The susceptibility to frost action in soil, an extremely important consideration in colder climates, can be predicted from the particle-size analysis described here and/or with the Hydrometer Analysis of Test 6.

Very fine soil particles are easily carried in suspension by percolating soil water, and underdrainage systems will rapidly fill with sediments unless they are properly surrounded by a *filter* made of appropriately graded granular materials. The proper gradation of this filter material can be predicted from the particle-size analysis. Particle-size of the filter

Table 5-1 (Continued)

U.S.[a]		BRITISH STANDARD[b]		FRENCH[c]		GERMAN DIN[d]	
Size or No.	Opening, mm or μm	No.	Opening, mm	No.	Opening, mm	Designation, μm	Opening, mm
20	850 μm	18	.853				
				30	.800	800	.800
25	710	22	.699				
				29	.630	630	.630
30	600	25	.599				
35	500	30	.500	28	.500	500	.500
40[f]	425	36[f]	.422				
45	355	44	.353	27[f]	.400	400[f]	.400
50	300	52	.295	26	.315	315	.315
60	250	60	.251	25	.250	250	.250
70	212	72	.211				
				24	.200	200	.200
80	180	85	.178				
				23	.160	160	.160
100	150	100	.152				
120	125	120	.124	22	.125	125	.125
140	106	150	.104				
				21	.100	100	.100
170	90	170	.089			90	.090
				20	.080	80	.080
200	75	200	.076			71	.071
230	63	240	.066	19	.063	63	.063
						56	.056
270	53	300	.053				
				18	.050	50	.050
325	45					45	.045
400	38			17	.040	40	.040

[a]ASTM E-11 (vol. 14.02).
[b]British Standards Institution, London BS-410.
[c]French Standard Specifications, AFNOR X-11-501.
[d]German Standard Specification, DIN 4188.
[e]For standard compaction test.
[f]For Atterberg limits.

material must be larger than the soil being protected so the filter pores permit passage of water but collect the smaller soil particles from suspension.

The particle-size analysis is an attempt to determine the relative proportions of the different grain sizes that make up a given soil mass. Obviously, to have significance the sample must be *statistically representative* of the mass. Actually it is not possible to determine the individual particle sizes—the test can only bracket the approximate size range between two sieves.

The size bracketing is done by stacking a series of sieves ranging in aperture size from the largest at the top to the smallest and sieving a known quantity of material through the stack. This is done by placing the material on the top sieve and shaking to separate the particles into ever smaller diameters from the top to the bottom.

The diameters of the soil particles in the mass of soil retained on any sieve have smaller diameters than the mesh openings of any of the sieves above this one. They are also larger than the smaller mesh openings of any of the sieves below. Statistically there may be some grains smaller that are trapped in this mass—particularly if the amount is large in relation to diameter of sieve—but almost none if the test is carefully done.

It should be evident that the material retained on any sieve in the stack contains a range of particle sizes. For example if the top sieve is a No. 10 (sieve openings = 2.00 mm) and the next sieve in the stack is a No. 20 (sieve openings 0.850 mm) the material retained on the No. 20 contains particles ranging from 0.850 up to 2 mm Because of this problem it is customary to use the term *percent passing* or *percent finer* to describe the relative amount retained on a given sieve.

In this context it is evident that all the material below any sieve in a sieve stack is "finer" than the material on that sieve—or any sieve in the stack above it. The computation of particle size versus percent finer thus becomes quite simple, as illustrated later.

Table 5-1 gives the latest ASTM and U.S. National Bureau of Standards official sieve designations. These sizes are slightly different from former (prior to 1970) designations and/or those that may be found on some older sieves still in service.

Sieves are made of woven wire with rectangular openings ranging in size from 101.6 mm (4 in) in the coarse series to the No. 400 (0.038 mm) in the fine series. The No. 200 sieve (0.075 mm) is the smallest practical sieve size. This mesh is about the finest size that will permit relatively free passage of water. Soil, of course, provides considerably more resistance than water to sieving; thus, sieve sizes smaller than No. 200 are more academic than practical.

All the soil classification systems use the No. 200 sieve as a dividing point; that is, classifications are in terms of the amount retained on or passing the No. 200 sieve. The No. 4 or No. 10 sieve is the usual division point between gravel and sand. Sometimes it is necessary to estimate particle sizes smaller than the No. 200 sieve. When this need arises (as it does in most university laboratories for student familiarization with the procedure), the method of Test 6, "Particle-Size Analysis—Hydrometer Method," is used.

The sieving process does not provide information on the shape of the particles, that is, whether they are angular or rounded. If you want an indication of particle shape you will have to look at some of the particles under magnification. A stereoscopic microscope of 10 to not over about 40x power is preferred so you can see the particles in three dimensions.

Sieving yields information only on particles that can pass, or with proper orientation do pass, through rectangular sieve openings of a certain size. Obviously, not all particles in larger samples that can pass a given opening do pass, since they may not ever become properly oriented with the square opening. In some cases smaller particles may be in the form of small clods which are not broken down to the elemental particle size in the pulverizing process. These apparently larger particle sizes are then incorrectly retained on a sieve with a larger mesh opening. Another problem with the very fine particles—in particular those in the (−) No. 200 sieve size (i.e., material that will pass the No. 200 sieve)—is that they may adhere to the larger particles as dust and be retained on larger sieves.

The particle-size analysis is presented as a semilog plot of percent finer versus particle size. This makes it easier to compare several soils; besides, certain computations may be made from the curve data. Additionally a curve allows one to estimate the percentage of particle sizes that are finer than select diameters which may not have been used in the sieve stack.

A semilog plot is used for the particle sizes to give both small and large diameters as nearly equal weight as possible. You might wish to verify this by making an arithmetic plot of the data in Fig. 5-3 (which were used to plot curve B of Fig. 5-4). Percent finer is always plotted as the ordinate using an arithmetic scale. In United States practice the grain sizes decrease from left-to-right in log decrements of 10 as in Fig. 5-4. A few authorities plot grain sizes in increasing log increments of 10, from left-to-right. The resulting curve is similar to Fig. 5-4 if one looks at it from the back side of the paper.

Particle-size distribution curves for two *fine-grained soils*[2] are shown in Fig. 5-4. One of the curves is obtained from the data shown in Fig. 5-3. A distribution curve for a coarser (larger soil particles) soil would be shifted to the left. Curve A is a combined plot that was obtained from both this test method and a Hydrometer Analysis on a soil.

[2]Fine-grained is used here in the context that more than 50 percent passes the No. 4 (sometimes the No. 10) sieve.

It should be evident that a particle-size distribution curve can only be approximate. This is for the several reasons considered here, including physical limitations on obtaining a statistically representative sample, the presence of soil lumps (or clods), the practical limitations of using square sieve mesh openings for irregularly shaped soil particles, and the limit on the number of sieves used in a stack for the analysis.

Oven-dry soil is commonly used for this test in order to start from a known mass. Some authorities suggest that one can use air-dry soil,[3] but one must first determine its water content using a representative sample of at least 30 percent[4] of the test mass. The water content is used to compute the "dry" sample mass used in the test.

There is an inherent error in the use of an air-dry sample because the distribution of hygroscopic water depends on particle size. The argument against using oven-dry soil is that the dimensions of some of the clay particles become altered (curl or otherwise distort from flat platelets) due to loss of hygroscopic moisture during oven-drying. Clearly, using oven-dry soil can alter the results, but from this discussion these alterations would seem to produce a more accurate particle size analysis.

Note: The U.S. Army Corps of Engineers' laboratory manual requires use of oven-dry soil. The U.S. Bureau of Reclamation's *Earth Manual* (page 414) specifies oven-dry soil. Both ASTM and AASHTO (usually a copy of ASTM) standards require air-dry soil. This means that the users who perform the greatest numbers of this test and Test 6 following are not in agreement. As will be pointed out later the ASTM "standard" method may be more academic than practical.

An air-dry soil that gives off dust may still have hygroscopic water absorbed from the atmosphere in the range of 3 to 15 percent. In most cases, if the soil *is mostly* sand and gravel the hygroscopic water is in the range of only 3 to 5 percent and is well within the precision of this test without doing a water content determination.

The oven-dry versus not oven-dry debate will not be easily resolved since a particle size test is not reproducible. However there are two other significant factors to address that are at least as important—if not more so—than whether to use oven-dry soil. One of these is that a representative sample be used; the other is that there be a complete reduction of clay lumps to elemental particles.

GENERAL PROCEDURE—An Overview

A particle-size distribution curve obtained by the following method is satisfactory for predicting the behavior of cohesionless soils and for obtaining the relative quantities of (+) and (−) No. 200 material for soil classification. The methodology below is based on guidance from ASTM/AASHTO as well as personal experience. No method of particle-size analysis is reproducible, but the procedure here will come as close as any and should be used when the most accurate results are desired. Washing should also be used when an unwashed soil is borderline in any of the classification methods.

a. Obtain a representative *oven-dry* quantity of material of adequate mass depending on the maximum particle size. Refer to the ASTM D 421 or your Data Sheet No. 5 for a suitable mass. In any case the amount used should be such that on any sieve in the stack the distribution over the screen is not over about 2 grains deep. A large accumulation of soil usually results in particle interference and some particle retention. Larger samples are statistically more representative of the soil mass but may give poorer results unless larger diameter sieves are used.
b. Using a mortar and pestle pulverize this mass as fine as possible if it is mostly (+) No. 200. If it appears to have substantial (−) No. 200 sieve material soak the mass for 2 to 24 h to dissolve the lumps. Strictly, it is not necessary to soak the mass to dissolve

[3]AASHTO suggests that samples oven-dried at not over 60°C can be considered "air-dried."

[4]Interestingly, ASTM currently allows using a water content sample of only 10 to 15 g of (−) No. 10 sieve material to be representative of the larger sample mass.

the lumps but it is easier to wash the sample through sieves if the lumps are already on the verge of breaking down. Omit soaking and washing if the soil only has 4 or 5 percent of (−) No. 200 material, for ordinary sieving will be adequate.

c. Visually inspect this soaked soil. If it appears a substantial amount is larger than either the No. 10 or No. 40 sieve use that appropriate sieve size and wash the sample through it first, but catch the washings in a large dish. If the dish tends to overfill, temporarily stop washing and decant (pour off) the clear top water. Continue washing until the water coming through the sieve is clear.

d. Backwash the residue on the sieve used in "c" into a large weighed dish. Decant as much of the backwash water as possible which will usually be nearly all.

e. From the dish with the washings from "c" above decant as much water as possible. Next carefully pour into and wash the soil-water mixture through a standard No. 200 sieve, or, if available, use one with deep sides made especially for washing. Exercise care in pouring the mixture into the sieve and continue until the wash dish is empty. Washing soil through a No. 200 sieve is illustrated in Fig. 5-2.

Figure 5-2
Washing material through a No. 200 sieve. Deeper sieves are available especially for washing; however, with care the regular No. 200 sieve can be used as shown here. Some persons suggest putting the soil on a larger sieve over the No. 200 (not shown) to protect the No. 200 sieve from damage. When the top sieve is washed remove it and its residue and continue washing on just the No. 200 sieve.

f. Wash the material on the No. 200 sieve until the wash water is clear. Pause as necessary to decant excess water from the washings dish if you are saving the (−) No. 200 material.

g. When the wash water is clear, backwash the material retained on the No. 200 sieve together with the material previously saved in step "d." Oven dry this material, then weigh and record the mass of dry soil. The difference in this and the starting mass is the amount of (−) No. 200 material.[5]

h. Now perform the sieve analysis on this coarse fraction. Remember, however, to use the initial starting mass from step "a" in your computations of percent finer and not the washed and oven dry mass from step "g."

i. If a hydrometer analysis is to be performed using only the (−) No. 200 soil particles, decant as much water from the washings dish as possible and put the dish in the oven. Obtain the mass of the oven-dry fines and add to step g above as a check since the total of these two oven-dry masses should equal the starting mass in step "a" within one or two grams.

[5]The author prefers doing the hydrometer test (Test 6) on the (−) No. 200 sieve fraction obtained in this manner.

COMMENTS

Sieve washing is not desirable nor practical for soils (gravels and gravelly sands) when less than 10 to 15 percent passes the No. 10 (2.00 mm) sieve. Sieve washing is usually unnecessary when only 5 to 10 percent passes the No. 100 (0.150 mm) sieve for the fine-grained soils. In most cases larger sample masses of 1 to 10 kg necessary with coarse-grained soils make washing impractical.

Due to the statistical distribution of particle sizes, even the most representative soil samples do not yield reproducible distribution curves. One has obtained a "reproducible" analysis if the percent finer of any size is on the order of ± 2.0 percent for (−) No. 4 soil and perhaps ± 3.5 percent for coarser grained soils. This is why the scale and log range of the enclosed semilog graph data sheet are satisfactory for general use.

A sieve stack generally consists in a series (usually 4 to 7) of sieves with the following control sieves used if the soil is to be classified.

Control Sieve Number	CLASSIFICATION METHOD	
	AASHTO Requires	ASTM/USC Requires
4		x
10	x	
40	x	
200	x	x

Additional sieves to make up a stack may be required, as you will see in Project 8.

Sieve sizes in the numbered series (No. 10, 20, . . . , 200, etc.) have a mesh opening that is approximately halved for each doubling of the sieve size, thus, a No. 20 is approximately 1/2 the mesh opening of a No. 10 (0.85 versus 2.0 mm), the No. 40 is approximately 1/2 that of the No. 20 (0.425 versus 0.850 mm), etc. The actual mesh opening is given by the equation shown in the caption of Table 5-1.

A halving ratio obtains the best spread of points for a particle size plot, however, with the control sieves that are usually required this can seldom be done. When control sieves are used, look at the sieve sizes and try to obtain a reasonable spread of points. For example, you might use 4, 10, 20, 40, 100 and 200 plus, of course, a lid and pan. Use of a large number of sieves in a sieve stack seldom improves the plot.

A No. 200 sieve is always used as the bottom sieve for fine-grained soils; a No. 100 sieve is always used as the bottom sieve for concrete sand. *A pan is always placed beneath the bottom sieve to catch the "fines."*

DATA OBTAINED FROM SIEVE ANALYSIS

For soil classification (as in Project 8) you directly obtain the percent passing the control sieves.

From the particle-size distribution curve, grain sizes such as D_{10}, D_{30}, and D_{60} can be obtained. The D refers to the size, or apparent diameter, of the soil particles and the subscript (10, 30, 60) denotes the percent that is smaller. For example, $D_{10} = 0.16$ mm from curve B of Fig. 5-4 means that 10 percent of the sample grains are smaller than 0.16 mm. The D_{10} size is also called the *effective size* of a soil.

An indication of the spread (or range) of particle sizes is given by the *coefficient of uniformity*, defined as

$$C_U = \frac{D_{60}}{D_{10}} \qquad (5\text{-}1)$$

A large value of C_U indicates that the D_{10} and D_{60} sizes differ appreciably. It does not ensure that a condition of gap grading, when sizes are missing or present in very small relative quantities, does not exist.

The *coefficient of concavity* C_C is a measure of the shape of the curve between the D_{60} and D_{10} grain sizes, and is defined as

$$C_C = \frac{D_{30}^2}{D_{10}D_{60}} \tag{5-2}$$

A value of C_C greatly different from 1.0 indicates particle sizes missing between the D_{60} and D_{10} sizes. Typical computations for C_U and C_C are shown on Fig. 5-4 for soil *B*. Values to the nearest 0.1 (as shown) are sufficiently precise. These values may be required when classifying certain soils by the ASTM soil classification method in Project No. 8 following.

The D_{15} and D_{85} sizes may be used in determining suitability of the soil for filter design in an earth dam or to surround the openings of a pipe laid in the ground for a subdrainage system.

STANDARDIZATION OF THIS TEST

Neither ASTM D 421 nor D 422 (as of 1991) provides much guidance in producing a particle size distribution curve. ASTM D 422 spends much time on the Hydrometer Test (Test 6 following). AASHTO tests T 87 and T 88 are essentially direct copies of D 421 and D 422. The text above and the test method outlined in the "Procedure" following represent the author's experience and interpretation of ASTM and AASHTO and borrows very heavily from the procedure given in the U.S. Army Corps of Engineers laboratory manual (cited reference).

PROCEDURE

This is suggested as a group project.

Note that there are two parts to this project:

a. *A sieve analysis of a relatively coarse-grained contaminated sand,*
b. *Preparation of a sample of cohesive soil, from Test 3, for use in Test 6.*

PART A:

1. Each group will obtain exactly 500 g (200 g at discretion of instructor) of oven- or air-dry coarse-grained soil supplied for this test. Be sure the sample is as *representative* as possible by using a sample splitter (if available) or by randomly spooning the soil from the container and regularly remixing the remaining stock until you obtain the necessary quantity.

   Weigh the sample mass and record on your data sheet, 5a as M_s.

   If the sample is to be washed, it is not necessary to pulverize the soil, but the washing process is considerably speeded by first pulverizing and dry-sieving as much as practical through the No. 200 sieve and discarding the soil that passes.
2. If the sample contains appreciable gravel, very few fines, or if, at the discretion of the instructor, washing is to be omitted, go to step 4. Otherwise, place the test sample on the No. 200 sieve and carefully wash the material through the sieve using tap water (see Fig. 5-2) until the water is clear. Be very careful in this process not to damage the sieve or lose any soil by splashing the material out of the sieve.

   Note: If there is a considerable amount of larger material and you use a 500 g sample, you should put either a No. 10 or a No. 40 sieve onto the No. 200 sieve and wash the sample through both sieves. This tends to reduce damage to the No. 200 sieve.

   Remove the top sieve when it appears that the soil retained on it is clean of fines. Check by separately washing into the No. 200 sieve. Backwash the residue from this sieve into a large weighed dish and save.
3. Complete washing of the soil on the No. 200 sieve and then backwash its residue into the same dish with that from the top sieve (if one was used). Let this dish sit for a short time until the top of the suspension becomes clear.

Pour off (i.e., decant) as much of the clear top water as possible, then place the dish and remaining soil-water suspension in the oven for drying. You will do your sieve analysis on this soil later when it is oven-dry.

4. On the following day, return to the laboratory and weigh the oven-dry residue. (Omit this step, of course, if you do not wash the sample.)

For sandy to fine-grained soils, two recommended sieve stacks (in the order from top to bottom) are as follows:

TYPICAL SIEVE STACK		ALTERNATIVE SIEVE STACK	
Sieve No	Opening, mm	Sieve No	Opening, mm
Lid		Lid	
4	4.75	4	
10	2.00	10	
20	0.850	40	0.425
40	0.425	60	0.250
60	0.250	100	0.150
140	0.106	200	0.075
200	0.075		
Pan		Pan	

Since this soil will be one of those later classified in Project 8, the control sieves necessary to classify this soil by either the AASHTO or ASTM method are included in both "stacks."

Now set up your sieve stack using the control sieves (No. 4, No. 10, No. 40 and No. 200) and any others that are available for use. After a sieve is used it usually will have soil grains wedged in some of the screen openings. Before you use one of these sieves clean as many of these particles from the screen as possible by using the brass wire brush provided in the lab for that purpose. Do not punch grains out with a screwdriver or other pointed object. If the particles do not all come out after brushing (from the bottom side) get what you can.

Next determine the mass of each sieve to the nearest 0.1 g and record it on a piece of paper (you might put these values beside the sieve number on your data sheet, as 4/240.2, 10/235.3, etc.).

When there appears to be as much as 4 to 10 percent gravel present [pieces larger than about 6 mm (1/4 in)] you should place a 12.5- and/or a 6.3-mm sieve before the No. 4 sieve in the stack. When the soil contains a large amount of gravel (material larger than 6 mm) you should use a larger sample mass and a sieve stack appropriate for the probable range of grain sizes. A typical stack[6] might be as follows:

75 or 50 mm (depends on maximum aggregate size)
25 mm
12.5 mm
6.3 mm
No. 4 (4.75 mm) and/or
No. 10 (2.00 mm)
Pan

Your sample mass should range from 1500 g for 20-mm gravel to at least 5000 g for 75-mm maximum size material. Note that you can use the above stack as a first part of a sieving operation. The material in the "Pan" shown would be finer than the last sieve so

[6]For sample masses over about 500 g you should use larger diameter sieves. The most common diameter is 200 mm (8 in); sieve diameters of 305 mm (12-in) and 460 mm (18-in) are available. If you are using a large sample mass be sure to check with the instructor if there are larger diameter sieves available.

you can start another stack using the No. 20, 40, 100 and 200 and pan. The starting material is that from the "pan" above which is poured onto the No. 20 sieve. The initial mass of 1500, 5000 g (or whatever) is used for all of the subsequent computations for percent finer—and for both sieve stacks.

5. Place the stack of sieves in a mechanical sieve shaker (if available) and shake for 5 to 10 min. Longer times are usually recommended, however, in a student laboratory there may only be one shaker for the class. If the sample was washed or you are using only 200 g of soil 5 min may be adequate; otherwise use about 10 min.

   If the entire stack of sieves will not fit into the mechanical shaker, perform an initial shaking operation by hand until the top few sieves can be removed from the stack; then place the remainder of the stack in the mechanical shaker.

   If a mechanical shaker is not available, shake by hand for 10 to 15 min. Alternate the mode of shaking so that the grains are being continually moved across the sieve screens—do not shake in a defined pattern.

6. Next remove the stack of sieves from the shaker, carefully separate the sieves and put each one on a paper towel. From the top down weigh each sieve with the soil retained on it to the nearest 0.1 g. If the sieving has not been well done some particles may drop out of the mesh onto the paper towel during handling or onto the balances. Remove these from the balances or paper towel and put into the next lower sieve before it is weighed. Now reweigh this sieve and record its total mass (sieve + soil retained).

   Obtain the mass retained on each sieve by subtracting the sieve mass from the sieve mass + retained soil. Record these values on your data sheet under column headed "Mass Retained".

   Now sum this column of masses (including that in the pan) and compare with the mass obtained in Step 4 (the oven-dry residue you started with). If you did not wash the sample compare this sum with your initial mass M_s to detect any soil loss or gain.

   A loss of more than 2 percent by weight of the residue weight (Step 4) is considered unsatisfactory, and the test should be repeated. *Question:* What might account for the discrepancy if you have a larger weight than you started with?

7. Compute the percent retained on each sieve by dividing the weight retained on each sieve by the *original sample mass M_s from Step 1 above.* This is valid, since any material passing the No. 200 sieve will pass any sieve above it in the stack.

8. Compute the percent passing (or percent finer) by starting with 100 percent and subtracting the percent retained on each sieve as a cumulative procedure. For example, from the displayed sieve-analysis data shown in Fig. 5-3, the amount of soil passing the No. 4 sieve is

$$\text{Quantity passing} = \text{Total mass} - \text{mass retained}$$

$$= 500 - 9.7 = 490.3 \text{ g}$$

The percent retained is computed as

$$\%\ \text{Retained} = \text{Mass retained/total mass}$$

$$\%\ \text{Retained} = 9.7/500 \times 100 = 1.9\ \%$$

From this the % passing = 100 − 1.9 = 98.1 %. For sieve No. 10

$$\text{Quantity passing} = \text{Mass arriving} - \text{mass retained}$$

$$= 490.3 - 39.5 = 450.8 \text{ g}$$

$$\%\ \text{retained} = 39.5/500 \times 100 = 7.9\ \%$$

$$\%\ \text{passing} = 100 - 1.9 - 7.9 = 90.2\ \%$$

In equation form the percent passing is always

$$\% \text{ Passing} = \% \text{ Arriving} - \% \text{ retained} \tag{5-3}$$

Using data for the No. 10 sieve, in this manner we have

$$\% \text{ Passing} = 98.1 - 7.9 = 90.2 \ \%$$

The rest of the data of Fig. 5-3 are computed similarly.

9. Each individual should make a semilogarithmic plot of particle size versus percent finer, using the graph on data sheet 5*b* for this test as in Fig. 5-4.

   If less than 12 percent passes the No. 200 sieve, compute C_U and C_C and show on your graph as in Fig. 5-4 (also list these values in your Discussion). Answer the question: Why are these values computed only if less than 12 percent passes the No. 200 sieve? *Hint:* Refer to Project 8: Soil Classification.

10. In your report "Discussion," give five uses for the particle-size distribution curve that has just been made.

    Computer Program: There is a computer program on the included diskette which does these computations. You input the sieve numbers or openings (in mm), the sieve openings in mm, and masses retained on each sieve in g. You also input the initial mass. The program outputs Sieve Number; Sieve Opening Diameter (in mm); Mass Retained, g; and the % Finer. You may output this data to a disk file to later plot using a Graphics (or CAD) program.

PART B:

1. In preparation for Test 6, Hydrometer Analysis, obtain a representative sample of the soil used for Test 3, Liquid and Plastic Limits. If the soil is very wet obtain about 250 g and if already quite air-dry about as follows:

   Sandy soil use 110 to 125 g
   Clay soil use 60 to 75 g

   Refer to the soil description you used for Test 3. The larger quantity of either type is recommended.

2. Do one of the following three alternatives as specified by the instructor.

   a. If the soil is very wet, pulverize the soil as much as practical and put it out in a flat pan for air-drying. After air-drying you should end up with at least the above amounts.

   b. If the soil is already air-dry, pulverize the soil and *then* take a representative water content sample of 10 to 25 g (so the amount remaining is on the order of either 100 g or 50 g). Put the rest of the soil in a closed container with a slightly damp paper towel (to maintain humidity) to save for the hydrometer test. Return to the lab the next day and weigh the oven-dry sample. Compute the water content of this soil and record for use when the Hydrometer Test is run.

   c. (This alternative was used to produce Curve A of Fig. 5-4.) Wash 200 to 500 g of air-dry soil through the No. 200 sieve. Save the material passing in one dish and the material retained in another. Oven-dry both dishes of soil and sum the two masses of soil for the total mass of dry soil. Next do a mechanical sieve analysis on the (+) No. 200 sieve fraction using a stack of sieves with the No. 200 on the bottom. Compute the % Finer for the composite grain size curve using the *total dry mass*. Save the oven-dry material passing the No. 200 sieve for the Hydrometer Test following.

GRAIN SIZE ANALYSIS—MECHANICAL Data Sheet 5a

Project Sieve Analysis Job No. —

Location of Project Soil Laboratory Boring No. — Sample No. —

Description of Soil Brown, Medium Coarse Sand Depth of Sample ~

Tested By JEB Date of Testing 7/12/19--

Soil Sample Size (ASTM D1140-54)

Nominal diameter of largest particle	Approximate minimum mass of sample, g
No. 10 sieve	200
No. 4 sieve	500
3/4 in.	1500

Mass of dry sample + dish	893.7
Mass of dish	421.2
Mass of dry sample, M_s	472.5

Washed 500 g
472.5
27.5 g = washed (-) 200

Sieve analysis and grain shape

Sieve no.	Diam. (mm)	Mass retained	% retained	% passing
4	4.75	9.7 g.	1.9 %	98.1
10	2.00	39.5	7.9	90.2
20	0.840	71.6	14.3	75.9
40	0.425	129.1	25.8	50.1
60	0.250	107.4	21.5	28.6
100	0.150	105.0	21.0	7.6
200	0.075	8.5	1.7	5.9
Pan	—	1.3	-	-
		Σ:472.1 (472.5 o.k.)		
#4: % Ret = $\frac{9.7(100)}{500}$ = 1.9				
% Pass = 100 − 1.9 = 98.1				

% passing = 100 − Σ % retained.

Figure 5-3
Mechanical analysis soil data used to construct curve B of Fig. 5-4.

GRAIN SIZE DISTRIBUTION Data Sheet 5*b*

Project Sieve & Hydrometer Job No. —

Location of Project Soil Laboratory Boring No. — Sample No. —

Tested By JEB Date of Testing 7/12/19-- (sand)

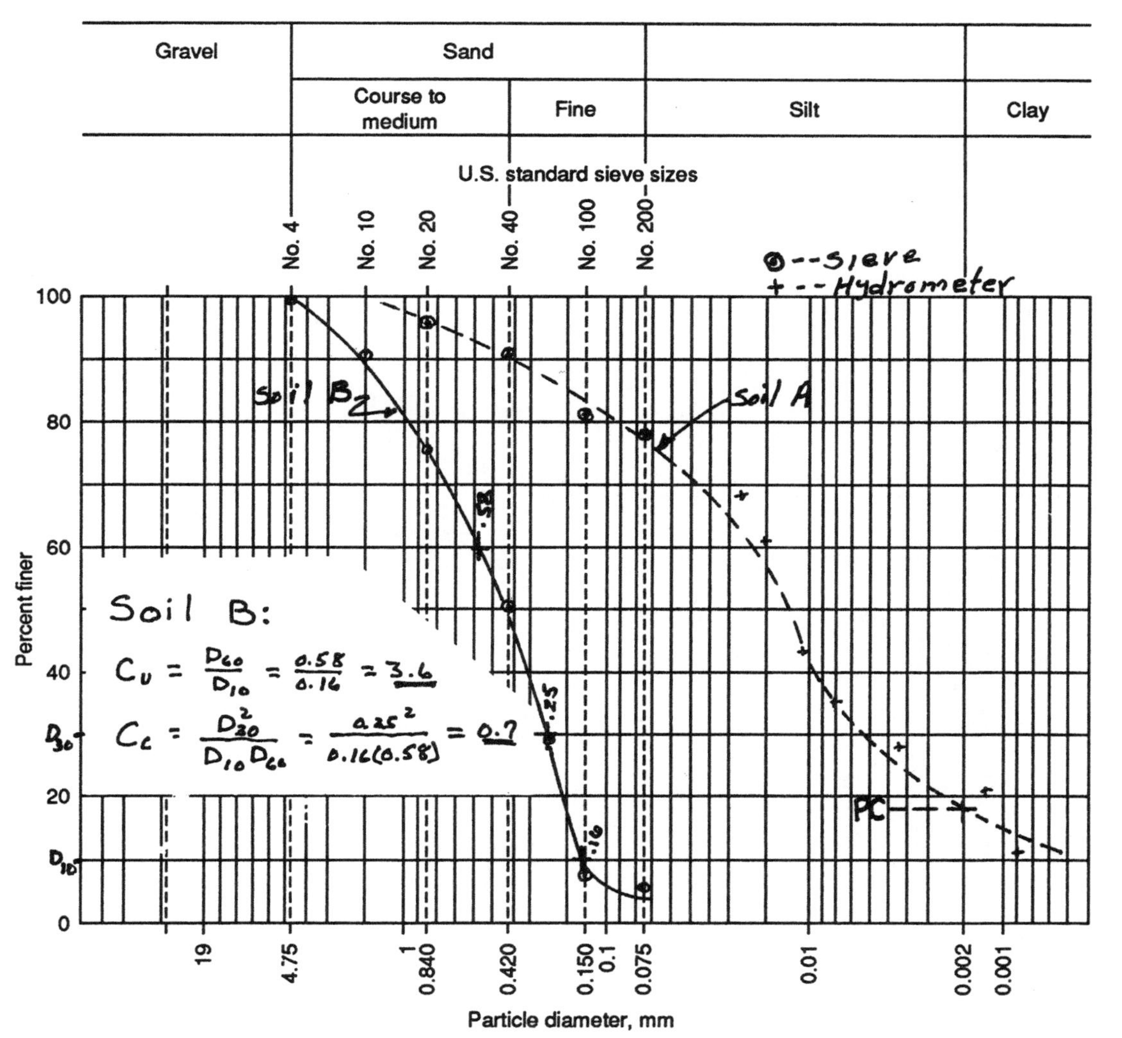

Visual soil description Soil B: See Data Sheet 5a

Soil A: Brown, silty clay; Hydrom. data from Fig. 6-3

Soil classification — sieve analysis data not shown

System —

Figure 5-4
Two grain size curves. Soil B curve is from data shown in Fig. 5-3. The hydrometer analysis part of Soil A curve is from Fig. 6-3*a* and is based on adjusting the % Finer based on 78% passing the No. 200 sieve as for Method C of Test 6 and using 50 g of (−) No. 200 sieve material.

7: SPECIFIC GRAVITY OF SOIL SOLIDS

References

ASTM D 854
AASHTO T 100

Objective

To familiarize the reader with a general method of obtaining the specific gravity of a mass of any type of material composed of small particles which has a specific gravity greater than 1.00. The project is specifically applicable to soil and fine aggregates (or sand) as used in concrete and asphalt mixtures.

Equipment

Volumetric flask, also called a pycnometer, preferably 250 or 500 mL (see Fig. 7-1)
Vacuum pump or aspirators for supplying a vacuum
Mortar and pestle
Balances weighing to 0.1 g
Supply of deaerated, temperature-stabilized water[1]

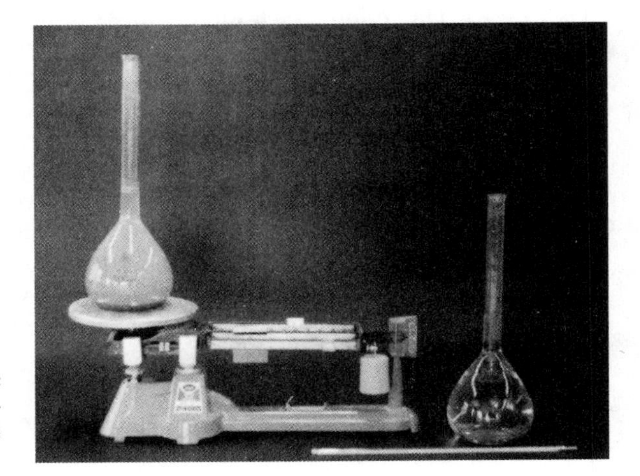

Figure 7-1
Balances and volumetric flask (500 mL shown) for the specific gravity test. Shown is the weighing to determine M_{bws} after carefully deaerating the soil-water mixture and filling the flask to the volumetric mark on neck.

General Discussion

The specific gravity G_s of a soil without any qualification[2] is taken to be the average value for the soil particles. If numerical values are given in a discussion where it may not be clear to what the specific gravity is referred, the magnitude of the values may indicate the correct usage since the specific gravity of the soil particles will always be larger than the bulk specific gravity based on including the soil voids in the computation [and either full of air (dry) or full, or partly full of water].

[1]Prior to class a supply of tap (or distilled) water should be collected and deaerated (about 1000 mL per group); use tap water, hot water, and/or ice water to produce a stabilized water temperature as close to 20°C as practical. Also check and oven dry the volumetric flasks as necessary.
[2]Some texts use the symbol G_s to indicate the specific gravity of the soil particles; others use the symbol G. The author prefers using G_s as in this laboratory text.

A value of specific gravity is necessary to compute the void ratio of a soil, it is used in the hydrometer analysis, and it is useful to compute the soil density ρ. Occasionally, the specific gravity may be useful in soil mineral classification, for example, iron minerals have a larger value of specific gravity than silicas.

The specific gravity of any substance is defined as the density of the material divided by the density of distilled water at 4°C. Thus, if one is considering only the soil particles, one obtains G_s as

$$G_s = \frac{\rho_{\text{material}}}{\rho_{\text{water at 4°C}}} \tag{7-1}$$

The same form of equation is used if bulk specific gravity is computed, the only difference being in the value of ρ_{material}. The specific gravity of a material can also be computed using any ratio of equal mass volumes of substance to water as

$$G_s = \frac{M_s/V}{M_w/V} \tag{7-2}$$

From Eq. (7-2) it is evident that this ratio is valid since the volume term V cancels. Note, however, that if one does not cancel the V term we have $\rho = M/V$ in Eq. (7-2) which reduces to Eq. (7-1).

The specific gravity determination reduces to obtaining the volume of a known mass of soil particles and dividing this by the mass of the same volume of water, that is, applying Eq. (7-2). This form is easier to visualize as well as to obtain the necessary laboratory test data.

The volume of a known mass of soil particles can be obtained by using a container of known volume and the Archimedes principle that a body submerged in a volume of water will displace a volume of water equal to the volume of the submerged body.

In this Test, the container is the volumetric flask, which holds a known volume of distilled water at 20°C. At temperatures above this value, the volume will be slightly more; below 20°C, the volume will be slightly less. Since the volume change is small for small temperature deviations and it is relatively easy to hold the test temperature close to 20° C, an approximate temperature correction for small temperature deviations can be applied in the required computations which produces satisfactory accuracy without recourse to experimentally determining the change in flask volume with temperature. Alternatively, one may develop a calibration curve for any given volumetric flask as follows:

1. Carefully clean the flask.
2. Fill with distilled, demineralized, or tap water at a known temperature.
3. Make a plot of mass M_{bw} vs. T°C (use about four points at, say, 16, 20, 24, and 28°C).

Tap water instead of distilled water is often used in this test (at least for routine work); again, relatively little error is involved. The error introduced from using tap water can be determined by filling the volumetric flask will tap water to the volume mark and obtaining the temperature and mass. From this data and the mass of the empty flask, one can compute the density of the tap water and compare with the density of distilled water in tables such as Table 6-1. Note that if the temperature is not exactly 20°C the volume of the flask will require calibration as previously outlined. Generally, if the density error is less than 0.001, it can be neglected.

Since the laboratory work to determine the specific gravity of the soil using the volumetric flask is somewhat indirect it is instructive to derive the expression to compute the specific gravity:

1. Let M_{bw} = mass of flask + distilled or tap water to the volume mark on the flask. Refer to the data sheet of Fig. 7-3 as a discussion aid. Now empty the flask.

2. Next place the mass of soil M_s into the flask and again fill the flask to the volume mark with water and weigh. Let this mass value be M_{bws}.
3. Recalling that one is dealing with a constant volume, if no water was displaced from the bottle when *soil mass* M_s was added the total mass would be

$$M_T = M_{bw} + M_s$$

However, since the volume of soil mass M_s occupies part of the space, the addition of M_{bw} to the flask would fill it above the volume mark. Since the volume mark is the reference point, the amount of water that cannot be put into the flask because of the soil volume already there is

$$M_w = M_T - M_{bws} = M_{bw} + M_s - M_{bws}$$

4. Equation (7-2) can be directly solved since we have equal volumes of masses M_s and M_w now known, or $V_w = M_w/\rho_w = V_s$.

Rewriting Eq. (7-2) with the mass values from Step 4 above into a form convenient to use in the order that the laboratory data is obtained gives

$$G_s = \frac{M_s}{M_{bw} + M_s - M_{bws}}$$

A slight increase in precision to account for temperature effects on the density of water can be obtained by rewriting this equation using a temperature correction α as

$$G_s = \frac{\alpha M_s}{M_{bw} + M_s - M_{bws}} \tag{7-3}$$

The temperature correction coefficient α is computed as

$$\alpha = \frac{\rho_T}{\rho_{20°C}} \tag{7-4}$$

and is the ratio of the density of water (or G_w) at the test temperature T and at 20°C formed such that the value of G_s obtained at temperature T (which will be too large if $T > 20°C$) is appropriately reduced.

Typical values of the correction factor α are as follows:

T,°C	α	ρ_w, g/cm^3
16	1.0007	0.99897
18	1.0004	0.99862
20	1.0000	0.99823
22	0.9996	0.99780
24	0.9991	0.99732
26	0.9986	0.99681

A source of error, which can be important, is the use of masses obtained from poorly adjusted balances or from not using the same balance for all weighings.

The most serious error in this test will occur from not properly deaerating the soil-water mixture. Any temperature error will be minor if the test is done between 18 and 22°C. The use of tap water also produces a minimal error unless it contains so much mineral hardness it is visible in a glass—in this case use distilled or demineralized water.

DEAERATION FOR THE TEST

Water normally contains dissolved air. The soil particles will also contain air, and if the air is not removed from both of these materials, the volume of air will result in a decrease in the mass M_{bws}. This in turn will result in too small a computed value of G_s that is, $M_{bw} + M_s - M_{bws}$ will be too large.

Deaerating the soil-water mixture is accomplished by applying a vacuum and/or heating. Use of a vacuum is usually sufficient for sands, silts, and clays. For organic soils it may be necessary to boil the soil-water mixture about 30 min, adding water as necessary to keep the flask about half full. If you use boiling, be sure the flask is a heat resistant type and using minimum heat.

The length of time the vacuum should be applied may range from a few minutes to

6 to 8 h for very plastic soils
4 to 6 h for soils of low plasticity

The efficiency of air removal from any soil may be improved by boiling for about 10 min, taking care not to boil the sample dry or lose any soil material. To check the deaeration.

1. Apply vacuum to a half-to-three-quarters-filled flask of soil-water mixture for a period of time.
2. Then fill flask to about 20 mm of volume mark with temperature-stabilized deaerated water.
3. Reapply the vacuum for several minutes and, with a colored grease pencil, mark the water level in the neck.
4. Carefully pull the stopper to break the vacuum; if the water level drops not more than 3 mm, the deaerating should be sufficient.

Typical values of G_s that can be used as a guide in determining whether the test results are correct are as follows:

Type of Soil	G_s
Sand	2.65 – 2.67
Silty sand	2.67 – 2.70
Inorganic clay	2.70 – 2.80
Soil with mica or iron	2.75 – 3.00
Organic soil	1.0+ – 2.60

Procedure

This is suggested as a group project.

1. Weigh a *representative* sample (exact weight is not important at this point) of *well-pulverized* air-dry soil. If you use a 500 mL volumetric flask this sample should be between 100 and 120 g. Put this sample into the flask, being careful not to lose any soil, for the sample at this point is "representative". If you are using a 250 or 500 mL flask you may wash the soil into the flask using a funnel and some of the stabilized water. If you do so be careful not to fill the flask more than half full of the soil-water mixture as you will need to wash soil out of the neck. Now add stabilized water to fill the flask about two-thirds full.
2. Attach the flask to a high vacuum for at least 10 min. During this time gently agitate the mixture by carefully shaking and turning the flask. Observe that the reduced

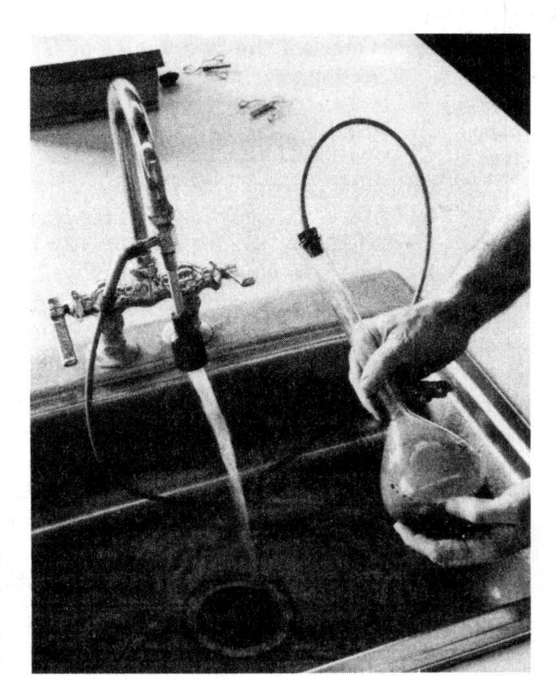

Figure 7-2
Using an aspirator connected to the sink faucet to produce deaerating vacuum. Vacuum intensity is determined by faucet flow. Flask is about two-thirds full, and contents are "boiling" (bubbles and foam inside flask). When using this vacuum device be very careful to unstopper the flask prior to shutting off the faucet so water in the system is not drawn into the flask from the pressure differential between the flask and exit piping.

air pressure in the flask causes the water to "boil" (see Fig. 7-2). Check the vacuum efficiency as outlined Step 4 of the "Deaeration for the Test" section. This step can take several hours, however; for student laboratories, check with the instructor after a reasonable amount of time for next step.

3. When the deaerating process is complete (or has been terminated) carefully add temperature-stabilized, deaerated water until the *bottom* of the meniscus is exactly at the volume mark. Be very careful not to reintroduce air into the flask when this water is added. Carefully dry the neck of the flask above the calibration mark with a rolled paper towel or some other means.
4. Weigh the flask and its contents to the nearest 0.01 g (by estimation) to obtain M_{bws}. Insert a thermometer and take a temperature reading to the nearest 1°C.
5. Empty the flask and its contents into a deep evaporating dish or other container. Let the soil-water mixture stand for a few minutes and decant as much water as practical, then put the dish with the remaining soil-water suspension in the oven for drying[3]. *Be very careful not to lose any soil* at this point.

   Repeat Steps 1 through this point for a second test.

   On the following day weigh the two oven-dry dishes of soil to obtain M_s for each test.
6. Next check the flask is clean and then fill it two-thirds full of deaerated water, apply vacuum for a short period then completely fill to the volume mark. Be sure the temperature is within 1°C of the temperature used for obtaining M_{bws} (unless a calibration curve is used). Weigh the flask with deaerated water as M_{bw} and record on your data sheet in the space provided.
7. Perform the necessary arithmetic on your data sheet entries to compute your two values G_s using Eq. (7-3).

   Compare the density of water used with that of distilled water as outlined in the "General Discussion" and comment in your report on the effect of using tap water (but only if you used temperature-stabilized, deaerated tap water).

[3]Avoid excess water in the dish as the 110°C oven temperature boils water. Heavy boiling may splash some of the particles out of the dish and ruin your test. With small very clayey samples it may be necessary to air-dry the soil dish to remove the most of the water before oven-drying.

8. Check if the two values of G_s just computed are within 2 percent of each other, defined as follows:

$$\frac{\text{Larger } G_2}{\text{Smaller } G_s} \leq 1.02$$

If these two test values are not in agreement you should do an additional test and use the two closest values.

When you have obtained two G_s values that agree sufficiently, take their average and round to the nearest 0.01. Report this value as the specific gravity of the solids for this soil.

The ASTM acceptance criterion for a single operator (your group) is that the difference in G_s between two tests should not exceed:

Cohesive soil	0.06
Non-cohesive soil	no criterion

Note that the author's suggestion for acceptance is almost exactly that of ASTM in the range of 2.40 to 2.80.

SPECIFIC GRAVITY OF SOIL SOLIDS (G_s) Data Sheet 7

Project Specific Gravity Test Job No. —

Location of Project Soil Laboratory Boring No. — Sample No. —

Description of Soil Brown Silty Clay Depth of Sample —

Tested By JEB Date of Testing 12/3/19

fl. = volumetric flask

Test no.	1	2		
Vol. of flask at 20°C	500 mL	500 mL		
Method of air removal[a]	Vacuum	Aspir.		
Mass fl. + water + soil = M_{bws}	753.66	754.69		
Temperature, °C	23°	23°		
Mass fl. + water[b] = M_{bw}	693.27	693.27		
Dish no.	5	4	(evap. dishes)	
Mass dish + dry soil	350.11	368.49		
Mass of dish	254.52	270.52		
Mass of dry soil = M_s	95.59	97.97 g		
$M_w = M_s + M_{bw} - M_{bws}$	35.20	36.55	(displ. soil)	
$\alpha = \rho_T/\rho_{20°C}$ $\frac{0.99757}{0.99823}$	0.9993	0.9993	(see Table 6-1)	
$G_s = \alpha\, M_s/M_w$	2.71	2.68		

[a]Indicate vacuum or aspirator for air removal.

[b]M_{bw} is the mass of the flask filled with water at same temp. ± 1°C as for M_{bws} or value from calibration curve at T of M_{bws}.

Remarks $\% \approx \frac{2.71}{2.68} \approx 1.01 < 1.02$ o.k.

Average specific gravity of soil solids (G_s) = (2.71 + 2.68)/2 ≈ 2.70

Figure 7-3
Data from two specific gravity test trials with average value reported.

8: CLASSIFICATION OF SOILS

References

ASTM D 2487-85 "Classification of Soils for Engineering Purposes"

ASTM D 3282-88 "Classification of Soils and Soil-Aggregate Mixtures for Highway Construction Purposes"

AASHTO M 145 in *AASHTO Materials: Part I: Specifications*

Casagrande, A. (1948), Classification and Identification of Soils, *Transactions, ASCE,* Vol. 113, pp. 901–991.

Highway Research Board (1945), Classification of Highway Subgrade Materials, *Proceedings,* Vol. 25, pp. 376–392.

Howard, A. K. (1987), "The Revised ASTM Standard on the Description and Identification of Soils (Visual-Manual Procedure)," *Geotechnical Testing Journal, ASTM,* Vol. 10, No. 4, Dec., pp. 229–234.

Objective

To introduce the user to two methods of soil classification. Secondary benefits will be derived from the additional practice obtained in performing the required soil tests to obtain necessary classification data.

Equipment

As required.

General Discussion

There are several methods of soil classification including:

The AASHTO System (M 145 and ASTM D 3282 noted above)
The Unified Soil Classification System (USC)
The ASTM Method of D 2487 (an extension of the USC)
U.S. Department of Agriculture (USDA)

There have also been several others used or proposed, but at present the above have substantial acceptance. The USC system is internationally used, and the AASHTO system, either directly or with minor modifications, is used by all the state departments of transportation (formerly state highway departments) in the United States. Since the USC and AASHTO systems are widely used, they will be the concern of this laboratory project.

Both the USC and AASHTO systems rely on the liquid and plastic limits of the soil and a sieve analysis. Depending on the soil, the sieve analysis may require only one sieve to determine the percent passing the No. 200 sieve, or a stack sufficient to plot a grain-size curve may be necessary. No other soil tests are required.

A part of the classification should include a description of the soil in addition to the symbols used for specific identification. This part of the soil classification is a recent addition to the original USC system. The AASHTO system is simpler and will be taken up first.

General classification	Granular materials (35 percent or less of total sample passing No. 200)							Silt-clay materials (More than 35 percent of total sample passing No. 200)			
Group classification	A-1		A-3	A-2				A-4	A-5	A-6	A-7
	A-1-a	A-1-b		A-2-4	A-2-5	A-2-6	A-2-7				A-7-5[a] A-7-6
Sieve analysis percent passing											
No. 10	50 max										
No. 40	30 max	50 max	51 min								
No. 200	15 max	25 max	10 max	35 max	35 max	35 max	35 max	36 min	36 min	36 min	36 min
Characteristics of fraction passing No. 40											
Liquid limit, w_L				40 max	41 min	40 max	41 min	40 max	41 min	40 max	41 min
Plastic Index, I_P	6 max		NP	10 max	10 max	11 min	11 min	10 max	10 max	11 min	11 min
Significant constituent materials	gravel and sand		fine sand	silty and clayey gravel and sand				silty soils		clayey soils	

[a]See Fig. 8-1*b*

(*a*) AASHTO Soil Classification System soil groups. A-8 (not shown) is peat or muck classified visually.

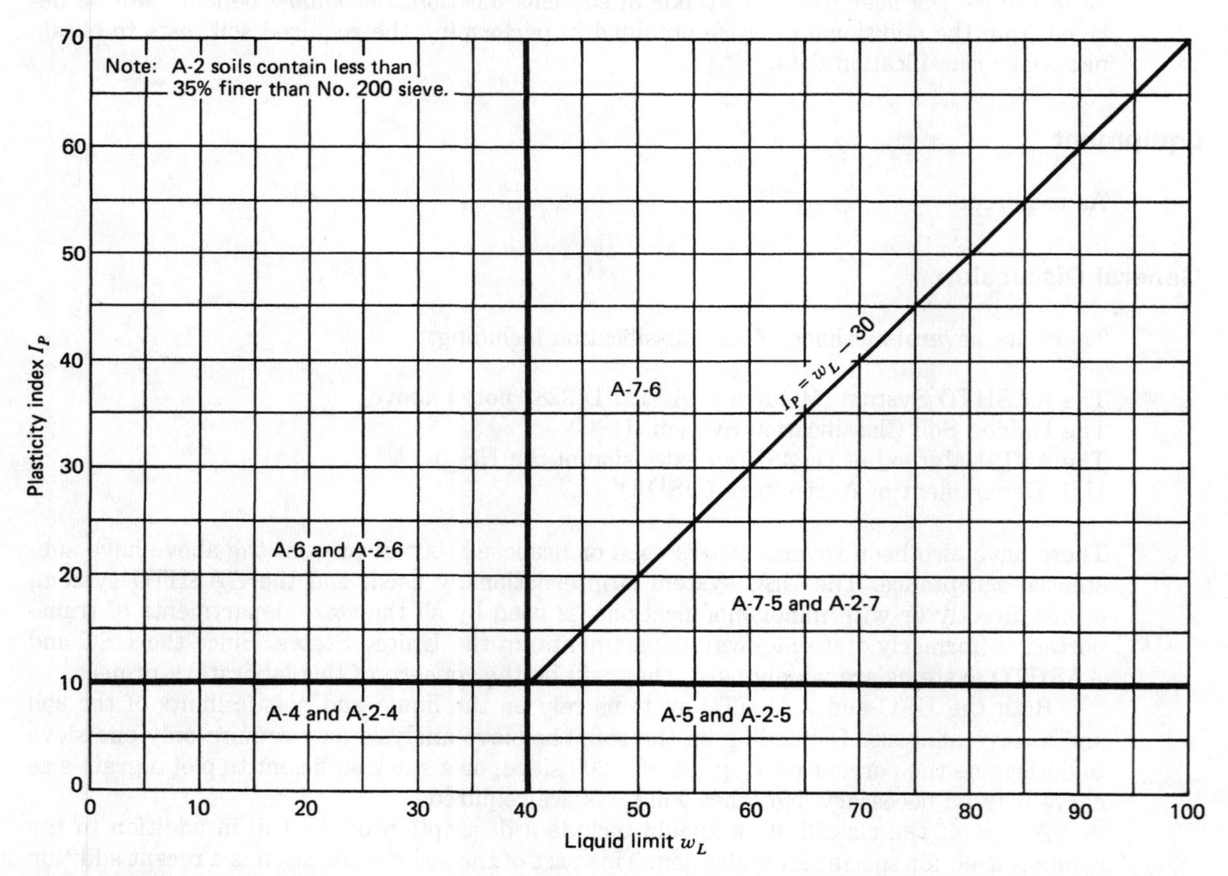

(*b*) Liquid-limit and plasticity index ranges for group classification of silt-clay materials. (*Standard Specifications for Transportation Materials and Methods of Sampling and Testing*, 15th ed., Washington, DC., American Association of State Highway and Transportation Officials, Copyright 1990. Used by permission.)

Figure 8-1
Charts for use in AASHTO Soil Classification System.

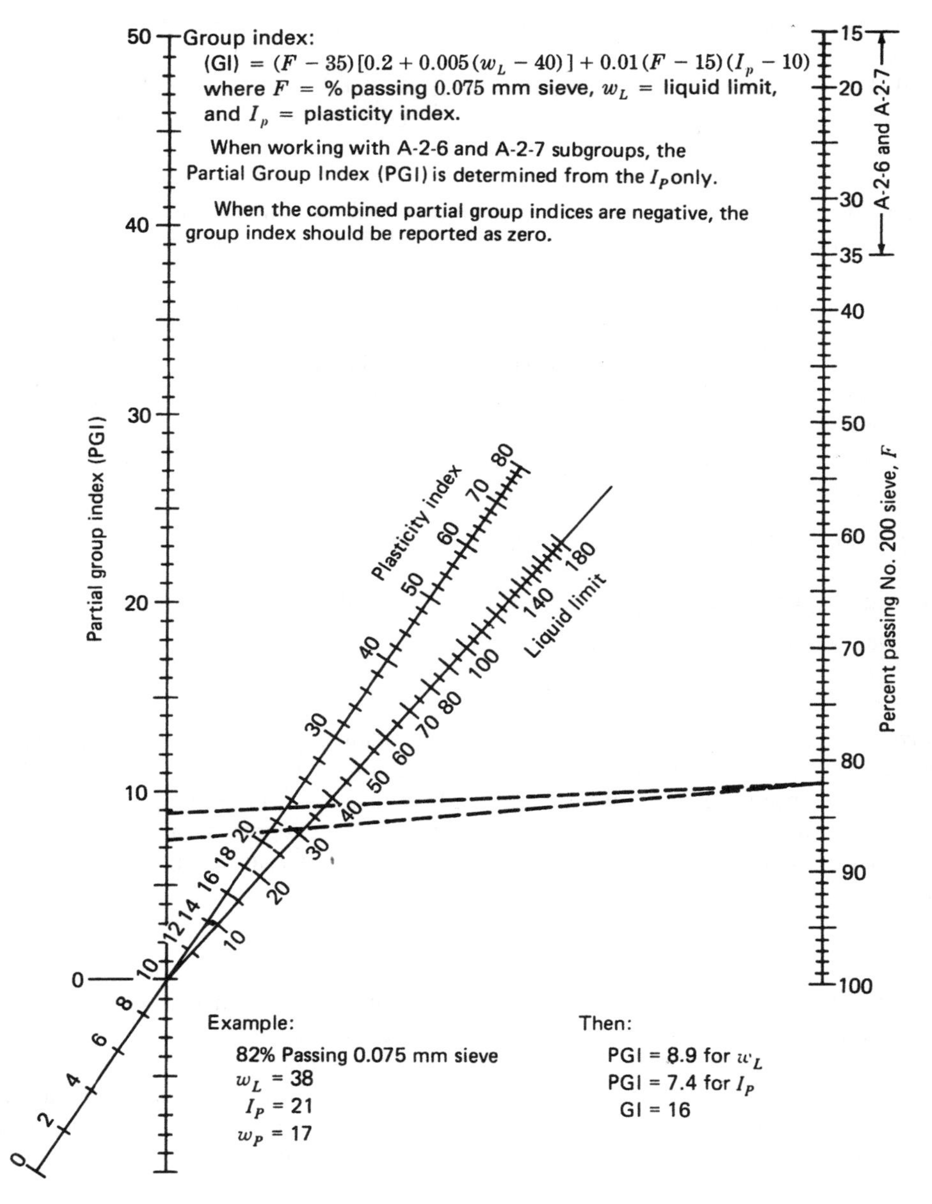

Group index:
(GI) = (F − 35)[0.2 + 0.005(w_L − 40)] + 0.01(F − 15)(I_p − 10)
where F = % passing 0.075 mm sieve, w_L = liquid limit,
and I_p = plasticity index.
When working with A-2-6 and A-2-7 subgroups, the
Partial Group Index (PGI) is determined from the I_p only.
When the combined partial group indices are negative, the
group index should be reported as zero.
Partial group index (PGI)
Plasticity index
Liquid limit
Percent passing No. 200 sieve, F
A-2-6 and A-2-7
Example:
82% Passing 0.075 mm sieve
w_L = 38
I_P = 21
w_P = 17
Then:
PGI = 8.9 for w_L
PGI = 7.4 for I_P
GI = 16

(c) AASHTO group index (GI) chart. (*Standard Specifications for Transportation Materials and Methods of Sampling and Testing*, 15th ed., Washington, DC., American Association of State Highway and Transportation Officials, Copyright 1990. Used by permission.)

A. AASHTO SOIL CLASSIFICATION

This method classifies a soil as belonging to one of eight groups A–1 through A–8 (see Fig. 8-1*a*).

1. Soil A-1 is a gravelly sand with only a small amount of "fines" (or material passing the No. 200 sieve). The fines are non-plastic—that is, they do not contain cohesive material such as clay or clayey silt.
2. Soil A-3 is a fine sand—with little gravel and the fines are non-plastic.
3. Soil A-2 is granular but the (−) No. 200 sieve fraction is plastic (has a w_L and w_P) and is further sub-divided based on plasticity characteristics.

4. Soils A-4 through A-7 depend on both plasticity and the amount passing the No. 200 sieve.
5. Soil A-8 (not shown on Fig. 8-1a) is *peat* or *muck* and is visually classified; peat is a woody material with a woody or organic odor and muck is a thin, watery, organic material.

This classification procedure can be programmed without much difficulty and a computer program is included with this manual for use if allowed or to check your personal classification technique.

Figure 8-1*a* displays the AASHTO soil classification system in a convenient chart form for direct use. Figure 8-1*b* provides a rapid means of using the liquid and plastic limits (and plasticity index I_P) to make determination of the A-2 subgroups and the A-4 through A-7 classifications. Figure 8-1b is based on the percent passing the No. 200 sieve (whether greater or less than 35 percent).

The AASHTO system further uses a group index (GI) to rate a soil qualitatively within a group. The equation for the GI is

$$GI = (F - 35)[0.2 + 0.005(w_L - 40)] + 0.01(F - 15)(I_P - 10) \tag{8-1}$$

where F = percent passing No. 200 sieve; if $F < 35$, use $F - 35 = 0$

w_L = liquid limit, percent (not a decimal)

I_P = plasticity index, percent

The group index equation can be directly solved using a programmable calculator or computer. The group index can also be solved using a nomograph such as Fig. 8-1*c*. The nomograph solution obtains the two parts of Eq. (8-1), which are then added, whereas a calculator or computer solution is direct. In either case there are no negative values of GI (use 0 when GI < 0). In both cases the GI is rounded to the nearest whole integer ≥ 0.

In use, the soil classification is reported as, say,

A-2-6(4); A-2-6(8); A-4(5), and so on

The soil within a group with the larger GI is of poorer quality for use as a subgrade material, for example, the A-2-6(8) above is not so suitable as the A-2-6(4).

Soil classification using Fig. 8-1*a* is done by proceeding from left to right in the chart until the *first soil group* that meets the properties of the soil under consideration is found.

To illustrate the method of soil classification the following three soils will be classified—first by the AASHTO procedure, then by the USC method:

	SOIL, % PASSING		
Sieve No.	A	B	C
4	–	91.4	69.3
10	68.5	79.5	59.1
20	54.9	–	48.3
40	36.1	69.0	38.5
60	–	–	28.4
100	22.5	61.0	19.8
200	18.1	54.3	5.1
Plastic properties			
w_L	34.8	54.5	Non-plastic (NP)
w_p	17.5	30.7	
Visual description	Light brown sandy, and silty clay	Dark brown silty clay, trace of gravel	Medium brown very gravelly coarse sand

AASHTO CLASSIFICATION OF SOIL A:

1. Compute the plasticity index, $I_P = w_L - w_P$

$$I_P = 34.8 - 17.5 = 17.3 > 10$$

   Therefore, from Fig. 8-1*a* the value of $I_P > 11$ minimum controls.
2. Since 18.1 percent (< 35 percent max) passes the No. 200 sieve, the soil is an A-2 with subgroup to be determined from I_P and w_L.
3. Proceeding across Fig. 8-1*a* from left to right with

$$w_L = 34.8 < 40 \qquad \text{and } I_p = 17.3 > 11$$

   the *first* soil found that satisfies these criteria is an A-2-6.
4. Next compute the group index GI (which is generally not more than about 4 for any A-2 soil). For this soil we will both compute the group index and obtain the value from Fig. 8-1*c*. Computing, and using Eq. (8-1) for terms:

$$F - 35 = 0 \qquad \text{since less than 35 percent passes the No. 200 sieve}$$

$$F - 15 = 18.1 - 15.0 = 3.1$$

$$w_L - 40 = 0 \qquad \text{since } w_L \text{ is less than 40}$$

$$I_P - 10 = 17.3 - 10 = 7.3$$

$$\text{GI} = 0[0.2 + 0.005(0)] + 0.01(3.1)(7.3) = 0.22$$

   and rounding to the nearest integer, obtain GI = 0.

   From Fig. 8-1*c*, the GI is approximately 0.0 from both parts using F, w_L, I_p, and % passing No. 200 sieve. The final classification of soil *A* is

Light brown, sandy, and silty clay, A-2-6(0)

AASHTO CLASSIFICATION OF SOIL B:

1. The plasticity index is

$$I_P = 54.5 - 30.7 = 23.8 > 10$$

   Therefore the value of $I_P \geq 11$ controls.
2. The percent passing the No. 200 sieve is 54.3 > 35; therefore, the soil is an A-4, A-5, A-6 or A-7 (the visual description eliminates A-8).
3. With $w_L = 54.5$ and $I_P = 23.8$ the soil is an A-7, but we must still find if an A-7-5 or A-7-6. Inspection of Fig. 8-1*b* displays that the scale is too small to use, so we will compute the coordinates using the equation shown on the figure:

$$I_P = 23.8 \qquad \text{(computed above for this soil)}$$

$$I_P = w_L - 30 = 54.5 - 30 = 24.5 \qquad \text{(computed)}$$

   Since 23.8 < 24.5 the $I_P - w_L$ coordinates plot in the A-7-5 zone, and thus, the soil is an A-7-5.

4. We next compute the group index from Fig. 8-1*c* and using % Finer $F = 54.3$ obtain:

approximately for $w_L = 54.5$ PGI $= 5.2$

approximately for $I_P = 23.8$ PGI $= 5.2$

Summing partial values, obtain GI $= 10.4$ round and use 10

The final classification of soil *B* is:

Dark brown, silty clay, trace of gravel, A-7-5(10)

AASHTO CLASSIFICATION OF SOIL C:

1. For soil C we find by a rapid elimination that the soil is either A-1 or A-3. With the percent passing the No. 40 sieve of $38.5 < 51$ but $38.5 > 30$, the soil must be A-1-b. Since the soil is non-plastic there is no group index GI for this soil. The final classification of soil *C* is:

Medium brown, very gravelly, coarse sand, A-1-b

B. UNIFIED SOIL CLASSIFICATION

The essential elements of this system of classifying soils were first proposed by Casagrande (1942) and were subsequently adopted by the U.S. Corps of Engineers for airfield construction. Currently, this system is used with slight modifications in many countries outside the United States. The system is rather widely used inside the United States by organizations such as the U.S. Corps of Engineers, Bureau of Reclamation, and with slight modifications by most consulting firms.

In 1985 ASTM approved a standard method for making this classification. The principal differences between the USC method as proposed by Casagrande and the ASTM method is identification and use of the percentages passing a 3-in (75-mm) sieve and retained on the No. 4 (4.75-mm) sieve and the percentages passing or retained on the No. 200 (0.075-mm sieve) as part of a soil description. Although the standard does not mention computer programming, the percentages retained and/or passing are made specific so the method can be programmed. The ASTM standard method is programmed on the enclosed computer diskette.

The procedure used in this manual is based on the original Casagrande (1942) Unified Soil Classification System. If the ASTM system is to be used the user should obtain a copy of ASTM D 2487 and use the computer program included here. If you obtain a good understanding of the USC system you only need make a few extensions to be able to use the ASTM 2487 standard.

Table 8-1 presents the primary factors to consider in classifying a soil according to the Unified Soil Classification system. Basically the soil is:

Coarse grained if more than 50% is retained on		No. 200 Fine-grained if more than 50% passes
	4.75 mm	0.075 mm
Gravel	Sand	Silt or Clay
If more than 50% of coarse fraction is retained on the No. 4 sieve	If more than 50% of coarse fraction passes the No. 4 sieve	Fine grained soil is: Silt (M) Clay (C) Organic (O)

We shall consider these subdivisions in more detail in the following:

1. Gravels and sands are

   GW, GP, SW, or SP

   if less than 5 percent of the material passes the No. 200 sieve; G = gravel; S = sand; W = well-graded; P = poorly-graded. The well- or poorly-graded designations depend on C_U and C_C as defined in Test No. 5 and with numerical values as shown in Table 8-1.
2. Gravels and sands are

   GM, GC, SM, or SC

   if more than 12 percent passes the No. 200 sieve; M = silt; C = clay. The silt or clay designation is determined by performing the liquid and plastic limit test on the (−) No. 40 fraction and using the *A* chart of Fig. 8-2. This chart is also a Casagrande contribution to the USC system, and the *A* line shown on this chart is sometimes called *Casagrande's A line*.

   The chart as presented in this manual has been slightly modified based on the Corps of Engineers findings that no soil has so far been found with coordinates that lie above the "upper limit" or *U* line shown. This chart and lines are part of the ASTM D 2487 standard.
3. Gravels and sands are (note using dual symbols)

   GW-GC SW-SC GP-GC SP-SC

   or

   GW-GM SW-SM GP-GM SP-SM

   if between 5 and 12 percent of the material passes the No. 200 sieve. If, however, you have 5 to 6 percent passing the No. 200 sieve and it is not possible to perform the plastic limit tests, do not use dual classification—instead use designations described in "1" above. Carefully note that the M or C designation is derived from performing plastic limit tests and using Casagrande's *A* chart.
4. Fine-grained soils (more than 50 percent passes the No. 200 sieve) are:

   ML, OL, or CL

   if the liquid limits are < 50 percent; M = silt; O = organic soils; C = clay. L = *Less* than 50 percent for w_L
5. Fine grained soils are

   MH, OH, or CH

   if the liquid limits are ≥ 50 percent; H = Higher than 50 percent. Whether a soil is a Clay (C), Silt (M), or Organic (O) depends on whether the soil coordinates plot above or below the *A* line on Fig. 8-2.

   The organic (O) designation also depends on visual appearance and odor in the USC method. In the ASTM method the O designation is more specifically defined by using a comparison of the air-dry liquid limit w_L and the oven-dried w'_L. If the oven dried value is

   $$w'_L < 0.75 w_L$$

   and the appearance and odor indicates "organic" then classify the soil as O. The above air-dry vs. oven-dry liquid limit check is in the included computer program.

Table 8-1 The Unified Soil Classification System

Major divisions			Group symbol		Typical names	Classification criteria for coarse-grained soils	
Coarse-grained soils (more than half of material is larger than No. 200)	Gravels (more than half of coarse fraction is larger than No. 4 sieve size)	Clean gravels (little or no fines)	GW		Well-graded gravels, gravel-sand mixtures, little or no fines	$C_U \geq 4$ $1 \leq C_C \leq 3$	
			GP		Poorly graded gravels, gravel-sand mixtures, little or no fines	Not meeting all gradation requirements for GW ($C_U < 4$ or $1 > C_c > 3$)	
		Gravels with fines (appreciable amount of fines)	GM	d u	Silty gravels, gravel-sand-silt mixtures	Atterberg limits below A line or $I_p < 4$	Above A line with $4 < I_p < 7$ are borderline cases requiring use of dual symbols
			GC		Clayey gravels, gravel-sand-clay mixtures	Atterberg limits above A line with $I_p > 7$	
	Sands (more than half of coarse fraction is smaller than No. 4 sieve size)	Clean sands (little or no fines)	SW		Well-graded sands, gravelly sands, little or no fines	$C_U \geq 6$ $1 \leq C_C \leq 3$	
			SP		Poorly graded sands, gravelly sands, little or no fines	Not meeting all gradation requirements for SW ($C_U < 6$ or $1 > C_C > 3$)	
		Sands with fines (appreciable amount of fines)	SM	d u	Silty sands, sand-silt mixtures	Atterberg limits below A line or $I_p < 4$	Limits plotting in hatched zone with $4 \leq I_p \leq 7$ are borderline cases requiring use of dual symbols
			SC		Clayey sands, sand-clay mixtures	Atterberg limits above A line with $I_p > 7$	
Fine-grained soils (more than half of material is smaller than No. 200)	Silts and clays (liquid limit < 50)		ML		Inorganic silts and very fine sands, rock flour, silty or clayey fine sands, or clayey silts with slight plasticity	1. Determine percentages of sand and gravel from grain-size curve. 2. Depending on percentages of fines (fraction smaller than 200 sieve size), coarse-grained soils are classified as follows: Less than 5%–GW, GP, SW, SP More than 12%–GM, GC, SM, SC 5 to 12%–Borderline cases requiring dual symbols	
			CL		Inorganic clays of low to medium plasticity, gravelly clays, sandy clays, silty clays, lean clays		
			OL		Organic silts and organic silty clays of low plasticity		
	Silts and clays (liquid limit > 50)		MH		Inorganic silts, micaceous or diatomaceous fine sandy or silty soils, elastic silts		
			CH		Inorganic clays or high plasticity, fat clays		
			OH		Organic clays of medium to high plasticity, organic silts		
	Highly organic soils		Pt		Peat and other highly organic soils		

$$C_U = \frac{D_{60}}{D_{10}}$$

$$C_C = \frac{D_{30}^2}{D_{10} D_{60}}$$

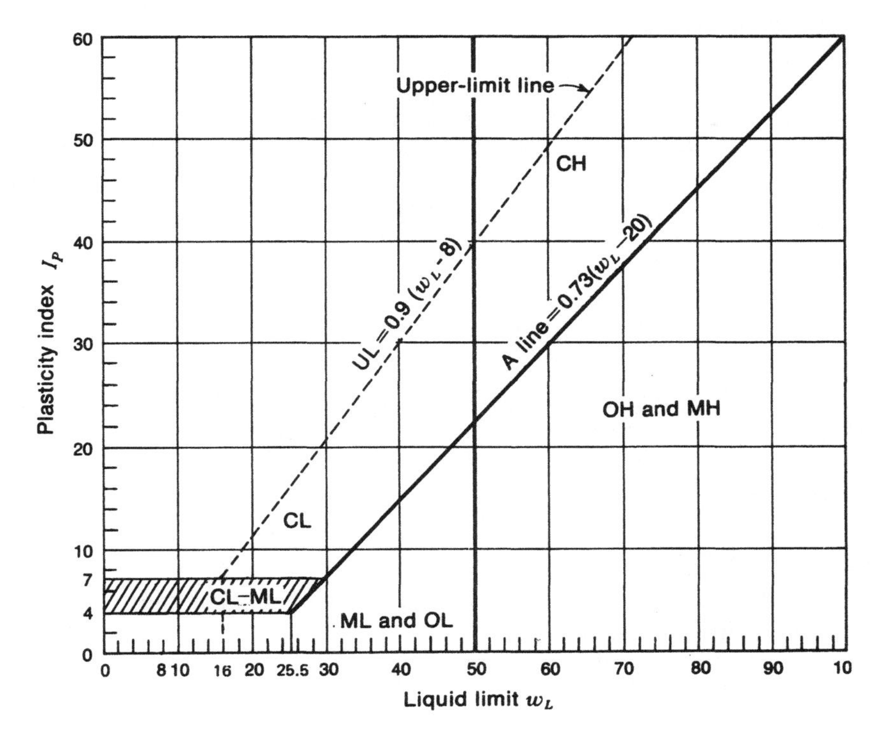

Figure 8-2
Plasticity chart (also called "Casagrande's A-chart") to use with Table 8-1 for the Unified Soil Classification System.

The liquid and plastic limits are performed on the (−) No. 40 sieve fraction of all of the soils, including gravels, sands, and the fine-grained soils using the procedures outlined in Test No. 3. Plasticity limit tests are not required for soils where the percent passing the No. 200 sieve ≤ 5 percent.

A visual description of the soil should accompany the letter classification. The ASTM standard includes some description in terms of sandy or gravelly, but color is also very important. Certain areas are underlain with soil deposits having a distinctive color (e.g., Boston blue clay, Chicago blue clay) which may be red, green, blue, grey, black, and so on. Geotechnical engineers become familiar with the characteristics of this material so the color identification is of considerable aid in augmenting the data base on the soil.

We will now reclassify the three soils previously classified by the AASHTO method. You can now see that the additional sieves in the stacks were necessary to allow this alternative soil classification. *You should also classify these three soils using the enclosed computer programs for both methods to develop confidence.*

USC CLASSIFICATION OF SOIL A

1. We have 18.1 < 50 percent passing the No. 200 sieve and more than 50 percent passing the No. 4 sieve (since 68.5 percent passed the No. 10); therefore the soil is either

   SM or SC

2. For $w_L = 34.8 < 50$ and $I_P = 17.3$ the coordinates on the plasticity chart of Fig. 8-2 indicate the soil is a CL. Taking C for "clay" the soil is classified as

   Light brown, sandy and silty clay, SC

USC CLASSIFICATION OF SOIL B:

1. Since 54.3 percent passes the No. 200 sieve, the soil is immediately fine-grained and one of MH, OH, or CH because $w_L = 54.5 > 50$ percent.
2. From the plasticity chart at $w_L = 54.5 > 50$ and $I_P = 54.5 - 30.7 = 23.8$, the soil coordinates are difficult to locate as above, below or on the A-line; we will compute I_P based on w_L and compare with 23.8 from using w_p

   From Fig. 8-2: $I_P = 0.73(w_L - 20)$

   $= 0.73(54.5 - 20) = 25.2 > 23.8$

   Since the computed value > 23.8 the soil point must plot below the A line and the soil is an MH: (O is eliminated from the given visual description) and soil is

   Dark brown clayey silt, MH

USC CLASSIFICATION OF SOIL C:

1. With 5.1 percent passing the No. 200 sieve and the soil being non-plastic (cannot obtain w_L, w_P) the soil must be one of

   GW, GP, SW, or SP

   Taking 5.1 as approximately 5 percent with sufficient accuracy and is strictly a matter of judgment.
2. It will be necessary to plot a grain-size distribution curve for this soil and obtain the following data:

   $D_{60} = 2.0$ mm $\quad D_{30} = 0.29$ mm $\quad$ and $\quad D_{10} = 0.086$ mm

   Use these values to compute

   $C_U = D_{60}/D_{10} = 2.0/0.086 = 23.3 > 6$

   $C_C = D_{30}^2/D_{10}D_{60} = 0.29^2/(2.00 \times 0.086) = 0.5 < 1$ (not between 1 and 3)

   The gradation criteria are not met for well-graded sand or gravel so the soil is poorly graded (P-suffix). Note that for some plots the C_i-values may be borderline. In these cases it will be necessary to recheck your curve and you may need to do a second sieve analysis to verify results. This is not the case here. However, if you plot the curve it is quite possible to compute slightly different values of C_C and C_U but probably not so different as to identify the soil as well-graded instead of poorly-graded.
3. With 94.9 percent retained on the No. 200 sieve, and with 69.3 percent passing the No. 4 sieve, the percent between the No. 4 and No. 200 is

   $69.3 - 5.1 = 64.2\%$

   % retained on No. 4 = $100 - 69.3 = 30.7\%$

   % passing No. 200 = 5.1%

   Total = 100.0%

   With 64.2 percent of total sample between the No. 4 and No. 200 sieves *more than half the coarse fraction is sand.* The soil is therefore

   Medium brown, coarse, poorly graded sand, SP

Procedure

This is suggested as an individual project

1. Each person will classify the soils assigned by the instructor[1] according to the USC and/or AASHTO systems.
2. Give some thought to the information required by a particular classification system so that the amount of soil testing is kept to a minimum. Use the one-point method of liquid limit determination (if allowed).
3. Be sure to obtain reasonably representative samples from the soil containers so that you can check your work if necessary.
4. Use the wash method to find the percent passing the No. 200 sieve; otherwise you may misclassify the soil. It is not unusual to dry-sieve a soil with considerable care and find only 40 to 45 percent passing the No. 200 sieve while with washing well over 50 percent passes.
5. Instead of a formal report, submit this work in a folder with a flyleaf, tabulation of results on the data sheets for this project and any additional data sheets used for sieve analysis, grain-size distribution curves, Atterberg limits, and so on. Be sure to provide a visual description for each soil along with the classification system symbols.

[1]Four to six soils should be classified using two laboratory periods plus whatever additional time the student requires to do the necessary laboratory work/classification. This can be done by keeping on hand 15 to 20 different appearing soils in containers holding 10 to 15 kg. Each student is given x numbers corresponding to containers + the laboratory soil(s) used in Test 3, 5, and 6. This tends to ensure individual work on the project.

9: MOISTURE-UNIT WEIGHT RELATIONSHIPS (COMPACTION TEST)

References

ASTM D 698-78 (standard) and D 1557-78 (reapproved 1990)
AASHTO T 99-90 (standard) and T 180-90 (modified)
Burmister, D. M. (1965), "Environmental Factors in Soil Compaction," ASTM STP No. 377, pp. 47–66.
Gordon, B. B., W. D. Hammond, and R. K. Miller (1965), "Effect of Rock Content on Compaction Characteristics of Clayey Soil," ASTM STP No. 377, pp. 31–46.
Johnson, A. W., and J. R. Sallberg (1962), "Factors Influencing Compaction Results," Highway Research Board Bulletin No. 319, 148 pages.
Lambe, T. W. (1960), "Compacted Clay: A Symposium," *Trans. ASCE*, Vol. 125, pp. 682–756 (also in *J. Soil Mech. Found. Div.*, ASCE SM 2, May 1958).
Lee, P. Y., and R. J. Suedkamp (1972), "Characteristics of Irregularly Shaped Compaction Curves," Highway Research Record No. 381, pp. 1–9.
Proctor, R. R. (1933), Fundamental Principles of Soil Compaction, *Eng. News-Record*, Aug. 31, Sept. 7, 21, and 28.
Ring, G. W., III, J. R. Sallberg, and W. H. Collins (1962), "Correlation of Compaction and Classification Test Data," Highway Research Board Bulletin No. 325, pp. 55–75.
Seed, H. B., and C. K. Chan (1959), "Structure and Strength Characteristics of Compacted Clays," *J. Soil Mech. Found. Div.*, ASCE, SM 5, Oct., pp. 87–128.
Wilson, S. D. (1950), Small Compaction Apparatus Duplicates Field Results Closely, *Eng. News-Record*, Nov. 2, pp. 34–36.

Objective

To familiarize the reader with the laboratory compaction test and to obtain the moisture-unit weight relationship for a given compactive effort on a particular soil.

Equipment[1,2]

Compaction mold[3] with base plate and collar (refer to Fig. 9-1)
Compaction rammer (24.5 N × 0.305 m drop or 44.5 N × 0.46 m drop) or mechanically operated rammer
10 to 12 moisture cans
Steel straightedge to smooth sample ends flush with mold
Sample extruder (jack) or mechanical pulverizer as in Fig. 9-4
Large mixing pan and large spoon for dispensing soil
Soil mixer (optional) or necessary soil-mixing tools such as trowel, spoon, spatula

General Discussion

In 1933, R. R. Proctor presented four articles in the *Engineering News-Record* [Proctor (1933)] which are the basis for the standard compaction test currently used (sometimes called Proctor tests or simply "Proctors").

[1]Instructor should obtain water content of soil for this project prior to laboratory period—also values of w_L, w_p from earlier laboratory tests so that the OMC can be estimated using Fig. 9-2.
[2]Soil may be used from Test No. 10 following, if done first.
[3]See the 2/e of this manual for method of fabrication of an "SI" mold of 1000 cm^3.

The original test consisted of taking about 3 kg of soil, passing it through a No. 4 sieve,[4] adding water, and compacting it into a 944 cm^3 mold in three layers with 25 blows per layer, using a 24.5 N compaction rammer dropping 0.305 m onto the soil. This provides a nominal compaction energy (kiloJoules, kJ) to the soil of:

$$\text{CE} = \frac{3(25)(24.5)(0.305)}{9.44 \times 10^{-4}(1000)} = 593.7 \qquad \text{kJ/m}^3$$

The compacted sample is then broken down to the No. 4 sieve size *as determined visually*, water-content samples are taken, more water is added, the soil is thoroughly remixed, and the process of compacting a mold of soil is repeated.[5] This sequence is repeated a sufficient number of times that a curve of dry unit weight vs. water content can be drawn which has a zero slope (a maximum value) with sufficient points on either side of the maximum unit weight point to accurately define its location. Dry unit weight is always the ordinate of this curve; however, in the original Proctor method the void ratio e was used instead of γ dry. The maximum ordinate value is termed *maximum dry unit weight* γ_d, and the water content at which this dry unit weight occurs is termed the *optimum moisture content* (OMC).

Several comments can be made at this point:

1. A standard test should be reproducible—this one is only by chance. Some of the factors causing this are the use of fresh samples for test points versus reusing samples,[6] mixing time, how well the water is distributed within sample, temperature, operator interest and plotting scale. Maximum dry unit weights can vary by as much as 0.8 to 1.2 kN/m^3 without some attention to test details.
2. When one is testing an initially air-dry soil, unless the sample is mixed with the first increment of water and allowed to cure (say, overnight), the results may yield a curve with erratic points (the locus of points is not a smooth curve) on the dry side of the optimum or peak point. Noncuring may also shift the OMC to the right (tend to increase it) on the unit weight vs. water content curve. Some soils may yield a somewhat erratic "dry side" curve as an apparent soil characteristic [see Lee and Suedkamp (1972)].

Figure 9-1
Essential compaction test equipment: Large 20-kg balance; compaction mold, base plate and collar; 24.5 N x 0. 305 m drop rammer; steel straightedge to smooth ends of compacted sample.

[4]Both AASHTO and ASTM include standard procedures for soils that contain materials larger than the No. 4 sieve.

[5]ASTM changed the initial test procedure in 1978 to require use of a fresh soil sample for each test point. AASHTO requires that the test sample be broken down to the No. 4 sieve, mixed with any soil left over, water added, the soil thoroughly re-mixed and the next test point compacted.

[6]Research by the author indicates that additional mechanical mixing, say, 8 to 10 min, will reduce the differences of using a fresh sample versus reused sample to around 0.2 to 0.4 kN/m^3. Soils with low plasticity tend to give almost no differences between fresh and reused soil. Differences are also less when the air-dry soil is mixed with water 10 to 12 h before the test.

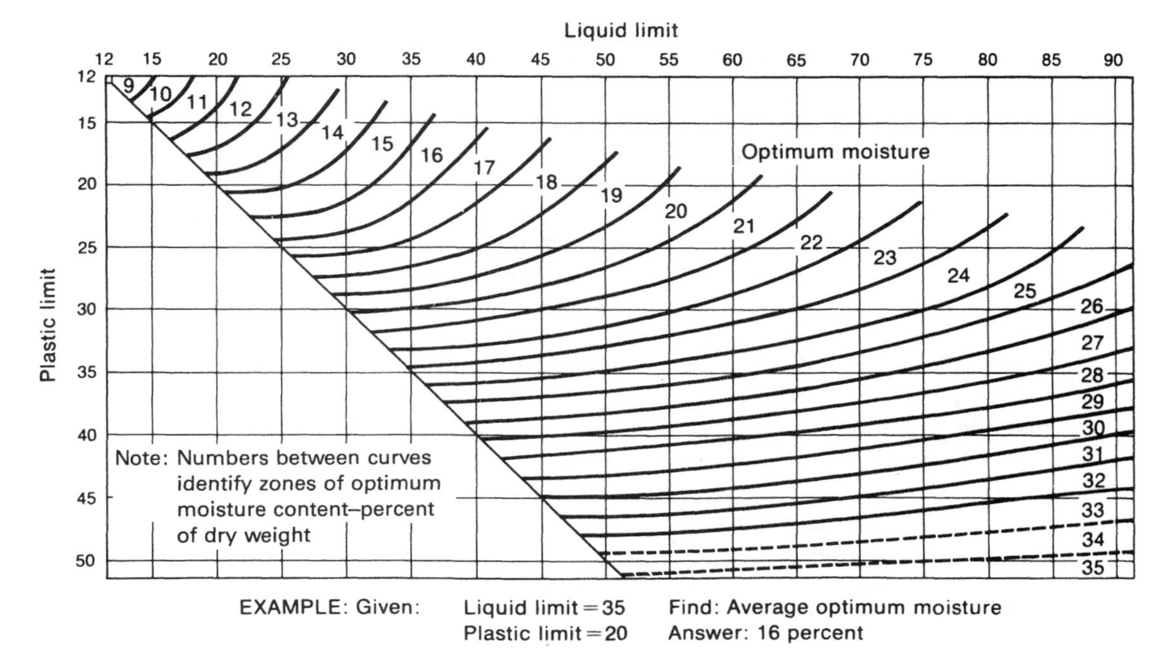

Figure 9-2
Chart to find the approximate optimum moisture content (OMC) for a soil using the standard compaction test. [From Johnson and Sallberg (1962), p. 125.]

3. If the cylinder of soil is not compacted in three approximately equal increments, the curve points will be somewhat erratic, that is, will not fall on a smooth curve on either side of optimum.
4. To avoid large numbers of compacted cylinders, and since about five well-placed points will define the curve very well, it is desirable to begin the first test at a water content about 4 to 5 percent from the OMC. Then adding 2 percent moisture (by weight) on successive trials, the last point should be about 4 to 5 percent on the wet side of the curve peak. It is not desirable to space the water content points over 3 percent apart, and 2 percent is preferred. If the vertical scale is adequate, this closer spacing more clearly identifies the OMC to the nearest 0.1 percent.

   Figure 9-2 may be used to estimate the starting-point water content. This chart should give the OMC within ± 2 percent if the liquid- and plastic-limit values are reasonably accurate.

   With practice, one may estimate the OMC by adding water until an average squeeze effort on a handful of soil produces definite finger imprints and the resulting molded soil has resistance to breaking apart. Dry this sample to obtain the water content as the OMC estimate. Since a highly accurate value is not necessary use the most rapid drying method available—particularly if time is critical.
5. The compaction mold should be placed on a surface that does not vibrate during the compacting process so that the rammer energy is fully transmitted to the soil as compaction effort. Both the AASHTO and ASTM standards specify that the mold should rest on a base consisting of a rigid cylinder or block of concrete with a mass of at least 90 kg (200 lb).

The original Proctor procedure specified "... 25 firm 12-in strokes" but this has been modified to "25 blows from a rammer dropping free from a height of 12 in." Currently the SI equivalent of a 0.305 m fall is used and the rammer has been put into a controlled height cylinder to assist in obtaining the correct fall.

Soil compaction is generally the cheapest method of soil stabilization available. Soil stabilization is improving the undesirable physical properties of a soil to achieve the desired shear strength, structure, and void ratio. There are many methods of stabilizing soil using chemicals such as lime, lime–fly ash admixtures, cement, and phosphoric acid compounds. These methods are usually more expensive and may utilize compaction in addition to the admixture since its incorporation into the soil mass may loosen the soil considerably. This is especially evident when adding lime, lime–fly ash, and cement since the soil is usually excavated, blended with the admixture and replaced, or the admixture is spread over the surface and then disced for blending.

Compaction achieves soil stabilization through the input of energy to produce a more dense soil structure than previously existed. The laboratory test supplies compaction energy to the soil in the mold by the impact of the rammer. In the field the energy comes from rolling with sheeps-foot, wobble-wheel, or smooth-wheel rollers. The smooth-wheel roller often has a vibratory device attached so both pressure and vibration are used to increase the soil density. The type of roller used depends on the soil. The smooth-wheel roller is best for sand; the sheeps-foot roller is best for cohesive soil. Note, however, that there is no field method that compacts the soil in a manner similar to this test.

Generally, the compaction effort imparts to the soil:

1. An increase in shear strength since this is a function of unit weight (the other variables are structure, ϕ and c)
2. An increase in swell potential
3. An increase in unit weight [$\gamma = f(e)$]
4. A decrease in shrinkage potential
5. A decrease in permeability [$k = f(e)$]
6. A decrease in compressibility [$\Delta H = f(e)$]

It has also been found that whether the field compaction water content is on the dry or wet side of the OMC also affects the end result—in particular Items 2, 4, and 6.

It is now recognized that the structure of the resulting soil mass (especially when fine-grained soils are present) is intimately associated with the compaction process and the water content at which the mass is compacted (whether on the dry or wet side of the OMC). This concept is extremely important for compacting the clay cores of dams (as one example), where settlements could cause cracks in the core. It has been found that the *dispersed* soil structure (parallel particle orientations) obtained by compacting these soils on the wet side of optimum results in a soil that has a somewhat lower shear strength but that can undergo larger settlements without cracking and causing a large leakage and/or actual dam failure. Compaction of the soil on the wet side of optimum also reduces the permeability, as compared with compacting the soil on the dry side of optimum. In this case the dispersed structure has fewer interconnected pores to produce continuous flow paths so is more resistant to water transmission.

Conversely, the *flocculated* structure (random particle orientations) resulting from compacting a clayey soil on the dry side of optimum is *less* susceptible to shrinkage on drying but is *more* susceptible to swelling on saturation. This is attributed to a combination of the flocculated structure, sensitivity of additional water at the contact points, and the lower reference (w at compaction) water content. For soil compacted on the wet side of the OMC the reference water content is already closer to saturation ($S = 100\%$) so swelling would be less and shrinkage as the soil dries would be more since the reference point is higher.

The ultimate strength of soils compacted on the dry side of the OMC is higher at low strains and the soil usually fails abruptly by fracture (also called *brittle* failure). If compacted on the wet side of optimum the failure is gradual and often not clearly defined (a *plastic* failure) and strength is ultimately lower. The residual strength of the soil compacted on the dry side of optimum is nearly the same as the ultimate strength of soil compacted on the wet side of optimum.

Thus, for highway work, where low strains beneath the pavement are desirable, the soil should be compacted at a water content from the dry side of optimum to optimum. The soil shell surrounding the clay core of an earth dam should also be compacted on the dry side of optimum since strength is more important than seepage. The clay core, which resists seepage, should be compacted on the wet side to produce a dispersed structure. This structure both reduces seepage and, since large settlements are possible, allows larger displacement strains without cracking. It should be apparent from this discussion that the soil behavior is being considered after either partial drying or complete saturation from the placing water content.

From this brief discussion, it is evident that compaction criteria should be based on the required soil behavior and not on some specified unit weight or percent of the laboratory determined unit weight of the soil. However, in many cases—especially where unit weight (and settlement control) is the only needed property—this does produce (as it has in the past) a satisfactory end product.

Relative compaction is the term used to compare the in situ (or field) compacted dry unit weight γ_{df} to the laboratory compacted maximum dry unit weight γ_{dL} given by the following equation:

$$RC = \frac{\gamma_{df}}{\gamma_{dL}} \times 100 \qquad (\%) \tag{9-1}$$

and may be more or less than 100 percent. For example, a particular project may specify that soil be compacted to 90 percent relative compaction; another project may specify 105 percent relative compaction. If the maximum dry unit weight from a laboratory compaction test is 18.50 kN/m^3 the field unit weight would be as follows:

At RC = 90 percent: 18.50(0.90) = 16.65 kN/m^3
At RC = 105 percent: 18.50(1.05) = 19.42 kN/m^3

Considerable data have been collected that indicate soil structure, unit weight, and OMC depend on the method of providing the compaction energy (kneading, impact, vibration, etc.). Despite this shortcoming, the standard compaction test as outlined here is widely used to establish the project quality control on field compaction unit weight. There is a considerable collection of data that indicates that this method works; also since the time the test was proposed, no one has developed a better method, so there is no valid reason at present to discard the procedure.

THE MODIFIED COMPACTION TEST

Just prior to and during World War II, the use of heavy military aircraft began requiring a subgrade density beneath airfield runways greater than that provided by the standard compaction unit weight. Rather than use relative compactions greater than 100 percent, a modified test was introduced (now termed *modified compaction test, modified AASHTO,* or *modified Proctor*). Details of this test and comparison with the standard test are in the following table based on Method A by either AASHTO or ASTM:

Using	Standard	Modified
Mold[7]	944 cm^3	944 cm^3
Rammer	24.5 N	44.5 N
Soil layers	3 @ 25 blows/layer	5 @ 25 blows/layer
Rammer fall	0.305 m	0.46 m
Compaction Energy	593.7 kJ/m^3	2710.5 kJ/m^3

[7] The compaction test uses the 944 cm^3 [diam approximately 102 mm (4-in)] mold for soil passing the No. 4 (4.75) sieve. If larger gravel is present, test modifications by both AASHTO and ASTM allow use of a 2123 cm^3 [diam approximately 152 mm (6 in)]. In these cases consult the appropriate standard for method B, C, or D.

The essentials of the modified test are the same as the standard compaction test, that is, use (−) No. 4 sieve material for Method A or consult ASTM or AASHTO for cases where larger gravel is present, develop the curve as for the standard test by adding increments of water, compact, break the sample down[8], and repeat the process until a curve is obtained.

The modified compaction test introduces a nominal compaction energy into the soil of 2710 kJ/m^3 which is about five times that of the standard test. For this extra input energy a 5 to 10 percent increase in unit weight is obtained together with a smaller value of OMC. The data for the modified compaction test are also presented as a curve of dry unit weight vs. water content.

The plot of dry unit weight vs. water content shows that the compaction process becomes more efficient at *any given compactive effort* as the water content increases—but only up to the OMC—then the efficiency decreases. This initial increase in efficiency is from a number of factors but predominating are the slaking and/or general breakdown of interparticle bonds of the clay lumps and possibly some particle contact point lubrication. On the wet side of optimum, the lumps have generally been slaked, but now the excess moisture causes high instantaneous pore pressures as the rammer strikes. The high pore pressures result in a greatly reduced shear strength, and the rammer energy simply shears (remolds) the soil rather than increasing the unit weight further.

The soil mass involved in the compaction process starts as a three-phase system of soil, water, and air. During the initial trials considerable air is present, but the process results in a change of state with more soil and water being present. Even at the OMC there is considerable air present. On the wet side of optimum, the principal effect is to displace more and more air with water. If the process were completely efficient, it might be possible to replace all the air in the voids with water to produce a two-phase system (a condition of *zero air voids*). Since it is never possible to get all the air out of the voids, which would result in a condition of S = 100 percent, any compaction curve *will always fall below* the zero-air-voids curve.

For any given water content w the zero-air-voids dry unit weight γ_{zav} is computed as

$$\gamma_{zav} = \frac{G_s \gamma_w}{1 + wG_s} \tag{9-2}$$

where G_s = specific gravity of soil solids
w = water content (usually arbitrarily selected for convenience as .05, .10, .15, .20, .30, etc.—and decimal, not percent
γ_w = unit weight of water in units of test (9.807 kN/m^3 or 62.4 lb/ft^3)

The locus of points from a compaction test often produces a slightly concave upward curve which may result partly from the initial soil state (whether air-dried and water added about 10 to 12 h before test or starting from air-dry soil). It is not usually possible to perform the test without substantial air-drying in order to sieve the soil through a No. 4 sieve and/or to achieve breakdown to elemental particles prior to performing the test.

Numerous factors influence soil compaction test results using either the standard or modified method, including:

Temperature
Size of molds (if the diameter/height ratio is approximately constant this effect is negligible)
Distribution of blows on any layer
Whether layers are of about equal thickness; Excess quantity of soil in mold (should not extend more than about 6 mm into collar)
Type of soil (only cohesive soil can be compacted using this test method)
Amount of processing (mixing, curing, manipulation)

[8]This is not a trivial task using the modified compaction method and a device such as developed by the author and shown in Fig. 9-4 is suggested.

The reader should use the reference list for additional study of the factors that can influence compaction results. These references are about the most recent available since it was fairly easy to do extensive research on this test and thus received early attention.

THE HARVARD MINIATURE COMPACTION EQUIPMENT/TEST

While doing graduate studies at Harvard University around 1949 Wilson (1950) developed the compaction apparatus shown in Fig. 9-3 to perform a compaction test using a kneading effort. The intent was to duplicate more closely field compaction using a sheeps-foot roller. This apparatus consists of a mold 1 5/16 in. in diameter × 2.816 in. long having a volume of 1/454 ft^3 (62.4 cm^3). During compaction the mold and its collar are attached firmly to the base. Compaction is accomplished by means of a tamper employing a spring-loaded plunger. There are two spring calibrations available: 89 N (20 lb) and 178 N (40 lb), that is, it takes 89 N to just break spring-plunger contact.

The soil may be compacted using any number of layers and tamps per layer. Commonly three layers at 25 tamps per layer may be used. After a sample is compacted the mold is removed from the base plate, the collar is removed, and the two ends are struck flush. The mold and sample are weighed for total mass and the mold mass subtracted. The sample mass in grams is the compacted unit weight γ in lb/ft^3; the sample mass in g divided by the mold volume of 62.4 cm^3 obtains the compacted density ρ in g/cm^3.

The principal miniature equipment test deficiency is that there is no direct correlation between the Proctor compaction density and that obtained by this method. The author's experience has been that the maximum density for some kneading effort (say 25 tamps/layer and three layers) occurs at an OMC that is on the dry side of the Proctor value (you may get 10 percent vs. 14 or 15 percent for the Proctor method). If you want to reproduce the Proctor density with this equipment you may have to try several compacting efforts (3 layers at 25 tamps, 4 layers at 30 tamps, etc.) to find which is the closest to the Proctor value.

The principal test advantages are that (1) only small amounts of soil are required (but must all pass the No. 4 sieve) and (2) one can obtain samples of dimensions suitable for testing in unconfined (Test 14) or triaxial (Test 15 or 16) compression. These compacted samples also may be suitable for falling-head permeability (Test 12) tests using triaxial apparatus.

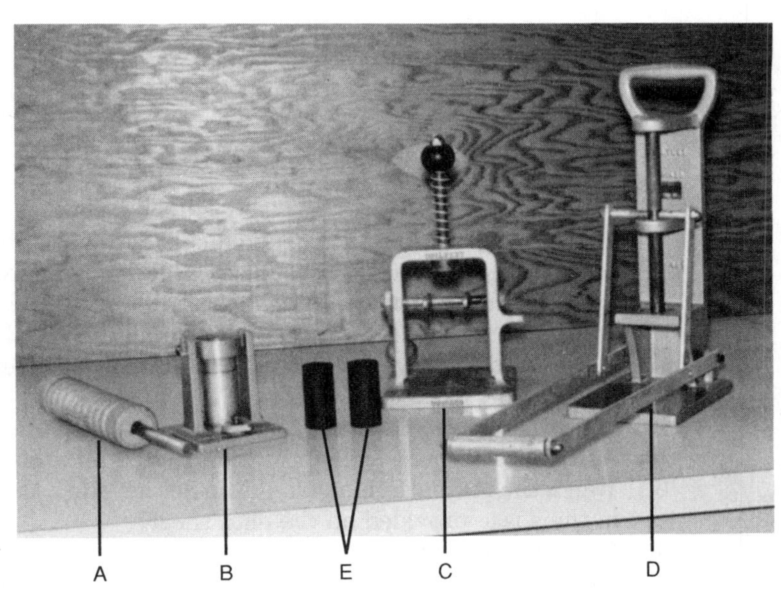

Figure 9-3
Harvard Miniature Compaction Apparatus. Letter codes as follows: A = tamper (calibrated spring inside gripping handle); B = mold with collar attached to base; C = device to remove mold collar without tearing last layer; D = sample extruder; E = two compacted samples.

Procedure

This is suggested as a group project.

NOTES

a. The initial water content is useful in this test. If rapid test equipment is available w can be obtained at the start of the project. If not, a water content sample should be taken the day before so that it will be oven-dry and w can be computed before the project begins. If a poor start is made it may take six to eight trials to obtain the compaction curve—especially for soils where the OMC is 17 to 22 percent.
b. It is suggested that the soil be reused in student laboratories (the AASHTO method) so that less soil is needed and there will be less to dispose of. This is also the "original Proctor" procedure.
c. Consider using a fabricated soil so the OMC is not over about 10 to 14 percent. A high OMC is difficult for students to obtain without using six to eight test points. Fabricate a "soil" by blending a cohesive stock material with sand. For first group use 85% sand with 15 percent clay for about 3.5 kg of soil; second group 80–20; third group 75–25, and so on. The students should then run necessary tests to classify this soil by either AASHTO or USC (or both.)

1. Each group should take 3 to 4 kg (nominal mass) of air-dry soil, pulverize sufficiently to run through the No. 4 sieve, then mix with an initial amount of water (mL). The initial amount of water should be based on the present water content; one should attempt to start with an initial moisture content 4 to 5 percent below OMC. Obtain the estimated OMC from Fig. 9-2 or as described earlier using the squeeze test.

   For example:

   estimated OMC = 18 percent

   present water content = 8 percent

   initial starting water content = 14 percent

   initial air-dry soil mass = 3 kg (3000 g)

   $$\text{Water to add} = \frac{3000(14 - 8)}{100} = 180 \text{ g} = 180 \text{ mL}$$

   Measure 180 mL into a graduated cylinder, add to the soil, and mix to a uniform color either using trowels and spoons or a mechanical mixer. Put in a large container, cover with a damp paper towel and seal to "cure" for 12 to 24 h. The curing step may be omitted for student laboratories but recall the initial part of the dry unit weight versus water content curve may have a characteristic convex upward shape if this is not done. If curing is omitted go to Step 3.
2. If the soil has been cured, add 1 percent moisture by mass [.01 x 3000 = 30 g (or 30 mL)] to account for evaporation losses. Mix this water into the soil carefully.
3. Weigh the compaction mold, but do not include collar or base plate and record its mass in the space provided on the data sheet.
4. Measure the compaction mold using calipers to determine its volume [or at the discretion of the instructor assume the volume is 944 cm^3 or 1/30 ft^3 (or 1000 cm^3 as appropriate)].
5. Using either the standard or modified compaction method as specified by the instructor, compact a cylinder of soil. If you use a 1000 cm^3 mold, use 26 blows per layer instead of 25 for either test to produce the same compaction energy.

6. Carefully strike both the top *and the base* of the compacted cylinder of soil with the steel straightedge. If the smoothing process pulls out pieces of gravel, backfill the holes using both smaller pebbles and soil.

   Note: If the last compacted layer in the mold is not above the collar joint, do not add soil to make up the deficiency—redo this test point. You can avoid this unpleasant situation by carefully watching and if, after about 10 blows on the last layer, the soil is below the collar joint, add enough material to fill above the collar joint and then continue with the remainder of the blows. On the other hand, you should try not to have more than about 6 mm of soil above the collar joint (see Fig. 9-4). If you have much more than this amount of excess and are not careful, you will remove the last layer of compacted soil cake when you remove the collar. If you do this, redo the test, since you can never replace the soil cake properly.

   If the collar is difficult to remove, do not risk twisting off the third (or last) layer of soil; take a spatula and trim along the sides of the collar until it comes off easily. Remember that you have an error multiplier of 1060 in this project; therefore, an error of 15 g of soil is about 0.15 kN/m^3 of compaction error—15 g of soil is not a very large quantity. In any case, an overfilled mold tends to produce a smaller density for the test point and does not plot a smooth curve.
7. Weigh the mold and cylinder of soil and record its mass.
8. Extrude the cylinder of soil from the mold, split it, and take two water-content samples—one near the top and the other near the bottom—of as much as the moisture cups will hold (60 to 80 g). Alternatively, use the sample reducer (Fig. 9-5) to remove the soil, and pulverize it directly from the mold. Again take two water-content samples from near the top and bottom of the sample as the soil emerges from the mold.

   When using the sample ejector (jack), you may find that some molds taper slightly so that the bottom is larger than the top. If this is the case, the sample will extrude much more easily if the sample exits from the larger end[9].
9. Break the sample to (−) No. 4 sieve size by visual appearance (unless the sample reducer device was used) and add 2 percent of water based on the original sample mass of 3 kg. Carefully remix and repeat Steps 5 through 9 until, based on wet masses, a peak value is followed by two slightly lesser compacted masses.
10. Return to the laboratory the following day and weigh the oven-dry water content samples. Compute the *average* water content at each test point using the two water-content samples.

Figure 9-4
Compacted soil specimen with collar removed and initial trimming started. Note that the soil projection is about 6 mm above the mold.

[9]If you use a fresh sample for each point, split the soil cylinder and obtain two water content samples—one near the top and the other near the base and discard the remainder of the sample.

Figure 9-5
A pulverizer developed by author for reducing compacted samples when the soil is reused as in the AASHTO standards.

(*a*) Pulverizer with sample attached for auguring from mold.

(*b*) Pulverizing a compacted sample. Soil is reduced to approximately No. 4 sieve size (visual) for adding water for next test point.

11. Compute the dry unit weight of each test and make a plot of γ_{dry} as the ordinate versus water content w as the abscissa (similar to Fig. 9-6). Also show a curve of zero-air-voids computed using Eq. (9-2) with about four assumed values of water content w.
    If G_s is not known, assume that the zero-air-voids unit weight at optimum moisture is 5 percent larger than the maximum dry unit weight. From this assumption you can compute a value of G_s for the soil and find the dry unit weight at other values of water content to plot the zero-air-voids curve. If the curve falls below the compaction curve at any point, the 5 percent assumption was not correct, so add 1 percent increments of density $G_s\gamma_w$ and recompute until you obtain a zero-air-voids curve that everywhere falls above the compaction curve. Try to use a scale for the compaction curve that allows you to obtain values to the nearest 0.1 for either γ_{dry} or w.
12. Show the liquid limit, plastic limit, G_s, and the AASHTO classification of this soil on the data sheet beside the compaction curve plot. If you used a different soil for this test than was used earlier for Tests No. 3, 5, and 7, you should perform these tests as part of this project.
13. In your sample computations, show the derivation, using a block diagram of Eq. (9-2) for zero-air-voids unit weight.
14. (*Optional*) Based on the value of specific gravity G_s from Step 11 or test value, compute the void ratio of the soil for each test point. On a sheet of graph paper from the data sheet section, replot the compaction curve; then, using void ratio for the ordinate (on right side), plot e vs. water content (use same water-content scale for both) and comment on which method displays the data better. Proctor's original method used e vs. water content.
15. (*Optional*) State what you recommend as relative compaction and the compaction moisture content for this soil in a 10-m highway fill.
16. If you used soil from Test No. 10, compute the relative compaction of the field soil.

10: DETERMINATION OF IN-PLACE SOIL DENSITY

References

ASTM D 1556-82 (Sand Cone Method)
ASTM D 2167-84 (Rubber Balloon Method)
AASHTO T 191-86 (Sand Cone Method)
AASHTO T 205-86 (Rubber Balloon Method)

Objective

To present to the reader two commonly used methods to determine the in-place density of a soil.

Equipment

Sand-cone apparatus (see Fig. 10-1)
Balloon-density apparatus
Digging tools
Friction-top cans [4000 mL (1 gal)][1]
Sack per group to salvage density sand

General Discussion

Once compaction criteria are established for the soil to be used at a particular site, generally with both moisture and density limitations, some means of verification of the results must be used. On all small projects and nearly all large projects, quality control is achieved by either the sand-cone method or the balloon density method (see Fig. 10-2). On a few large projects, nuclear devices have been and are being used. The nuclear method is beyond the scope of this manual and will not be considered further.

Basically, both the sand-cone and balloon-density methods use the same principle. That is, one obtains a known mass of damp-to-wet soil from a small excavation of somewhat irregular shape (a hole) in the ground. If the volume of the hole is determined, the wet density is simply computed as

$$\rho_{\text{wet}} = \frac{\text{mass of soil}}{\text{volume of hole}} \tag{10-1}$$

If the water content w of the excavated material is obtained, the dry unit density of the material is

$$\rho_{\text{dry}} = \frac{\rho_{\text{wet}}}{1 + w} \tag{10-2}$$

[1]Standard 3-lb (or 39-oz) coffee cans with plastic lids make excellent field-sample cans to replace the 1-gal friction-top cans available from laboratory supply houses. Coffee cans hold between 2500 and 3000 mL and are easy to open, close and empty. The plastic lid fits tightly enough that moisture can be retained for several days. Spray the inside with a quality paint to increase rust resistance. Also excellent for use are the 1-gal all plastic ice cream containers with metal bail. The friction-top plastic cover is tight enough to retain moisture for an adequate period of time. They are particularly easy to clean and are rust-proof.

Figure 10-1
Equipment for determining field soil density.

(a) Sand-cone equipment: Sand cone, plastic jug of sand, hole template, digging tools (spoon and screwdriver), 3-lb coffee can with plastic lid (painted inside to control rusting) for collecting soil, and a 64-mm paint brush for sweeping soil crumbs for full excavated sample recovery.

(b) Balloon density equipment: Balloon device equipped with an optional pressure gauge, hole template, digging tools, commercial friction-top can for collecting soil, a paint brush for sweeping soil crumbs for full excavated sample recovery.

where the water content w is in decimal, not percent.[2] Obtain the in situ dry unit weight as

$$\gamma_{\text{dry}} = \rho_{\text{dry}}\gamma_w$$

where $\gamma_w = 9.807$ kN/m^3 for SI and 62.4 lb/ft^3 for fps units.

The sand-cone and balloon methods are used to find the volume of the excavated hole used in Eq. (10-1).

With the *balloon method*, the volume is found as a direct measurement of the volume of water pumped into a rubber balloon that fills the hole. This volume is read directly from a graduated cylinder that forms the reservoir for the balloon. This rapid means of finding the volume of the hole is often a distinct advantage in terms of test time over the sand-cone method, which is more indirect.

The principal precaution with the balloon density test is to excavate a hole with regular sides so gravel protrusions do not rupture the balloon. Also there should be no voids that the balloon can bridge. A zero reading must be obtained prior to using the device, after which, unless excessive evaporation occurs in the reservoir or the balloon ruptures, many hole-volume readings can be taken before a new zero reading is required.

[2]This equation was used in Test 9 to convert the wet *unit weight* for plotting the curve of γ_d vs. w.

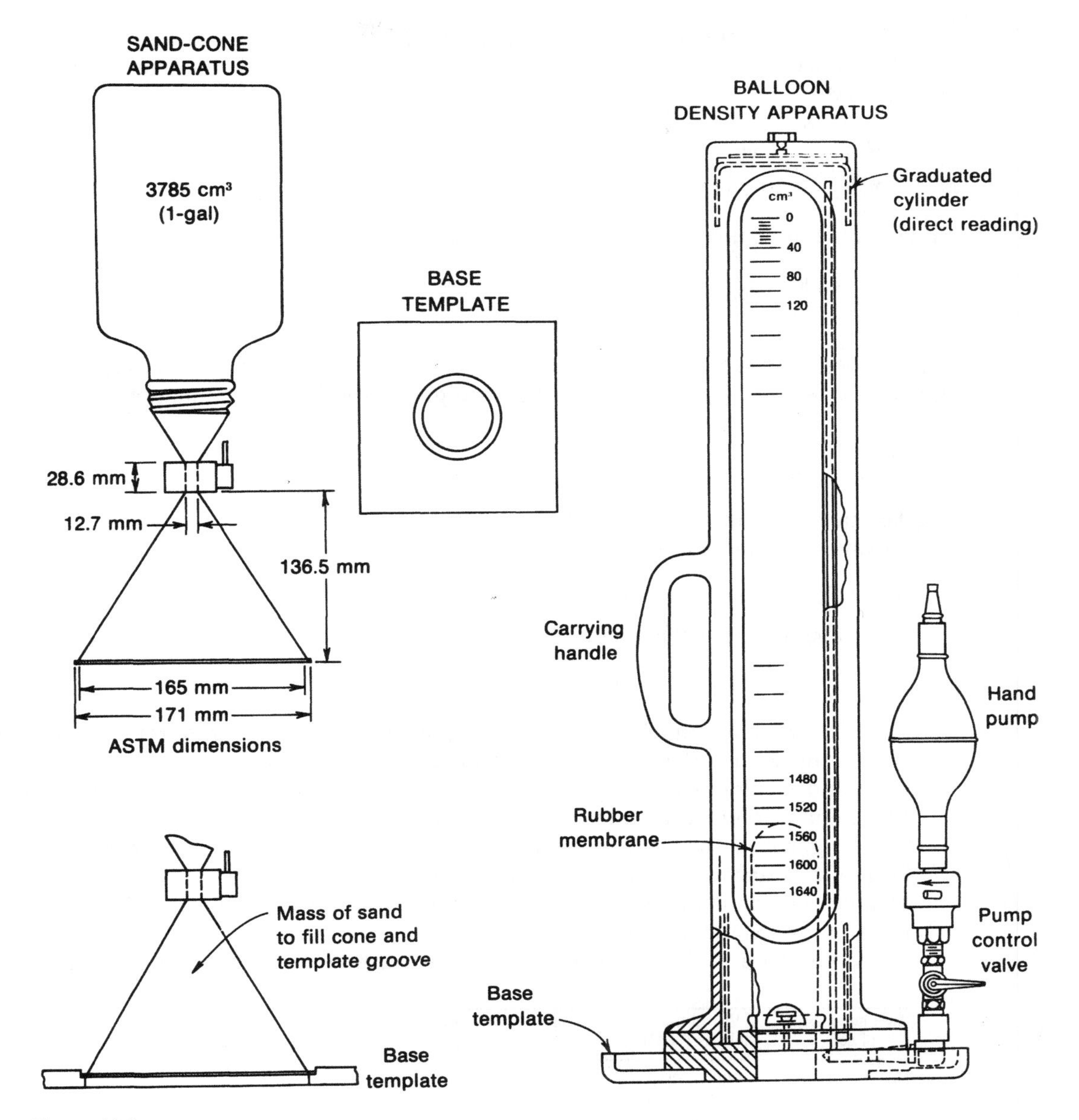

Figure 10-2
General line details of the sand-cone and balloon apparatus.

The *sand-cone method* is an indirect means of obtaining the volume of the hole. The sand used (often Ottawa sand) is generally material passing the No. 20 sieve but retained on the No. 30 sieve.[3] Although (−) No. 30 and (+) No. 40 or (−) No. 30 and (+) No. 50 sieve material can be used, it is generally desirable to have a uniform or "one-size" sand with rounded grains to avoid segregation problems (a volume of fine sand may weigh more than the same volume of coarse sand, and a volume of the mixture may weigh still more).

The use of rounded instead of angular particles reduces particle packing. Sand characteristics should be such that if sand was poured through the cone apparatus into a hole and then completely recovered and then a second sand container is used the volume of the hole would be approximately the same.

[3]ASTM only requires that the sand be free-flowing with a $C_U < 2.0$ with a maximum particle size < 2.00 mm (No. 10 sieve) and less than 3 percent by weight passing the No. 60 (0.25 mm) sieve.

If the sand has a known, constant density of, say, $\rho = 1.60 \text{ g/cm}^3$ and one pours 4800 g of this material into an irregularly shaped hole, the volume of the hole can be found by proportion[4]:

$$\frac{V}{4800} = \frac{1 \text{ cm}^3}{1.60 \text{ g/cm}^3}$$

Solving for the volume of the hole, one obtains

$$V = \frac{4800}{1.60} = 3000 \text{ cm}^3$$

and in general, the hole volume is computed as the mass of material to fill the hole M_{mat} divided by the unit density ρ of the material giving the following equation:

$$V_{hole} = \frac{M_{mat}}{\rho_{mat}} \tag{10-3}$$

For the sand-cone method the material is sand.

Several comments are in order at this point:

1. The balloon-density apparatus is available in only two sizes[5]

   $$1596 \text{ mL (or cm}^3) = 1/20 \text{ ft}^3$$

   $$2230 \text{ mL} = 1/13 \text{ ft}^3$$

   The most commonly used sand-cone apparatus uses a 3785-mL (1-gal) glass or plastic jug with sufficient material to fill a hole (and cone combined) of not over 3700 mL—and results depend on how full the jug is prior to the test. Two-gallon (7500-mL) and ½-gal (1700-mL) jugs are available but not widely used.

   In general, the field test holes will be quite small, thus the error multiplier is large. It is absolutely essential that no soil be lost during excavation and that the volume determination not be done in any way that gives an apparent hole volume that is too large (or too small).

   As a guide, ASTM suggests that for reasonable results the test be as follows:

Max. sieve size of soil	Vol. of test hole, cm^3 (ft^3)	Size of water-content sample, g
No. 4	700 (0.025)	100
12 mm	1400 (0.050)	250
25	2100 (0.075)	500
50	2800 (0.100)	1000

   The water content of the soil from the hole is required to compute the dry density or unit weight. The largest possible water content sample should be used to improve test reliability. In laboratories where oven space is available, the best water-content results are obtained by drying the entire amount of soil excavated from the test hole. When this is not done the water content samples should be (see ASTM D 2216) as follows:

90% soil passes No. 10	100 to 200 g
90% soil passes No. 4	300 to 500 g
90% soil smaller than 20 mm	500 to 1000 g

[4]It is definite computational convenience to use g and cm^3 in these computations rather than the basic SI units of kg and m^3.

[5]Soiltest, Inc., Lake Bluff, IL, sells replacement balloon cylinders graduated in both mL and cu ft units.

2. The excavation should be as rapid as possible to maintain the representative moisture content of the natural soil in the holding can. As soon as the excavation is complete, the lid should be put on this container so no further drying occurs.
3. When using the sand-cone method, avoid vibrating either the ground in the area or the sand jug, as this will introduce too much sand into the hole thus causing an apparent increase in the hole volume.

Procedure

This is suggested as a combination group and lab section project.

1. Each group will perform a sand-cone density test in an area designated by the instructor.
2. Each laboratory group will (as a joint project) test a single hole, using the balloon apparatus to measure the volume of the hole.
3. The entire amount of soil collected from the test holes should be oven-dried to determine the water content of the in-place soil.
4. Use data sheets from the data sheet section of manual for project. Refer to Figs. 10-4*a* and 10-4*b* following for typical data entry.

A. FIELD WORK

The field work will be done first as follows:

1. Prior to going to the field: (a) Each group will weigh its sand-cone jug full of sand to obtain M_j and the sample recovery cans, friction-top or other, with lid to obtain M_c. (b) Each group will place the balloon-density apparatus *on its template* on a flat surface and obtain a zero reading. Write the zero reading on the blackboard so it so it can be averaged for the class—but only if using a single apparatus.
2. Each group will then proceed to the field and be directed by the instructor where to locate a site to excavate the test hole. Carefully smooth an area and position the cone template (see Fig. 10-3). Be sure there are no visible voids between the inner template ring and the ground (minimal backfilling is allowed)

   Now carefully excavate the hole and place all of the removed soil into your previously weighed recovery can. Leave the can open while digging the hole. After finishing, put the lid on the can to avoid any further loss of sample moisture.
3. Next check and close the sand cone valve then turn the sand cone upside down and place into the grooved ring of the template. Open the valve and allow the sand to fall into the hole (and fill the cone). When the sand ceases to pour, close the valve and lift off. Salvage as much of the sand from the hole and template as possible (Ottawa sand is relatively expensive) and place in the sacks provided.
4. One group, as designated by the instructor, will place the balloon template at a site and excavate a hole the same shape as that for the sand-cone method but of not nearly so large a volume.
5. Each group will then place the balloon apparatus on the template, pump the balloon into the hole for a volume measurement, and record the final reading of the water reservoir. If the device is equipped with a pressure gauge, use a pressure of not over 0.2 kg/cm^2 (20 kPa or 3 psi)
6. For on-campus student work, be sure to landscape your hole so that the site is left as neat as you found it.
7. This completes the field work. Return to the soil laboratory. Check to make sure you are returning with all the equipment you brought out.

Figure 10-3
Field density test using sand cone.

(*a*) Digging tools and site preparation. Note that the site has been smoothed for template placement; holes produced by removal of small surface pebbles have been back-filled.

(*b*) Digging the hole. Use the screwdriver or spoon to loosen soil which is then spooned into sample can. Remove any large pebbles and save into can. If stones are much larger than 35 to 40 mm the site may not be suitable for testing with this size sampling equipment.

(*c*) Hole is now dug. It is carefully cleaned of loose soil crumbs and should be on the order of 100 to 150 mm deep.

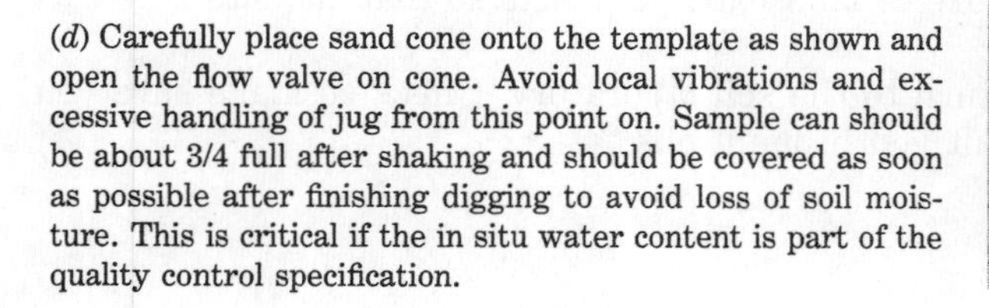

(*d*) Carefully place sand cone onto the template as shown and open the flow valve on cone. Avoid local vibrations and excessive handling of jug from this point on. Sample can should be about 3/4 full after shaking and should be covered as soon as possible after finishing digging to avoid loss of soil moisture. This is critical if the in situ water content is part of the quality control specification.

B. IMMEDIATE LABORATORY WORK

1. Each group will weigh the can of damp soil obtained from the field and record the mass on the data sheet in the appropriate line. Pour the soil into a weighed container and place in the oven for drying. Alternatively, obtain a water-content sample as previously described based on a visual inspection of percent passing any of the 20 mm, No. 4 or No. 10 sieves.
2. Each group will weigh its partially empty sand-cone jug and record the mass on the appropriate line of the data sheet.
3. Place the balloon volume readings on the blackboard in the soil laboratory, average, and subtract the averaged zero readings for a class average of the hole volume. Record the averaged final reading and the initial readings on the data sheet together with the averaged hole volume V_h'.
4. The group taking the soil sample from the balloon test hole will record to the nearest 0.5 g the wet mass of the soil from that test on the blackboard.

C: LATER LABORATORY WORK

1. Return to the laboratory the next day and weigh the oven-dry samples to find the dry mass of soil removed from the holes. Record this in the appropriate location on the data sheets.
2. Each individual will compute the water content and the wet γ_{wet} and dry γ_{dry} unit weights of the test hole of the individual's own group after Step *D* following.
3. Record on the blackboard for the rest of the laboratory group the mass of the dry soil for the balloon test hole.
4. Each individual will compute the water content and the wet and dry unit weight of the balloon test hole after Step *D* following.

D: EQUIPMENT CALIBRATION

Each group must determine three items of information:

1. The density ρ (g/cm^3) of the sand used in the field density test.
2. The mass of sand required to fill the sand cone and template groove (see Fig. 10-2).
3. The calibration of the balloon-density apparatus.

The following procedure can be used to determine the density of the sand used in the sand-cone apparatus:

1. Weigh a standard (944 or 1000 cm^3) compaction mold with base plate attached. Strictly, the volume should be determined also, but it can be assumed to have the correct volume within laboratory accuracy. Note that any available container of known volume of reasonably regular shape can be used in lieu of the compaction mold.
2. Carefully pour sand into the mold from the same approximate height as the sand falls into the hole in the field. Strike off the top of the mold with a steel straightedge and weigh. Determine the mass of sand in the mold by subtracting the mold mass from total mass.
3. Repeat Step 2 until you obtain two mass values that agree within about 10 g. Average these two masses and divide by container volume to get the average sand density ρ and write the value on the blackboard.
4. Average the values obtained by all the groups as displayed on the blackboard, except that any group whose value is substantially out of agreement with the other groups will have to do this part of the project over. The *average value* of density is to be recorded on the data sheet in the appropriate location and used by all groups for hole-volume determinations.

To determine the weight of sand to fill the sand cone and template groove:

1. Place the template on a flat surface. Weigh the sand cone with attached jug full of sand and record the mass.
2. Turn the sand cone upside down with the valve closed and place on the template. Open the valve and let the sand pour until it stops; then close the valve.
3. Reweigh the sand jug and the remaining sand. The difference between masses obtained in Steps 1 and 2 is the mass of sand to fill the cone and template groove. Repeat this series of steps for a second mass value. Average the two values and put the average on the blackboard.
4. Now determine an overall average of the several average values of sand-cone weights (*but only if all of the groups have the same type of sand cone*) for a value to use the data sheet.

To calibrate the balloon apparatus, the following procedure is recommended:

1. Place the template on a flat surface and obtain a zero reading by pumping the water against the flat surface.
2. Next, place the template over a standard compaction mold or other container of known volume that will not rupture the rubber balloon.
3. Place the balloon apparatus on the template and pump the balloon into the mold at least three times. Take the average of the three final readings (which should be very nearly identical). Try to use the same balloon pressure as used in the field (either by gauge or by estimation).
4. The difference between the zero and final averaged readings is the measured volume of the container ΔR or:

$$\Delta R = \text{final reading} - \text{zero reading}$$

5. The correction factor CF is computed by proportion as

$$\Delta R \times \text{CF} = \text{known volume}$$

from which the correction factor is

$$\text{CF} = \frac{\text{known volume}}{\Delta R} \tag{10-4}$$

All groups should show their correction factor on the blackboard so that a class average value can be obtained. If the CF is too small to affect the field density beyond 0.01 neglect it.

The Report

The discussion should comment on possible limitations of these two procedures. In the "Conclusion" show a tabulation of sand density and the weight of sand to fill the cone. List the balloon density CF. Tabulate the wet and dry unit weights γ obtained in the field along with water content data. Compare the unit weight obtained by the two methods. In the "Discussion," answer the following two questions:

1. What material(s) other than sand can be used to find the volume of the hole?
2. What problem may develop in using the balloon density apparatus in a loose soil or one with a low plasticity?

FIELD DENSITY TEST (Sand-cone, Balloon) Data Sheet 10*a*

Project In-place soil density Job No. —

Location of Project Construction site

Description of Soil Dark brown clay

Test Performed By JEB Date of Test 6/20/19--

Laboratory Data from Field Test

Sand-cone method

Mass of wet soil + can 3282 g

Mass of can 387 g

Mass of wet soil, M' 2895 g

Mass of wet soil + pan 3473 g

Mass of dry soil + pan 3142 g (W wt = 331)

Mass of pan 578 g

Mass of dry soil 2564 g

Water content, w% $\frac{331}{2564} \times 100$ = 12.9%

Balloon method

Mass of wet soil + can ______

Mass of can ______

Mass of wet soil, M' ______

Mass of wet soil + pan ______

Mass of dry soil + pan ______

Mass of pan ______

Mass of dry soil ______

Water content, w% ______

Field Data

Sand-cone method

Type of sand used Ottawa

Density of sand, ρ_{sand} = 1.62 (Sht 11b) g/cm³

Mass of jug + cone before use 7394 g

Mass of jug + cone after use 2850 g

Mass of sand used (hole + cone) 4544 g

Mass of sand in cone (from calib.) 1898 g

Mass of sand in hole, M 2646 g

Vol. of hole, $V_h = M/\rho_{sand}$ = 1633 cm³

Balloon method

Correction factor CF = ______

Final scale reading ______ cm³

Initial scale reading ______ cm³

Vol. of hole, V'_h ______ cm³

Vol. of hole = V'_h (CF) ______ cm³

$\rho_{wet} = M'/V_h = \frac{2895}{1633} = 1.773$ g/cm³

Unit Weight of Soil: Wet $\gamma_{wet} = \rho_{wet} \times 9.807$ = 1.773 × 9.807 = 17.39 kN/m³

Dry $\gamma_{dry} = \gamma_{wet}/(1 + w) = \frac{17.39}{1.129}$ = 15.40 kN/m³

Figure 10-4*a*
Field density data using the sand cone.

Name JEB Date of Testing 6/20/19--

Calibration Data

I. *Sand-cone method*

A. Sand density determination

Sand used ottawa

Type of vol. measure Stand. Comp. mold Vol., V_m 944.0 cm³ (1/30 c.f.)

Mass of sand to fill vol. measure: Trial no. 1 1531

Trial no. 2 1528

Trial no. 3 1530

Average mass M_a = 1529.6

Density of sand, $\rho_{sand} = M_a/V_m$ = 1529.6/944 = 1.62 g/cm³

B. Mass of sand to fill cone

Mass of filled jug + cone = 7391 g

Mass after trial No. 1 = 5491 Mass used = 1900 g

Mass after trial No. 2 = 3589 Mass used = 1902

Mass after trial No. 3 = 1697 g Mass used = 1892

Average mass to fill cone = 1898 g

II. *Volumeasure (balloon apparatus) calibration*

Type of container used ______

Vol. of container, V_c = ______ cm³

Initial reading ______

Reading after trial No. 1 ______; Change in vol. ______ cm³

Reading after trial No. 2 ______; Change in vol. ______ cm³

Reading after trial No. 3 ______; Change in vol. ______ cm³

Average ΔV ______ cm³

Correction factor CF = $V_c/\Delta V$ = ______
[Note, if correction factor is less than ±0.002, neglect it.]

Figure 10-4*b*
Calibration of sand cone and density sand.

11: COEFFICIENT OF PERMEABILITY—CONSTANT-HEAD METHOD

References

ASTM D 2434-68 (last changes 1984)
AASHTO T 215-70 (last changes 1984)
Bowles, J. E. (1973), "Permeability Coefficient Using a New Plastic Device," Highway Research Record No. 431, pp. 55–61.
Johnson, A. I., and R. C. Richter (1967), "Selected Bibliography on Permeability and Capillarity Testing of Soil and Rock Materials," *ASTM STP* No. 417, pp. 176–210.

Objective[1]

To introduce the student to a commonly used method of determining the coefficient of permeability of a cohesionless (granular) soil.

Equipment

Permeability device[2] (Fig. 11-1 or Fig. 12-1)
Timer (stop watch or stop clock)
Thermometer
Beaker for collecting water (500 to 1000 mL)
Graduated cylinder (to measure water collected in beaker)

General Discussion

The coefficient of permeability[3] is a constant of proportionality relating to the ease with which a fluid passes through a porous medium. Two general laboratory methods are available for determining the coefficient of permeability of a soil directly. These are the *constant-head method,* here described, and the *falling-head method* of Test 12 following. Both methods use Darcy's law given as:

$$v = ki$$

The corresponding flow rate (or quantity per unit time) is

$$q = kiA$$

where $q =$ quantity (cm^3, m^3, ft^3) of fluid flow in a unit time, s, h, day, yr, etc.
$k =$ coefficient of permeability, or hydraulic conductivity, in velocity units (cm/s, m/s, ft/day, m/yr, etc.)
$i =$ hydraulic gradient $= h/L =$ head loss across a flow path of length L (dimensionless).
$h =$ total head difference across the flow path of length L.
$L =$ length of sample or flow path that produces the head difference (and in units of) h.
$A =$ cross-sectional area of soil mass through which flow q takes place in units consistent with k.

[1]Consider doing Tests No. 11 and 12 during the same lab period on the same soil sample.
[2]The plans for the permeability device of Fig. 11-1 are available from the author for the cost of reproduction and mailing.
[3]This is also called *hydraulic conductivity* by a number of researchers. ASCE categorizes this property under the key words of both permeability and hydraulic conductivity.

Figure 11-1
A permeability device for granular materials (designed by the author).

(*a*) Permeability device disassembled; two end pieces clamp together with the three threaded rods shown attached to base. Round piece in foreground restricts the top of the sample with No. 200 wire mesh screen and includes the overflow wier shown in "c" below. Base part also includes a No. 200 mesh screen together with a diffuser to evenly distribute the inflow across the sample base.

(*b*) Building a sample with dry sand (wet sand can also be used). With care, a sandy sample can be compacted to a reasonably high density without damage to the filter screen.

(*c*) Sample saturated with upward flow of water. Sample confinement prevents sample expansion during saturation. Upward flow tends to produce S → 100%.

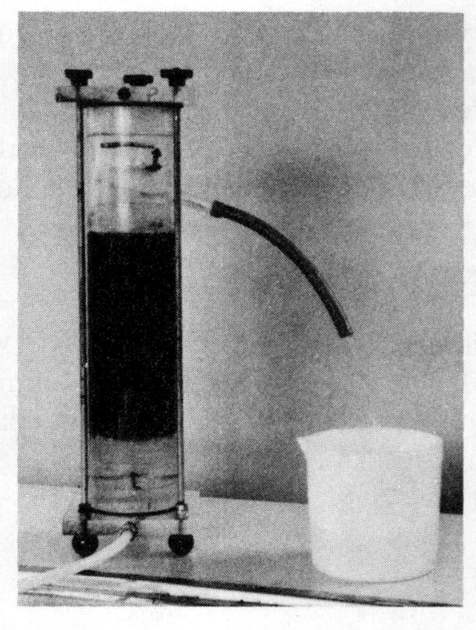

(*d*) Water flows through sample and exits across the overflow weir and is collected from the output tube into a beaker so Q in some time t can be measured.

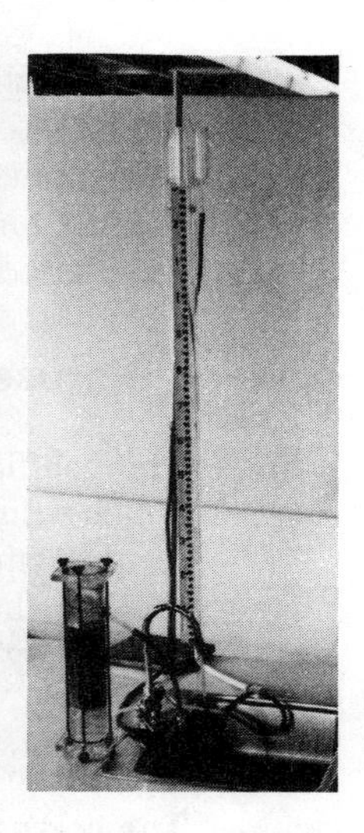

(*e*) A constant head setup. There is an overflow container at the top of the graduated standpipe to produce a constant hydraulic gradient (h/L) across the sample. This particular configuration using the graduated standpipe also allows a falling head test to be done. This type equip ment varies widely between laboratories.

The constant-head test is usually used for cohesionless materials since a large amount of water is wasted into the overflow weir, which controls the constant head, unless the test is of short duration (on the order of a few minutes). The falling-head method of Test 12 is usually used for cohesive materials (where the computed k is on the order of 10^{-4} cm/s and smaller). With a small k it may take several days to perform the constant-head test and a large amount of water would flow across the head control weir into the sewer.

Neither the constant-head nor the falling-head laboratory method provides a reliable value for the coefficient of permeability of a soil. One is fortunate if the value obtained is correct within one order of magnitude. Reasons for this are varied, but the major ones are as follows:

1. The soil in the permeability device is never in the same state as in the field—it is always disturbed to some extent.
2. Orientation of the in situ stratum to the flow of water is probably not duplicated. In sands, the ratio of horizontal flow to vertical flow is on the order of

$$k_h / k_v \geq 3$$

This ratio is impossible to duplicate in the sample—even where the void ratio may be duplicated by careful placement and compaction.

3. Conditions at the boundary are not the same in the laboratory. The smooth walls of the permeability mold make for better flow paths than if they were rough. If the soil is stratified vertically, the flow in the different strata will be different, and this boundary condition may be impossible to reproduce in the laboratory.
4. The hydraulic head h is usually 5 to 10 times larger in the laboratory test than in the field. Field values of $i = h/L$ are often less than 1.5 unless there is flooding or water impoundment (reservoirs or dams). The high laboratory head may produce turbulent flow and wash some of the fine materials out—or to the boundary and produce a "filter skin." If fines wash out, an increase in k results as pores enlarge and interconnect; a boundary filter skin reduces k as it inhibits flow.
5. Considerable evidence indicates that Darcy's law is nonlinear—at least at large values of hydraulic gradient i so that

$$v = k i^n \qquad \text{and not} \qquad v = k i$$

6. The effect of entrapped air on the laboratory sample will be large even for small air bubbles since the sample is small.

The *coefficient of permeability* of a homogeneous, isotropic soil mass depends primarily on the following factors:

1. The viscosity of the pore fluid (usually water). As the temperature increases, the viscosity of water decreases and the coefficient of permeability increases; that is, the flow rate increases. The coefficient of permeability is standardized at 20°C, and the coefficient of permeability at any temperature T is related to k_{20} by the following ratio:

$$k_{20} = k_T \frac{\eta_T}{\eta_{20}} \tag{11-1}$$

where η_T and η_{20} are the viscosities of the fluid at the temperature T of the test and at 20°C, respectively. Either absolute or kinematic fluid viscosity may be used in Eq. (11-1). Table 6-1 is a convenient source of absolute viscosity values for use in this test and also in Test 12 following. As an example:

Test temperature, $T = 16$°C
From test at this T obtain $k_{16} = 0.01000$ cm/s
Required: compute k at the standard $T = 20$°C (or k_{20})

Obtain from Table 6-1 $\eta_{16} = 0.01111$ and $\eta_{20} = 0.01005$.

Now substitute directly into Eq. (11-1) and compute

$$k_{20} = 0.01000\frac{0.01111}{0.01005} = 0.01105 \text{ cm/s}$$

Note that k_{20} *is larger than* k_{16} as one should expect since at the higher temperature the viscosity is lower and allows a larger water flow rate.

Temperatures recorded to the nearest degree are sufficiently precise for this test since it is quite approximate and there is not a large viscosity change with 1 to 2°C variations. Other test limitations are far larger than the temperature error. Realistically, the above correction is not necessary but is required in most "standard" test procedures and in student laboratories.

2. The void ratio e of the soil. Various attempts have been made to relate the coefficient of permeability of the soil at one void ratio to the same soil at a different void ratio [$k = f(e)$] by expressions such as

$$k_2 = k_1\left(\frac{e_2}{e_1}\right)^2 \qquad (11\text{-}2)$$

$$k_2 = k_1\frac{e_2^3(1+e_1)}{e_1^3(1+e_2)} \qquad (11\text{-}3)$$

In both these equations the coefficient of permeability k_1 is the value determined by test at sample void ratio e_1 and the permeability at void ratio e_2 is desired. In current practice it is strongly recommended that one perform permeability tests at various void ratios and use curve fitting to obtain the relationship between e and k.

3. The size and shape of the soil particles. Angular and platy particles tend to reduce k more than when rounded and spherical particles predominate in the soil. Allan Hazen studied filter sands for use in water works filters (ca. 1890) and concluded that for clean sands and gravels the coefficient of permeability could be expressed using the D_{10} grain size (as identified in Test 5—*but in cm*) approximately as

$$k = 100D_{10}^2 \text{ cm/s} \qquad (11\text{-}4)$$

in the following particle size (*in centimeters*) range

$$0.01 < D_{10} < 0.3 \text{ cm}$$

4. The degree of saturation S (as defined in the Introduction). As the degree of saturation increases, the apparent coefficient of permeability also increases. Part of this increase is due to the breaking of surface tension. The remainder is an unknown quantity, since it is difficult to determine k unless one considers the continuity of flow through the medium.

The flow through the medium can be measured only by considering the quantity going into and coming out of the soil mass. As an extreme case one could, in a dry soil, have a considerable flow into the sample and no flow at all out—an apparent $k = 0$ condition. After saturation flow in $=$ flow out and k becomes non-zero. Saturated samples are generally used in the laboratory to avoid this problem, although a few researchers have considered non-saturated soil samples—but with inconclusive results.

LABORATORY EQUIPMENT AND TEST DETAILS

A widely used permeameter is the standard compaction mold with a special cover and base plate as shown in Fig. 12-1. The base includes a porous stone, and the cover has a valve for connecting the inflow tube and a petcock for deaeration. Rubber gaskets are used to seal the mold to the base and to the cap.

The device of Fig 12-1 has the advantage of being able to compact a sample to the required project density (as for a dam core or the clay liner of a landfill) then putting the mold on the permeability base and replacing the collar with the special cap piece.

The device of Fig 12-1 has several shortcomings. First, for sand there can be a large head loss across the 12 mm porous stone in the base. Second, there is no provision for diffusion of the inflow across the sample area. Third, with down-flow and no potential for sample observation it is difficult to get all the air out of the sample when trying to saturate it. This last shortcoming *may* be somewhat reduced by submerging the assembled mold of soil in a basin of water (after putting an exit tube on the inflow valve) so it is saturated from the bottom up. This method of saturation tends to produce sample expansion since the top is not well confined.

These were the principal reasons for the author's developing the permeameter device shown in Fig. 11-1 for testing sands. This type of device is of particular value in a student laboratory since it is easy to observe sample length, head, drainage, saturation—and sample expansion is controlled. Head loss across the two No. 200 mesh end screens is negligible compared with that in the base of the compaction mold.

Laboratory tests to determine the coefficient of permeability often require the use of deaerated (and sometimes distilled) water. The use of deaerated water may have some merit, since the influence of a bubble of air coming out of solution in the small laboratory sample may be very important compared with the effect of a bubble of the same size in the field soil.

In the author's opinion, except for precise research, the use of deaerated (or distilled) water introduces a questionable increase in the precision of k determinations for routine laboratory work, considering the inaccuracies of this test and the fact that in situ water is neither deaerated nor distilled. The cost of using distilled water in a constant head test far exceeds any potential improvement in test quality.

Two fairly simple procedures may be used to reduce the dissolved-air problem. One technique is to use water for the test that is warmer than the soil sample so that the water cools as it percolates through the soil; this will attract air from the sample into solution. The other technique is to use a large enough gradient h/L that the pressure holds or forces the free air into solution.

Procedure

This is suggested as a group project.

1. Weigh the source container of cohesionless material to be used for the test by your group.
2. Determine the permeameter volume V_p in cm^3 if not already known. Next assemble as necessary so soil can be placed in it for the test.
3. Each group should make a test sample to a density different from any other group. This may be done by placing the soil loosely, placing it with some vibration, or placing it with considerable effort. After building your sample again weigh the source container to obtain the sample mass M_s. Compute the sample density as

   $$\rho = M_s/V_p \text{ g/cm}^3$$

   Write your group number and sample density on the blackboard so each group uses a different value of ρ. Try to vary the *unit weight* by about 0.4 to 0.6 kN/m^3 between groups (but in sand this may present some difficulty).

For the compaction permeameter (of Fig. 12-1) do Steps 4 through 7 below in order.

4. Place a piece of filter paper on top of the sand in the mold; then carefully clean the rim of the mold, place a rubber gasket on the rim and then firmly seat the cover. The cover should have a transparent piece of plastic tubing about 150 mm long connected to the inlet valve over which the water inlet tube can be slipped. Attach a 150 to 200 mm length of rubber tubing to the outlet pipe.
5. Place the permeameter in a sink (or other container) in which the water is about 50 mm above the cover. Be sure the outlet tube is open so that water can back up through the sample to produce saturation with a minimum of entrapped air. When water in the plastic inlet tube on top of the mold reaches equilibrium with the water in the sink, the sample may be assumed to be saturated. A soaking period of 24 h might provide better results but involves too much time for the objectives of this test. Note, however, that this procedure may expand the sample somewhat inside the mold. The trade off is sample expansion with reasonably good deaeration versus large amounts of trapped air inside the mold.
6. With the water level stabilized in the inlet tube, take a hose clamp and clamp the exit hose (reason for using a rubber tube). Remove the permeameter from the sink and attach the inlet tube to a rubber hose from the constant-head standpipe. The inlet hose should be clamped at this point so no water can go into the sample.
7. Remove air from the tubing at the top of the soil sample by opening the petcock in the mold cover (refer to Fig. 12-1) then remove the clamp from the hose slowly so water can trickle into the sample and any air trapped either under the cover or in the entry line is flushed out through the petcock. When no more air comes out, close the petcock. Now measure the hydraulic head across the sample.

If you are using the Bowles type of permeability device (Figs. 11-1 and 11-2),

6a. Complete the assembly of the device. Orient the inlet and exit tubes conveniently for collection of water and for the initial saturation drainage.

7a. Connect the water inlet to the water supply using a rubber hose. Next slowly (squeeze down the rubber tube so flow is restricted) allow the sample to saturate (visual observation) and the flow to stabilize by allowing some outflow that is not recorded. Now clamp the inlet hose.

Now for either device:

8. Using a 500- or 1000-mL container (larger is preferable), record the time to collect Q = 450 to 900 mL of water. Take the water temperature in the collection container to the nearest 1°C. Repeat the test two or three times using the same time (t = constant). The amount of water collected on successive trials usually decreases and the cause of this phenomenon should be considered in the "Discussion" part of your lab report. Record all the data from the test on the data sheet provided.
9. Each group should compute its value of k for the test temperature using the following equation:

$$k = \frac{QL}{Aht} \text{ cm/s} \tag{11-5}$$

where all terms have been previously identified.

Each group should also compute k_{20} using Eq. (11-1) for its test. If time t is constant and temperature T constant (or within about 1°C) you can average Q to reduce the computations to a single value for k_T. If all of these conditions are not satisfied or Q is not relatively constant you must compute individual values of k_T and reduce them to k_{20} and only then average them for your k_{20} test value.

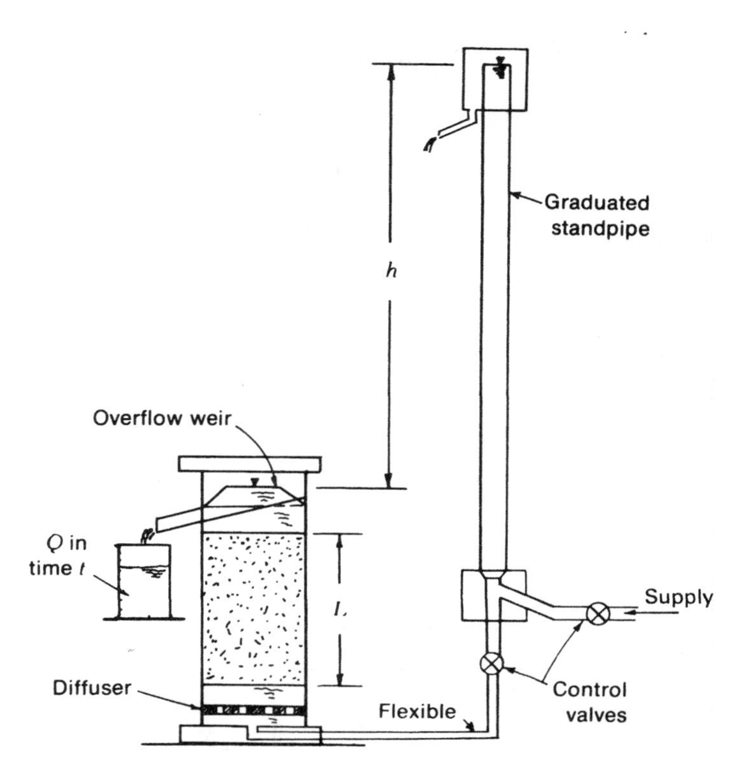

Figure 11-2
Line details of the constant-head permeability test using equipment illustrated in Fig. 11-1.

Place the average k_{20} value of your group on the blackboard alongside your sample density ρ previously written there with those of the several other lab groups (and identified by your group number). Alternatively, the instructor may require submission of these values on a sheet of paper so they can be combined and copies given to each member of the laboratory section.

Be sure the computations are checked and that the units of k_{20} are in cm/s.

10. Using a value of G_s furnished by the instructor (or estimated by you), compute the void ratio e for each density used by the lab section and make a plot of k_{20} versus e using data from your group and those of the other groups, obtained from the blackboard.
11. Fit an equation through the curve plotted in Step 10. If you have access to a regression curve program use that to fit the equation. Plot the curve equation on this plot using a color (but not red) so the instructor can compare your "fit" to the actual data. Also (and using still another color) plot a curve using Eq. (11-2) and *your* test values for k_1, e_1 and arbitrary values of e_2 between 0.4 and 1.0.
12. Compute the approximate actual water velocity for your test as

$$v_a = \frac{1+e}{e} v \text{ cm/s}$$

13. In your report (Refer to Fig. 11-3):
    a. Compare equations derived from using the laboratory test data. Also discuss how well Eq. (11-2) compares with equations derived from your test data.
    b. Do you think the actual water velocity from the hydraulic gradient of your test produced turbulent flow?
    c. Is the collection of a lesser quantity of water in the same time due to formation of a filter skin or to air coming out of solution? What can be done to produce a constant flow quantity Q (if one was not obtained)?
    d. Discuss any changes you think could be made to improve this test as a practical possibility.
    e. Derive the equation used in Step 12 for v_a.

COEFFICIENT OF PERMEABILITY (Constant Head, Falling Head) — Data Sheet 11

Project: k for Lab manual — Job No. —

Location of Project: Soil Laboratory (using permeameter of Fig. 11-1)

Description of Soil: Brown, medium coarse sand w/trace of gravel

Tested by: JEB — Date of Testing: 6/7/19--

Sample Dimensions: Diam. 7.62 cm; Area 45.6 cm²

Mass soil + pan Init. 2984.2 g — Ht. 20.3 cm

Mass soil + pan Final 1427.3 g — Vol. 925.7 cm³

Mass of Sample 1556.9 g — Density, ρ 1.682 g/cm³

Constant Head

$h =$ 87 cm

Test data / *Test data used*

Test No.	t, s	Q, cm³	T, °C	Test No.	t, s	Q, cm³	T, °C
1	180	775	23		180	775	
2	180	772	22		180	772	
3	180	761	22		180	761	
4							
				Average[a]	180	769	22°C

$k_T = \frac{QL}{Aht} = \frac{769(20.3)}{(45.6)(87)(180)}$ = 0.022 cm/s

Table 6-1: $\alpha = \eta_t/\eta_{20} = \frac{0.00958}{0.01005} = 0.9532$

$k_{20} = \alpha k_T =$ 0.9532(0.022) = 0.021 cm/s

Falling Head

Standpipe = [burette, other (specify)] ______ Area standpipe, $a =$ ______ cm²

Test data[b] / *Test data used*

Test no.	h_1, cm	h_2, cm	t, s	Q_{in}, cm³	Q_{out}, cm³	T, °C	Test no.	h_1, cm	h_2, cm	t, s	T, °C
1											
2											
3											
4											
							Average				

$\alpha = \eta_T/\eta_{20} =$ ______

$k_T = \frac{aL}{At} \ln \frac{h_1}{h_2} =$ ______ = ______ cm/s

$k_{20} = \alpha k_T =$ ______ = ______ cm/s

[a]Use averaged values only if there is a small difference in test temperature, say, 1–2°C.
[b]Simplify by using the same h_1 and h_2 each time, so you can average k.

Figure 11-3
Data from a constant-head permeability test using the equipment in Figure 11-1. Data using other types of equipment would be identical in format.

12: COEFFICIENT OF PERMEABILITY—FALLING-HEAD METHOD

References

See Experiment No. 11.
This test has not been standardized by ASTM (or AASHTO) as of 1991.
Tavenas, F., et al. (1983), "The Permeability of Natural Soft Clays. Part I: Methods of Laboratory Measurement," *Canadian Geotechnical Journal*, Vol. 20, No. 4, Nov., pp. 629–644.
Peirce, J.J., et al. (1987), "Parameter Sensitivity of Hydraulic Conductivity Testing Procedure," *Geotechnical Testing Journal, ASTM*, Vol. 10, No. 4, Dec., pp. 223–228 (has excellent reference list on p. 228).

Objective

To introduce the student to a method of determining the coefficient of permeability of a fine-grained soil (such as fine sand, silt, or clay). The test may also be used for coarse-grained soils.

Equipment

Permeability device[1]
Timer
Thermometer
Ring stand with test-tube clamp or other means to develop a differential head across soil sample
Buret to use (with ring stand or other means of support) as a standpipe[2]

General Discussion

The general discussion of Test No. 11 is also applicable to this test. The limitations of the constant-head test are inherent in this test, and, in addition, tests of long duration require some way of controlling evaporation of water in the standpipe (see Fig. 12-1).

The equation applicable to this test can be derived (making reference to Fig. 12-2) but is left as part of the exercise for the student report and will be merely given here:[3]

$$k = \frac{aL}{At} \ln \frac{h_1}{h_2} \qquad \text{(cm/s)} \tag{12-1}$$

where a = cross-sectional area of buret or other standpipe (Fig. 12-2 or 11-2), cm^2

A = cross-sectional area of soil sample, cm^2

h_1 = hydraulic head across sample at beginning of test ($t = 0$)

h_2 = hydraulic head across sample at end of test ($t = t_{test}$)

[1]A triaxial cell may be used for this test but may not be suitable for student laboratories due to time restraints or insufficient cells for a lab section.
[2]If the falling-head standpipe of Test No. 11 is graduated you may be able to use that by filling to a height in the graduated zone and closing the supply valves.
[3]Note the non-SI unit of cm/s. This is a common laboratory unit for k but other units can be reported. For example units of m/yr may have more physical meaning for laypersons.

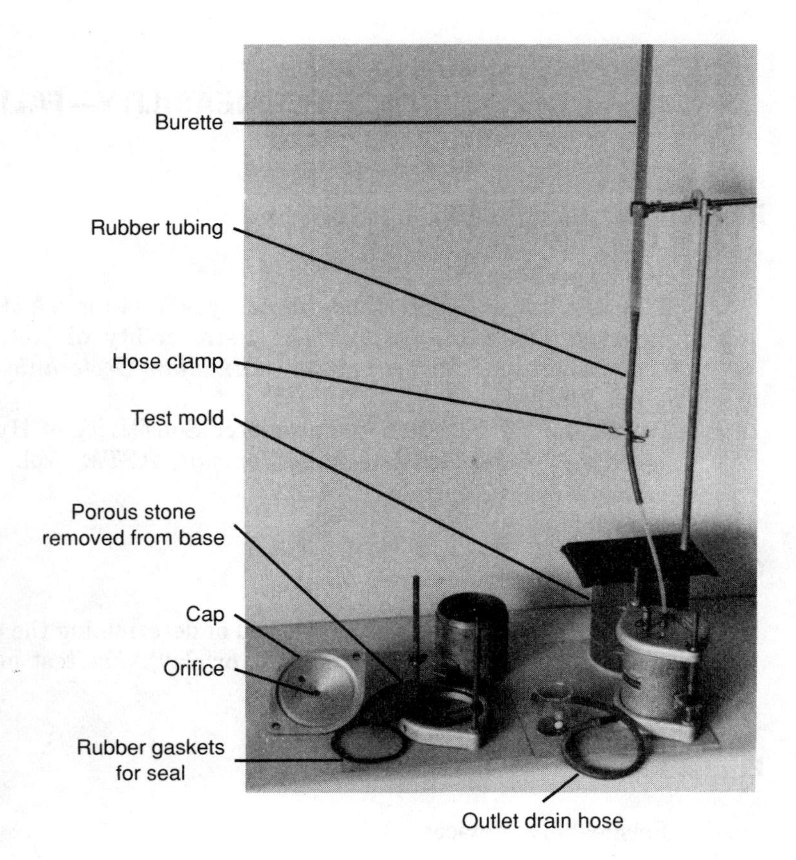

Figure 12-1
Falling-head permeability test using the standard compaction mold permeameter. Shown are both a disassembled device and a test setup using a 100-mL buret. A substantial head loss can occur through the thick porous stone in the base. The small water-entry orifice through the cap may produce a sample cavity form local flow conditions. Care is required to produce a watertight system. Use a meterstick to obtain the hydraulic heads h_1 and h_2.

L = sample length, cm

t = elapsed time of test, s (but min, days can be used)

ln = natural logarithm to base 2.7182818. . . .

It should be noted that this method of determining the coefficient of permeability k is primarily for economy, since a test to determine k for a fine-grained soil (k on the order of 10^{-4} to 10^{-9} cm/s) may take many hours or even several days or weeks. To maintain the constant-head of Test No. 11 could waste a large quantity of water over several days—depending on the equipment configuration.

For long duration tests and where the quantity of flow through the soil sample is small, one should make some provision to control evaporation of the water if an open standpipe is used and to avoid sample drainage and/or evaporation through the sample exit tube. One solution for this problem is to do the test in a controlled-humidity room. Another is to keep the standpipe reservoir covered with a small rubber balloon that has been partially inflated. Do not plug the standpipe, as a vacuum will eventually form, stopping the flow of water.

To control sample drainage and ensure that the exit tube flows full (or to control tail-water evaporation), submerge the exit tube in a container of water. Obtain the tailwater elevation for h_1 and h_2 based on your particular laboratory setup. Use judgment and ingenuity to control any sample leaks.

The coefficient of permeability is necessary to determine the time for fluid to travel between two known points and/or the amount of fluid that travels between these two points according to the following equations previously given in Test No. 11:

$$v = ki \qquad \text{and} \qquad Q = kiAt$$

There is still no exact method of determining k but improved methods are under investigation. Current methods are accurate within about one order of magnitude (the difference

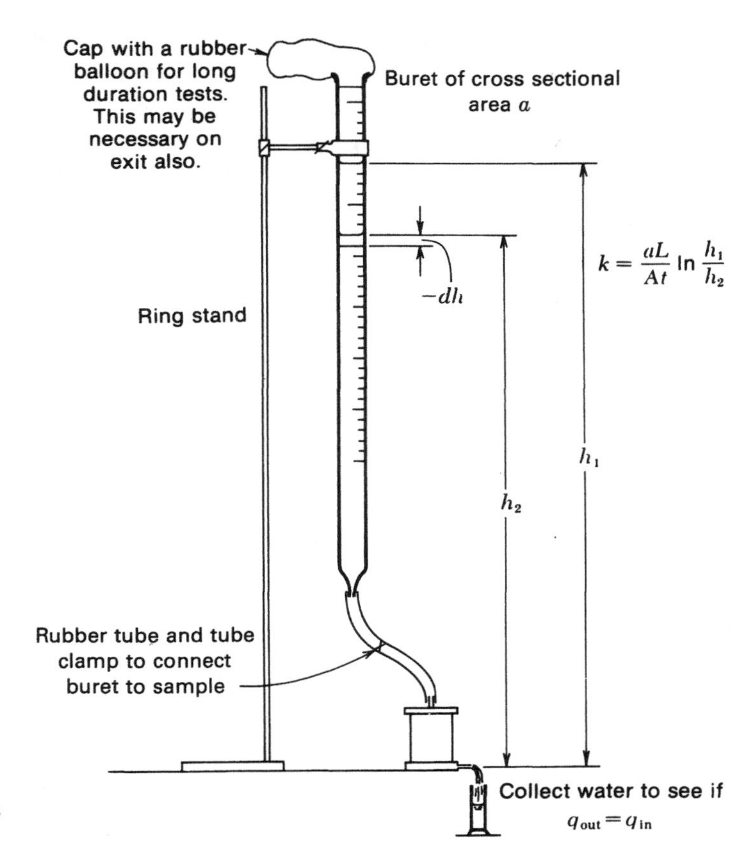

Figure 12-2
Line details of the falling-head test equipment shown in Fig. 12-1.

between 10^{-6} and 10^{-5} is one order of magnitude). Clearly, more refinement is necessary in determining k so that water detention basins (dam cores) and clay liners for landfills can be designed with some statistical confidence.

Unfortunately serious problems exist in determining k for clay liners in landfills—particularly landfills which develop leachates from penetration of surface water + any liquids already present in the wastes. Some of these leachates react with the clay liner and produce an increase (sometimes a decrease) in k over time. This requires the use of geotextiles together with the clay layers as well as monitoring pipes placed at strategic points beneath and around the landfills. The amount and time for any escaping leachate to show up in the monitoring pipes also depends on obtaining "accurate" values of k in these underlying materials.

Clay fills or clay liners are compacted to produce a material with both greater strength and a reduced k. It is necessary to find the coefficient of permeability (the term *hydraulic conductivity* is widely used for landfill work but uses symbol k) for these compacted materials.

In landfill liner tests, it is more nearly correct—but may not be practical—to determine k using a test liquid of similar chemical composition to that expected in the landfill leachate. The problems with this are (1) exposing the test personnel to hazardous and corrosive materials and (2) predicting the chemical composition of a leachate where the landfill materials are of unknown composition.

The compaction mold permeameter can be directly used to produce compacted clay samples. There are some problems with this; for example, the mold may have little resistance to corrosion, the smooth walls may produce alternative flow paths of less resistance than through the sample, and the method of compacting may produce a different layering than on site. If the layering is different there may be a substantial difference between k_h and the k_v actually measured in the laboratory test. Unfortunately there is little that can be done about potential differences between k in the horizontal and vertical directions.

There are some alternatives to the compaction mold permeameter that can be used. For permeability tests on soils of low permeability (say, 10^{-4} to 10^{-9} cm/s), it may be more practical to use a thin sample about 1 in (2 to 3 cm) thick in a specially constructed permeameter. Alternatively, if the results of the test will be used to plan for water retention one might be able to obtain the coefficient of permeability using a consolidation test.

For landfill leachates it may be preferable to use a triaxial cell constructed of corrosion-resistant materials. Both the consolidation test and the triaxial cell will be briefly considered below. However, they may be too time consuming for a first course in geotechnical engineering.

Permeability tests on soils of low permeability must be performed very carefully for the results to have meaning. Any leakage from the permeability device, or evaporation from the standpipe or from the collection container, or flow along the less-resistant interface of the permeability device and soil will materially affect the computed value of k. As an example, let us assume a constant-head test, a soil sample with $A = 81\ \text{cm}^2$ (approximate area of standard compaction mold), a hydraulic gradient of 20 (which is very large even for laboratory testing), and a real coefficient of permeability of 1×10^{-6} cm/min. How much water Q will be collected in 1 h for determination of k?

Since $Q = kiAt$ we have by direct substitution

$$Q = (1 \times 10^{-6}\ \text{cm/min})(20)(81\ \text{cm}^2)(60\ \text{min/h})$$

$$= 0.0972\ \text{mL/h}$$

This is a negligible quantity and could easily evaporate nearly as fast as it is collected unless special precautions are taken. Then the question is: How reliable is this k?

1. If water travels along the wall of the permeability device at a rate of 1×10^{-2} cm/min, about 20 to 40 times as much water will be collected as should be.
2. A hydraulic gradient of 20 (in this example) is far in excess of any likely in situ value (except possibly beneath dams) but is in the upper range of values used in some soil laboratories.

In any case these two factors indicate substantial difficulties in obtaining reliable values of k in the laboratory.

k FROM A CONSOLIDATION TEST

The consolidation test of Test No. 13 can also be used to estimate the coefficient of permeability using c_v from Eq. (13-9) as

$$k = \frac{a_v \gamma_w C_v}{1 + e}$$

with a_v identified in Eq. (13-13) of Test No. 13.

Referring to Fig. 13-2 and using a *fixed-ring* consolidometer with the extreme right piezometer tube connected to the sample base, one may use a graduated buret, as in this test, attached to that piezometer and, at the end of primary consolidation for any load, add water to some level and observe the fall and elapsed time. It may, of course, be necessary to cover this reservoir with a rubber balloon to control evaporation. At the end of the test, disconnect (or drain) the buret and continue the consolidation test with the next load increment.

A particular disadvantage of the consolidation method is that the sample is only about 20 mm thick. Also there is a substantial change in void ratio during consolidation that may cause invalid field results.

***k* FROM A TRIAXIAL CELL**

The triaxial cell apparatus can be used to estimate the coefficient of permeability k. This has the considerable advantage of being able to use a tube sample recovered from the site. It is also possible to use the "Harvard miniature compaction equipment" shown in Fig. 9-3 (see the instructor if the laboratory has a set of equipment or refer to the Wilson (1950) reference of Test 9). The Harvard equipment allows you to produce a sample whose density depends on the number of layers and tamps per layer, similar to the standard compaction test. This sample is about 33.3 mm diam × 71.5 mm in length.

One prepares a sample using the procedures of Tests No. 15 and 16. The sample is then saturated (a vacuum assist using about 50 to 75 mm of Hg will speed the process without unduly consolidating the sample inside the rubber membrane). When the sample is saturated the pore pressure apparatus can be used to produce either a constant-head test or a falling-head test (force water into one of the sample drainage lines and collect flow from the other). It may not be necessary to collect the flow but rather observe a buret tube in the pore pressure apparatus to measure the flow quantity. Exact test procedures depend on the equipment available and are beyond the scope of this manual.

The triaxial test determination with pore pressure apparatus gives substantial control over the hydraulic gradient h/L across the sample. This may allow one to obtain a curve of h versus k which will show if there is linear flow and if there is some threshold value of h/L necessary to produce flow. There is at present some conflicting evidence of a threshold gradient.

Procedure

This is suggested as a group project.

1. Build the soil sample following the instructions given in Test No. 11 or those of your instructor. If you use a sample prepared using instructor's instructions, ignore the following procedure which is predicated on your using a coarse-grained (primarily sand) sample for test illustration and for performing both tests on this sample during the same lab period.
2. Fill the buret (or other standpipe) to a convenient height and measure the hydraulic head h_1 across sample. Record this in the appropriate column of your data sheet.
3. Commence the flow of water and simultaneously start timing the test. Allow water to flow through the sample until the buret (or standpipe) is almost empty or to a convenient mark. Simultaneously stop the flow and timer. Obtain the final head h_2. Take the temperature of the test. Record these data in the appropriate columns of your data sheet.

   If it is necessary to obtain the area a of the standpipe you should collect the outflow in a beaker.
4. Refill the buret (or standpipe) to the same mark so h_1 is a constant and repeat the test two additional times. Use the same h_2 for each test but the times t may be different. Take a temperature reading for each trial.

   If the standpipe area a is to be determined, accumulate the outflow from all of the tests in a beaker. Record the required data in the data sheet columns. After the last test run compute the cross-section area of the standpipe as

$$a = \frac{q_{\text{total}}}{\sum(h_1 - h_2)} \qquad \text{cm}^2$$

If this test is done in combination with Test No. 11 and the constant-head data have not been obtained, take that data set next.[4]

[4]Either Test No. 11 or 12 can be done first. If there is limited equipment for these two tests do either test as soon as the equipment becomes available.

5. Each person should compute the coefficient of permeability at the test temperature k_T and for k_{20} (at 20°C). Obtain the viscosity values from Table 6-1 and use Eq. (11-1) to reduce k at the test temperature T to k_{20}. Use a data sheet from the data sheet section. Average the results for k_{20} for this soil in its current state.

Refer to Fig. 12-3 for a typical set of falling-head data.

6. In your report:
   a. Discuss test limitations (specifically for your test setup).
   b. Can you propose a better (and practical) way of doing the test?
   c. Compare the k_{20} values between Tests No. 11 and 12 (if both are done on the same sample). What might be the cause of any differences between the two values—if there are any?
   d. Show the derivation of Eq. (12-1) in sample computations.
   e. How long will it take for head h_2 to become theoretically (not visually) zero?

Project: K by Falling-head Job No. ________

Location of Project: Soil Laboratory (stand. Compaction mold)

Description of Soil: Light brown uniform fine sand

Tested by: JEB Date of Testing: 7/9/19--

Sample Dimensions: Diam. 10.2 cm; Area 81.7 cm^2

Mass soil + pan Init. 3074.9 g Ht. 11.6 cm

Mass soil + pan Final 1525.1 g Vol. 947.7 cm^3

Mass of Sample 1549.8 g Density, ρ 1.635 g/cm^3

Constant Head

h = ——— cm

Test data / *Test data used*

Test No.	t, s	Q, cm^3	T, °C	Test No.	t, s	Q, cm^3	T, °C
1							
2							
3							
4							
				Average[a]			

$k_T = \frac{QL}{Aht}$ = ________ $\alpha = \eta_t/\eta_{20}$ = ________

= ________ cm/s $k_{20} = \alpha k_T$ = ________ cm/s

Falling Head

Standpipe = [burette, ~~other (specify)~~] 100 mL Area standpipe, a = 1.71 cm^2

Test data[b] / *Test data used*

Test no.	h_1, cm	h_2, cm	t, s	Q_{in}, cm^3	Q_{out}, cm^3	T, °C	Test no.	h_1, cm	h_2, cm	t, s	T, °C
1	51.1	24.3	54.1		45.8	21°	1			54.1	
2	51.1	24.3	54.7		↓	21	2			54.7	
3	51.1	24.3	55.3		↓	21	3			55.3	
4											
							Average	51.1	24.3	54.7	21°

$a = \frac{45.8}{51.1 - 24.3} = 1.71\ cm^2$

Table 6-1

$\alpha = \eta_T/\eta_{20} = \frac{0.00981}{~~0.01005~~} = 0.9761$

$k_T = \frac{aL}{At}\ln\frac{h_1}{h_2} = \frac{1.71(11.6)}{81.7(54.7)}\ln\frac{51.1}{24.3}$ = 0.0033 cm/s

$k_{20} = \alpha k_T = 0.9761(0.0033)$ = 0.0032 cm/s

[a]Use averaged values only if there is a small difference in test temperature, say, 1–2°C.
[b]Simplify by using the same h_1 and h_2 each time, so you can average k.

Figure 12-3
Data from a falling-head permeability test using the equipment of Fig. 12-1 *or* of Fig. 11-1.

13: CONSOLIDATION TEST

References

ASTM D 2435-90

AASHTO T 216-83

ASTM (1986), *Consolidation of Soils: Testing and Evaluation,* ASTM Special Technical Publication No. 892, 750 pages.

AASHTO (1976), "Estimation of Consolidation Settlement," Transportation Research Board, Special Report No. 163 (with many references).

Becker, D. E., et al. (1987), "Work as a Criterion for Determining In Situ and Yield Stresses For Clays," *Canadian Geotechnical Journal,* Vol. 24, No. 4, Nov., pp. 549–564.

Casagrande, A. (1936), "The Determination of the Pre-Consolidation Load and Its Practical Significance," *Proc. 1st Int. Conf. Soil Mech. Found. Eng., Harvard University,* Vol. 3, pp. 60–64.

Crawford, C.B. (1964), "Interpretation of the Consolidation Test," *J. Soil Mech. Found. Div., ASCE, SM5* September, pp. 87–102.

Gorman, C. T., et al. (1978), "Constant-Rate-of-Strain and Controlled-Gradient Consolidation Testing," *Geotechnical Testing Journal, ASTM,* Vol. 1, No. 1, March, pp. 3–15.

Leonards, G. W., and A. G. Altschaeffl (1964), "Compressibility of Clay," *J. Soil Mech. Found. Div., ASCE, SM 5,* September, pp. 133–156.

Lo, K. Y. (1961), "Secondary Compression of Clays," *J. Soil Mech. Found. Div., ASCE, SM 4,* August, pp. 61–87.

Samarasinge, A. M., Y. H. Huang, and V. P. Drnevich (1982), "Permeability and Consolidation of Normally Consolidated Soils," *J. Geotech. Div., ASCE,* Vol. 108, GT 6, June, pp. 835–850.

Taylor, D. W. (1948), *Fundamentals of Soil Mechanics,* John Wiley and Sons, New York, 700 pages.

U.S. Army Corps of Engineers: *Laboratory Soils Testing,* EM 1110-2-1906, Appendix VIII: Consolidation Test.

Equipment

Consolidometer[1,2], and sometimes called an oedometer (refer to Fig. 13-1)

Dial indicators[3] reading to 0.01 mm (or 0.001 in) or a displacement transducer reading to at least 0.01 mm

Loading device

Stop watch or timer

Thermometer

Sample trimming equipment (see Fig. 13-1*a*) as available or necessary.

[1] There are a number of versions available. The one shown is particularly convenient but may no longer be available.

[2] The standard 2½-in consolidometer can be converted to SI [to use the existing load device ¼ tons per ft^2 (tsf) = 25 kilo Pascals (kPa); ½ = 50 kPa; 1 = 100 kPa, etc.] as follows: (a) Obtain a piece of brass (non-corrosive) tubing of sufficient OD and ID that a machinist can turn a 62.14 mm ID ring with wall thickness equal to that of the consolidometer you now use. Make the height 24 mm. Also obtain a sharp-edged sample trimmer. (b) Turn a piece of solid brass or aluminium to make a spacer disk with 2 mm and 4 mm shoulders to produce a final sample thickness of 20 mm. (c) Use a grinding wheel to carefully grind the porous stones to fit inside the 62.14 mm ID ring. (d) Machine a piece of solid brass tubing for a new load head (or turn down existing load heads). Counter-drill small holes to adjust load head mass consistent with load masses.

[3] ASTM specifies dial indicators that read to 0.0001 in (0.0025 mm). These are difficult to read—particularly the mm precision. This is excess precision when considering the initial sample thickness is on the order of 20 ± 0.01 mm at best. The author recommends using dial indicators reading to 0.001 in or 0.01 mm.

Figure 13-1
Floating ring consolidometer equipment and test details.

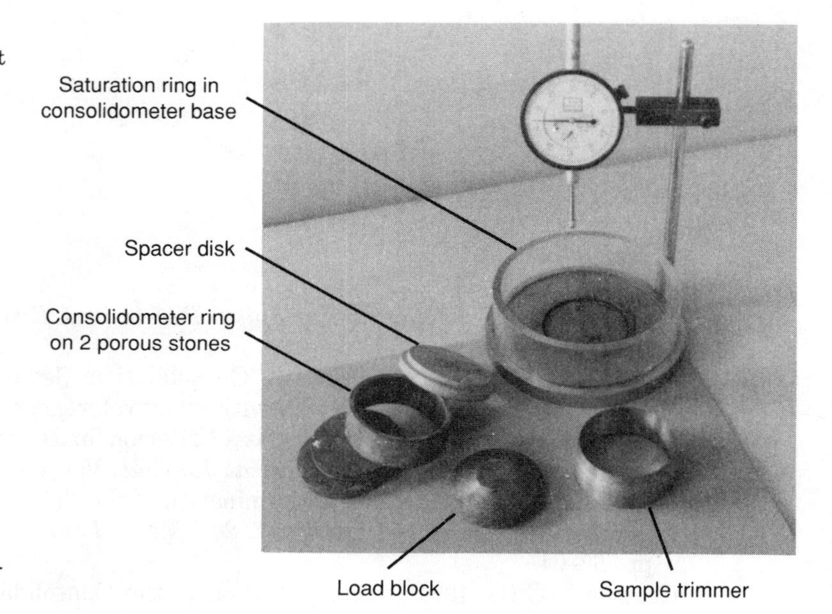

(*a*) A typical set of consolidometer equipment—refer to figure for identification.

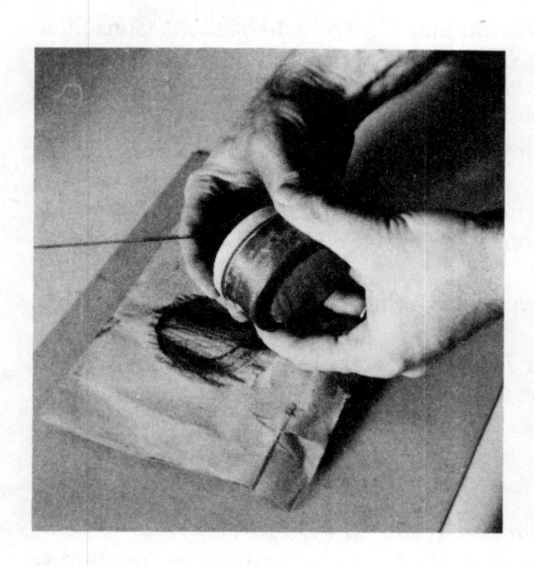

(*b*) Trim sample and place a piece of filter paper on both soil faces to protect porous stones from excessive contamination. Sample is trimmed flush with both ring edges and a piece of filter paper placed on one face. Next the spacer disk with the 2-mm shoulders is used to extrude 2 mm of sample for final trimming (using filter paper side to reduce sticking). A piece of filter paper is placed on this cut face and the 4 mm side of spacer disk used to push the sample back through to trim 2 more mm from other face and the filter paper. Now put on another piece of filter paper and use the 2 mm shoulders to push sample back. Now both faces of sample are approximately 2 mm below ring rims so porous stones can be fitted.

(*c*) Put a wet porous stone on one sample face with 4 mm protruding and place the stone and sample in consolidometer base in stone recess. Put other wet porous stone on top and cover with load block.

(*d*) Grease saturation ring slot and seat saturation ring. Next place consolidometer in loading device. You may want to delay adding saturation water until the second load. When water is added fill above level of top porous stone.

(*e*) Test in progress using a hydraulic loading device. Dial gauge is used here to measure the sample compression at select elapsed times. An LVDT with electronic readout could be used instead of a dial gauge. Other possible load devices include compressed air, or load hangers. The load hanger may be direct, or use a lever to obtain a load advantage.

General Discussion

Note at the outset: There is not complete agreement among the several authorities (including ASTM and AASHTO) on how to reduce the data from this test. The ASTM allowed options are such that two laboratories would not obtain the same results (within normal deviations from within the soil deposit) unless they did both the test and data reduction exactly the same way.

When any soil is subjected to an increase in pressure (or load), a readjustment in the soil structure occurs that may be considered as consisting primarily of *plastic deformation* with a corresponding *reduction in void ratio e*. A small amount of *elastic deformation* may also take place, but considering the magnitude of loads (or contact pressures) involved and that the modulus of elasticity of the soil particles is on the order of 20 MPa, the elastic deformation—recoverable on removing the load—is negligible.

When a load is applied to a dry, partially saturated, or fully saturated coarse-grained soil, or to a dry or partially saturated fine-grained soil, the process of plastic deformation with void-ratio reduction takes place in a short enough period of time for the process to be considered instantaneous. This can be explained by the fact that for dry and partially saturated soils the pore fluid (being primarily air) has almost no viscosity or flow resistance. Similarly, if the soil is coarse-grained the coefficient of permeability k is large and the pore water can quickly flow out. A consolidation test is not required in these cases.

When a load is applied to a fine-grained soil that is either nearly or completely saturated the time for all the plastic deformation and void-ratio reduction to take place is much longer.

The length of time for this process to take place will depend on several factors, of which the primary ones are

1. Degree of saturation
2. Coefficient of permeability of the soil
3. Viscosity and compressibility of the pore fluid
4. Length of path the expelled pore fluid must take to find equilibrium

Note that the pore fluid viscosity may be significant and depends on the temperature. Because of this it is suggested to limit temperature variations to no more than ± 2°C for laboratory tests. Inspection and interpolation of Table 6-1, on the properties of distilled water, indicate the laboratory time versus field time for consolidation can vary by up to 25 percent. In the lab T is in the range of 20-22°C where in the field it is in the range of 10-15°C. The resulting ratio for the viscosity of water at 10 and 20°C is approximately 0.01263/0.01005 = 1.25+.

Consolidation will now be defined as that plastic deformation with void-ratio reduction (generally termed *settlement* ΔH) which is a function of time, or

$$\Delta H = f(t)$$

Time-dependent settlements require data from a consolidation test. A consolidation model which can be used with reasonable confidence is only possible for the case of a fully saturated ($S = 100\%$) soil state. Elastic theory can be used with some confidence for the other saturation cases.

When a load is applied to a fine-grained saturated soil confined in a metal ring with peizometers inserted in the sample (as shown in Fig. 13-2*a*), the water level in all the piezometers will move to a new height of $h + \Delta h$ immediately after the load Δp is applied, as shown in Fig. 13-2*b*.[4] The reason for this is that the low coefficient of permeability of these soils (clays and silty clays) does not allow the pore water to drain instantly. Instead during drainage part of the load increment is carried by intergrain soil contact pressure and part by the pore water. The stress on the pore water produces the flow pressure but this gradually dissipates (the piezometric head decreases as shown in Fig. 13-2*c*) as the water drains. During this time of drainage the process called "consolidation" is occurring.

Since water flows from high potential (high pressure) to a lower potential, and obviously the water surface is the lowest potential for this system, the water in the soil mass always flows toward the exterior of the soil mass. The length of the flow path is shorter for water in the soil pores near the free surfaces of the soil sample where the water quickly flows out, causing a more rapid head reduction in the top piezometer of Fig. 13-2*c* than in the lower piezometer which has a longer flow path to the free water surface. This is also evident from Darcy's equation,

$$v = ki = k(\Delta h/L)$$

The water must travel some distance L in the soil sample. To travel this distance at the velocity of Darcy's equation requires some time interval Δt, as:

$$\Delta t = \frac{L}{v} = \frac{L}{ki} = \frac{L^2}{k\Delta h}$$

[4]This is very idealized since there is often some time lag before the pressure increase develops in the lower piezometer. The soil particles must displace at least a minute amount before any pressure develops. Also as water drains to produce the rise in the tube the pressure falls.

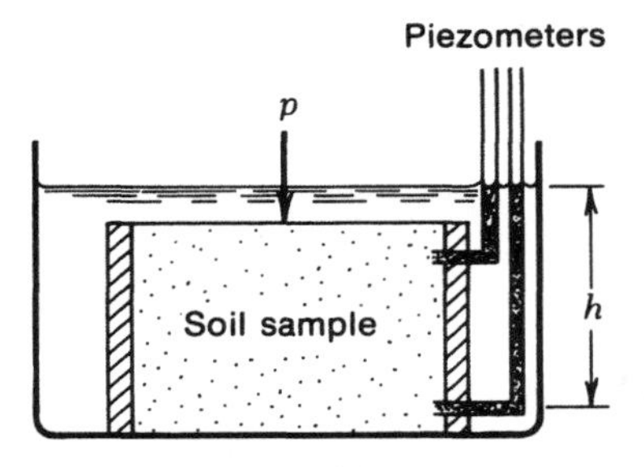

(*a*) Piezometers have reached static equilibrium under some compressive load p.

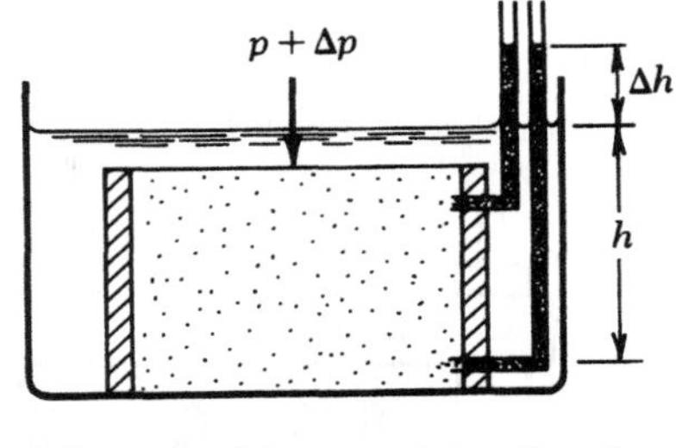

(*b*) On adding a load increment Δp the piezometers record a hydrostatic pressure increase of Δh as shown. There is often some finite time lag before Δh is fully developed but Δh is considered instantaneous in any idealization.

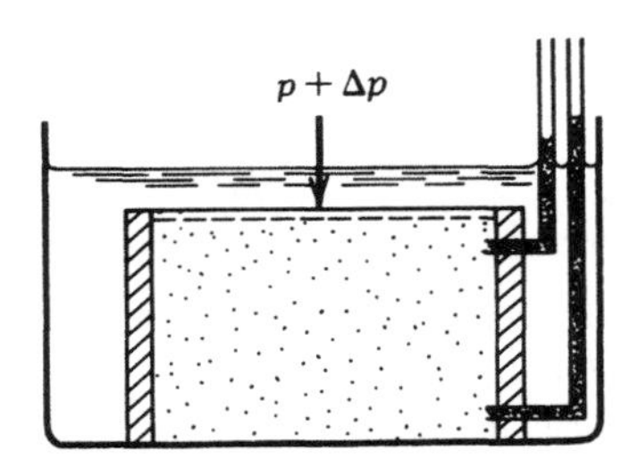

(*c*) Relative positions of piezometers some time after application of Δp. With shorter flow path the upper piezometer loses head faster than the bottom one.

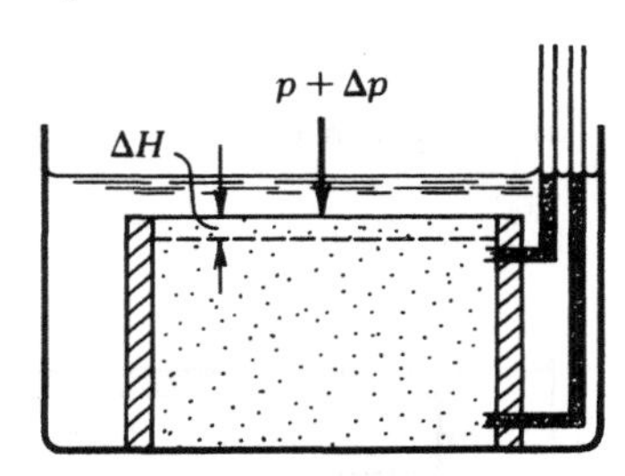

(*d*) At end of primary consolidation under pressure increment Δp both piezometers have returned to the initial position of "*a*." Secondary compression will continue under Δp for some additional time so that the final total settlement will be larger than the ΔH at the end of primary consolidation.

Figure 13-2
Stages in primary consolidation (idealized).

Consolidation (also termed *primary consolidation*) is considered to be complete when the water level in all the piezometers returns to the static water level, as shown in Fig. 13-2*d*. This can take a very long time since v becomes less with h; that is, rearrange Eq. (12-1) to solve for t as $h_2 \to 0$ or let $\Delta h \to 0$ in the above equation. Note that doubling L *will require four times as long for the water to flow out.*

We can now redefine consolidation as that plastic deformation with void ratio reduction of a soil mass which is a function of time and excess pore pressure.

The conventional laboratory test is *one-dimensional* in that, with a metal ring confining the sample no lateral soil or water movement takes place—all water flow and soil movement are in the vertical direction. Whether this duplicates field conditions may be subject to speculation but observations tend to support continued use of this test.

The laboratory consolidation test is performed on a specimen between 20 and 40 mm thick placed in a confining metal ring ranging in diameter from about 45 to 113 mm (100 cm^2 area). Most commonly, a diameter of 63 or 113 mm with a height $h = 20$ mm is used. The apparatus may be of the *fixed-ring* or *floating-ring* type. The floating-ring is more common since it is easier to set up and takes less time to run. If the soil to be tested is very soft the ring may require some support to keep it centered on the sample.

Ring-to-soil friction is indeterminate but can be reduced by limiting the sample thickness. It can also be reduced by spraying the inner ring wall with teflon powder, or using a teflon-lined ring. If neither is available friction can be reduced by polishing the inner ring face. Do not use a ring if it is pitted from corrosion. One of the advantages claimed for the floating-ring device is that it reduces the friction close to 50 percent as idealized in Fig. 13-3*c*.

The consolidation test is usually performed on "undisturbed" soil recovered from the site using tube samples. Great care is necessary to extract the sample from the tube, cut

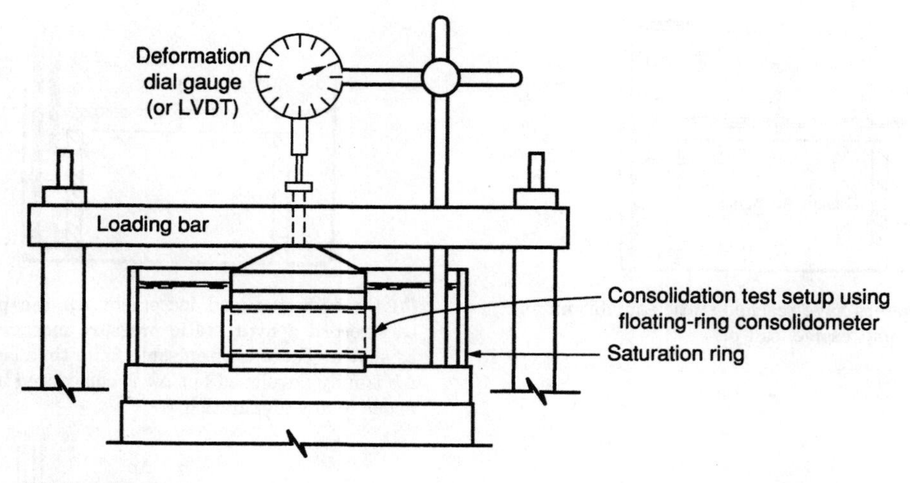

(*a*) Consolidometer.

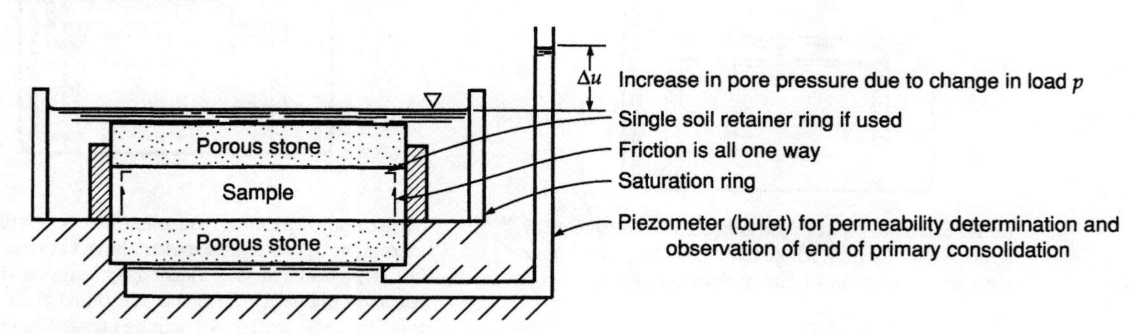

(*b*) Fixed-ring consolidometer. May be used to perform a falling-head permeability test by attaching a buret to the piezometer outlet at base. Fill buret with water to level h_1 and at some time interval measure h_2 as in Test 12. Remove and drain buret for next load increment.

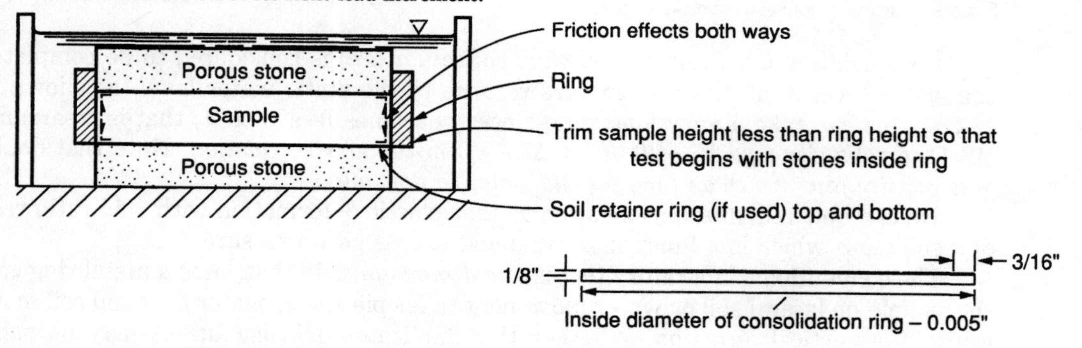

(*c*) Floating-ring consolidometer.

Figure 13-3
Select line details of a consolidation test.

a short length for the test sample, trim it to the ring diameter, put it in the ring and then trim the ends. For example, extruding the sample from the tube produces some disturbance (sawing the tube lengthwise to split it open heats the sample). Cutting the short length may produce some disturbance. Trimming it to the ring diameter certainly produces some disturbance—even if you can use the sharpened edge sample trimmer of Fig. 13-1*a* which cuts the tube sample to the correct ring diameter.

Finally, putting the sample into the ring and flush-cutting one end, then pushing it through to flush cut the other using the spacer disk, then pushing it back through again to retrim the previously flush-cut end does not improve the sample quality. Just do the best you can but try not to drop the sample on the floor.

Note that remolded samples can be used for this test but are primarily either for student laboratories to illustrate the procedure or as a check on the so-called "undisturbed" test. If the undisturbed test parameters are very nearly those of the remolded sample the "undisturbed" test is suspect. Probably the better check is to do at least two "undisturbed" sample tests + a remolded sample test.

The fixed-ring device can be used to measure the coefficient of permeability[5] of the sample at the same time the consolidation test is conducted since sample drainage is limited to one direction. The two-way drainage of the floating ring test, however, reduces the test time to about one-fourth that of the fixed-ring test (for the same sample thickness).

The porous stones should have a k about two orders of magnitude larger than that of the soil sample. They should fit closely (but not bind) inside the sample ring so that none of the sample is squeezed out between the stone and ring. Sample retainer rings (see Fig. 13-3*c*) may be used (if available) to avoid soil squeezing—a problem when testing very soft soil.

On balance and since the sample preparation effects are about the same for any size sample, larger diameter samples provide the highest confidence level. In any case, the consolidation sample ring ID should be at least 6 mm smaller than the tube sample and the diameter/height ratio > 2.5. The author recommends a minimum sample diameter of 62+ mm—requiring a tube sample of at least 75 mm diam.

The strain rate and internal pore pressure conditions are not controlled in the usual test procedure outlined here. Equipment is available that can produce a specified rate of strain and uses a closed sample cell so that internal pore pressures can be developed and/or controlled (termed using *backpressure*) to simulate pore pressure conditions in situ. Some of these procedures have been standardized by ASTM (D 4186). This equipment and methodology are beyond the scope of an introductory laboratory course.

The consolidation test proceeds by applying loads in a geometric progression with a load ratio $\Delta p/p = 1$ with a typical load sequence as follows:

25, 50, 100, 200, 400, 800, 1600, 3200 (sometimes 6400) kPa

An alternative loading sequence, depending on equipment available, might be:

5, 10, 20, 40, 80, 160, . . . etc., kPa

Other load increments can be used *and at least one must be larger than the preconsolidation pressure of the sample.* The load increment ratio is critical. It appears that the increment ratio should be $\Delta p/p \simeq 1$ so the soil does not build up internal resistance to the loads. If this happens the total deformation of the sample is usually less than obtained if the load increment ratio of $\Delta p/p = 1$ (as illustrated above) is used.

EQUIPMENT CALIBRATION

The load block and porous stones—as well as the soil sample—compress under the load increments and produce larger dial readings than just from soil compression. To obtain the amount the equipment compresses place the one or two stones, filter paper, and load block that you will use for the test in the consolidometer and attach the same dial indicator (you may have to use a stem extension). Now set the dial to 0.000 and apply the load sequence and record the deformation. Unload and also record the deformations. If these

[5]This is not a trivial procedure and requires setting up a falling-head permeability test as in Test 12. Use a 100 ml buret with a rubber balloon to control evaporation and run a falling-head test either before applying a load increment or at the end of a load increment. You cannot collect the drain water so the head change in the buret is your control.

are not in reasonable agreement do a second load sequence. Now record these values of ΔH_e to use in the test on the data sheet in the column provided. You may do this test as fast as loads can be changed since this deformation is nearly instantaneous.

These equipment deformations are used to correct the actual dial readings as illustrated in the sample data set at the end of this test discussion.

DATA FROM THE CONSOLIDATION TEST

Consolidation-test data (refer to Fig. 13-5) are obtained and used as follows:

1. Mass, dimensions (initial height H_i and diameter), and water content of the original consolidation test specimen are obtained so that the initial void ratio e_o can be computed. The initial height H_i is a nominal value obtained by trimming soil projections from the ring using a spacer disk to produce a sample height less that the ring height. This lesser height is necessary so the porous stones used to spread load over the sample and facilitate drainage penetrate inside the ring to avoid uneven loading or a hangup on one side of the ring edge so the load is tilted.

   The total volume of the sample can be determined from H_i and the ring area A and can also be computed as

$$V_t = V_s + V_v \tag{13-1}$$

   where V_s and V_v = volume of soil solids and volume of voids (or water if S = 100 percent), respectively. The volume of water can be computed from the water content if it is reliably known at the beginning of the test, using relationships from the "Introduction" preceding Test No. 1:

$$V_w = \frac{wM_s}{G_w \rho_w} = wM_s \tag{13-2}$$

   since the specific gravity and density of water can be taken as 1.0 with negligible error.

   The volume of water at the end of the test V_{wf} can be obtained directly as the loss of water from oven-drying the sample cake. At this time the saturation S should be 100 percent (even if it is less[6] at the beginning of the test). From initial sample dimensions, change in sample height, and final water content, a back-computation can be made for V_w, which is usually more reliable than using Eq. (13-2).

   If the soil cake is to be oven-dried, be sure to remove the water from the saturation ring used to maintain sample saturation before disassembling the test apparatus. You may use a battery bulb or similar device for the water removal. This is so the sample does not absorb water if it should expand as you remove the load in order to take the apparatus apart. To minimize absorption from the porous stones you should disassemble the equipment and remove the stones from the sample ends in under 30 s.

   Special Note: ASTM requires unloading the sample in increments prior to removing it. However, even if the sample expands, if you have removed access to water there is none for it to absorb (take in by suction). Also sample expansion does not necessarily mean water absorption so the procedure of the ASTM leaves S indeterminate at the end of the test.

   The height of the soil solids (solid mass in the block diagram of Fig. I-1) can be readily computed after ovendrying the soil cake using the total change in sample height ΔH *corrected for equipment calibration* (see Eq. (13-11) as:

$$H_f = H_{\text{initial}} - \Delta H$$

[6]Due to both samples expansion from loss of overburden pressure and loss of hydrostatic water pressure causing air to come out of solution.

where ΔH = difference between final dial reading D_f and initial dial reading $D_i + \Delta H_e$. The final sample volume is

$$V_{\text{final}} = H_f A = H_s A + V_{\text{wf}} \tag{13-3}$$

where V_{wf} = volume of water in the sample from assuming S = 100% and is also the mass M_w in g from oven-drying the soil cake. Solving for the height of soil solids H_s one obtains

$$H_s = H_f - \frac{V_{\text{wf}}}{A} \tag{13-4}$$

where A = area of consolidometer ring (cm^2 recommended).

The initial height of voids can be computed using the initial sample height H_i as

$$H_v = H_i - H_s \tag{13-5}$$

and the initial void ratio e_o can be computed as

$$e_o = \frac{H_v}{H_s} \tag{13-6}$$

An additional check on the computed value of H_s is obtained if the specific gravity of the soil solids is known; alternatively, we may compute G_s from H_s and the dry mass of soil solids M_s. If the value is not reasonable we may attempt some method of reconciliation among the several data items from the test.

2. Time versus settlement data. For each load increment the amount (or dial reading) that the sample has compressed at the end of a series of elapsed times (in minutes) is recorded as part of the data. This information is obtained by attaching a dial indicator[7] to take readings of sample ΔH. A total time for a sample to consolidate under a load increment may be 24 to 48 h or more. To give all the readings in this time span equal importance, these data are usually presented as a semilogarithmic plot of dial reading (DR) vs. time (time on the log scale) in minutes for each load increment.

   There are three methods for taking time versus dial readings. One method is to take dial readings until there is no appreciable change between the last two readings. In the second method readings are taken for a specified duration of time 12 h, 24 h, or longer. In the third method one uses a plot of dial reading vs. $\sqrt{\text{time}}$ (again in minutes).

   This last method of data presentation is credited to Taylor (1948), who found that the method provided reasonably reliable results for clay soils in the Boston, Massachusetts, area. Currently this method is used in other areas as well with claimed advantage that the test is more rapid.

   Note that a plot of dial readings is the same as a plot of ΔH vs. time since the difference between any two dial readings multiplied by the dial sensitivity is the numerical compression.

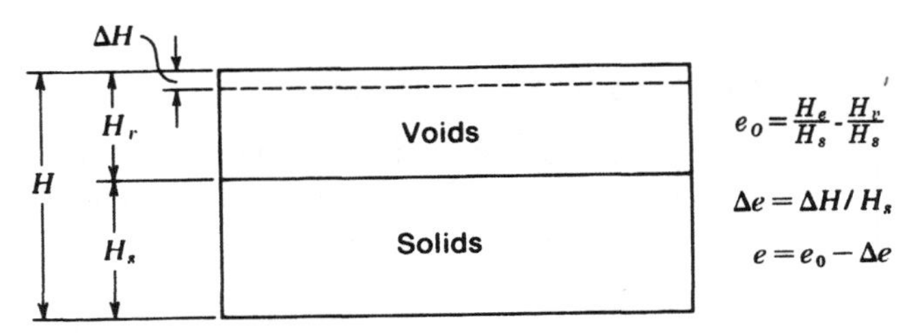

Figure 13-4
Relationship between change in sample height and void-ratio reduction.

[7] "Dial indicator" is used in this discussion as they are most commonly used. If a displacement transducer (LVDT) is used the deformations are directly obtained and used instead of "dial readings."

From a plot of dial reading vs. log time (or $\sqrt{\text{time}}$), as shown in Fig. 13-6*a* and *b*, one may obtain the dial reading corresponding to the *end of primary consolidation* D_{100}. This is also called the "end of 100 percent consolidation" or $U = 100\%$ for that load increment. Time elapsed when this occurs is termed t_{100}. It is also necessary to obtain the dial reading at the beginning of the test, called D_0. While one can always use the actual initial dial reading for that load increment for D_0 a semilog or a $\sqrt{\text{time}}$ plot may indicate that there is an apparent D_0 that should be used instead.

Data reduction usually requires using a D_{50} (dial reading at $U = 50\%$) and t_{50} (time in minutes when D_{50} occurs). How to obtain these follows:

A. USING A SEMILOG PLOT

a. First plot the curve using a vertical scale (the arithmetic ordinate) such that you can get at least two and preferably three time-settlement curves on the graph sheet. This allows some comparison of data quality. The principal requirement is that the scale be sufficiently large that the middle and end branches can be reasonably well identified. This cannot be done very well if you use a small scale and there is not much change in the dial readings for that load. You might first plot with a pencil using soft lead and see if the scale is adequate before using a more permanent marking material.

b. To obtain D_{100}, t_{100} proceed as follows:
   1. Through the midportion of the curve produce a best-fit tangent to its steepest part.
   2. Through the end portion produce a best-fit tangent.
   3. At the intersection of these two tangents draw a horizontal line to the ordinate and read D_{100}. Project this horizontal line to the actual curve then down to the log axis and read t_{100}. The t_{100} time value is not used in this test.

c. To obtain D_0 (the theoretical dial reading at $t = 0$, since one cannot plot $t = 0$ on a log scale) proceed as follows:
   1. Inspect the initial curve branch and if it appears that projecting the curve to the left towards $\log 0 = \infty$ would give the actual initial dial reading use that for D_0.
   2. If "1" does not apply then select a time t_1 and a time t_2 that is $4t_1$ (e.g., 2 min and 8 min). Do this only in the initial branch—do not extend into the mid-portion previously used. Also do this only if this portion of the curve appears nearly parabolic; if it is not parabolic use "1" above for D_0. Measure the ordinate between t_1 and t_2 and lay off this distance above t_1. Draw a horizontal line through this point and read D_0 where it intersects the DR ordinate. You may try one or more additional time sets and put your horizontal line through their average but this may be excessive refinement.

d. Obtain t_{50} as follows:
   1. First find D_{50} as

$$D_{50} = \frac{D_0 + D_{100}}{2} \tag{13-7}$$

   2. Project D_{50} to the curve and drop a vertical to the log axis and directly read t_{50}. *This value should be shown on the DR vs log t curve* for easy access later.

B. USING A DR VS. $\sqrt{\text{TIME}}$ PLOT

a. As the test proceeds and you have taken a DR and time reading, immediately plot the point. Note that it is most convenient if you take the time readings in the following initial time lapse sequence:

2, 4, 9, 16, 25, 36, 49 (or 64), 100 . . . min (as convenient)

This makes an easy plot since $\sqrt{2} = 1.414$, $\sqrt{4} = 2$; $\sqrt{9} = 3$, and so on usually fall on defined grid points.

b. When you have four or five points draw a best-fit straight line through them and extend it back to the ordinate as in Fig. 13-6*b*. The intersection of this line can be taken as D_0 (or you can use the actual initial dial reading). Also extend this line down to the abscissa.

c. Continue plotting the points as you obtain them and note that the points begin to deviate from the straight line. Use a drafting curve and draw a smooth best-fit curve when you have about three points accumulated. Next take a scale and measure a 15 percent offset from the abscissa value from step "b" above when that straight line was extended. Again refer to Fig. 13-6*b*. Draw a straight line from D_0 to the 15 percent offset. When the curved part of the settlement curve intersects this offset line that dial reading DR is arbitrarily taken as D_{90}.

   At this point the next load increment can be applied (or test terminated if this is the last load increment). This time for D_{90} is usually much less than using a specified time increment and is why this method is used by many commercial laboratories.

d. With D_{90} now known, obtain D_{100} as

$$D_{100} = D_0 + \frac{10}{9}(D_{90} - D_0) \tag{13-8}$$

   With D_{100} computed we can obtain D_{50} and from that obtain $\sqrt{t_{50}}$ and from this obtain t_{50}, as shown on Fig. 13-6*b*.

Both of the above procedures can be used to find times at any other percent consolidation U. Clearly t_{25} occurs at D_{25} and so on, where the subscript 25, 50, 100 refers to U.

The time at $U = 50$ percent (t_{50}) is most commonly used to estimate the *coefficient of consolidation* c_v from a test. This value is used to estimate rate of settlement. For example, the total settlement of a structure is estimated at 100 mm; we may want to know about when 20 mm, 40 mm, and so on of settlement will occur so we can plot a curve of settlement versus time. Use the consolidation test to compute c_v as follows:

$$c_v = \frac{T_i H^2}{t_i} \tag{13-9}$$

where T_i = time factor given in Table 13-1 for two assumed pore pressure cases; Case I is usually used and gives for $i = 50(t_{50})$, $T_{50} = 0.197$; for t_{90} use $T_{90} = 0.848$.

t_i = corresponding time for T_i as previously illustrated.

H = *average length of the longest drainage path* during consolidation for the given load increment. For a porous stone on both faces of the sample in the floating ring test $H = \frac{1}{2}$ the average sample height for that load increment. The average total sample height for the load increment i is computed as

$$H = H_i - D_{50} + \Delta H_e$$

if you use t_{50}; ΔH_e = equipment correction.

The coefficient of consolidation is plotted as c_v versus log pressure p, generally on the same graph plot as the void ratio vs. log pressure but at a different ordinate (arithmetic) scale. ASTM and AASHTO require using the *average* pressure; however, the curve is usually so erratic—because of limitations in the theory and because of the method of obtaining H in Eq. (13-9), that it is adequate to plot c_v against the corresponding load increment (both plots are shown in Fig. 13-9).

Generally one can use t_{50} for computing c_v but other convenient time values can also be used. You may reduce the vertical scale to obtain a more reliable-appearing curve. Since

Table 13-1 Time factors for indicated pressured distribution

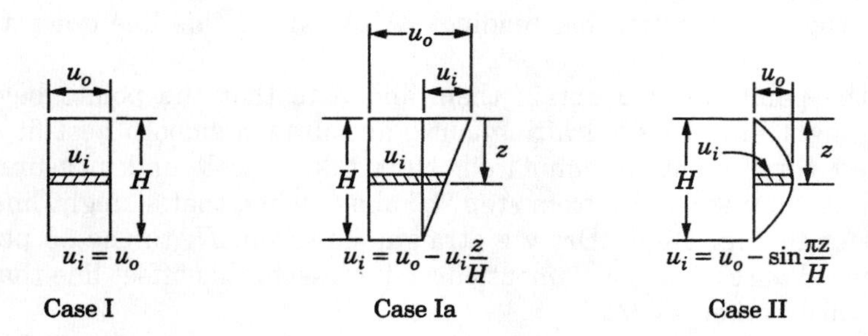

Pore-pressure distribution for case I usually assumed for case Ia.

U, %	Case I	Case II
0	0.000	0.000
10	0.008	0.048
20	0.031	0.090
30	0.071	0.115
40	0.126	0.207
50	0.197	0.281
60	0.287	0.371
70	0.403	0.488
80	0.567	0.652
90	0.848	0.933
100	∞	∞

the thin sample undergoes far more void ratio reduction during testing than in the field you probably should select a value near the preconsolidation pressure for time-of-settlement computations.

C: SECONDARY COMPRESSION

Observations indicate that additional soil deformation occurs after the end of primary consolidation. This additional settlement is termed *secondary compression* and is believed to be a continuing soil-particle readjustment to the applied load. The amount and time involved seems to depend on the type of soil (organic, inorganic, overconsolidated, etc.). In some heavily organic soils the secondary compression component is larger than primary settlement but in most soils primary consolidation predominates.

There is no mathematical model currently available to compute this additional compression with much confidence. However, we may be able to make a crude estimate from the laboratory test. That end part of a plot of DR vs. log time is called the secondary branch of the curve and is defined by the end slope that intersected the middle slope to locate the D_{100} reading. The additional settlement from the D_{100} reading to the end of the test is *secondary compression in the laboratory test.*

The slope of the secondary compression branch is approximately constant for a given remolded soil. It may not, however, be constant for natural soils and there is some evidence that the slope in these cases is stress-dependent.

The slope of the secondary branch of the curve can be used to compute a secondary compression coefficient C_α as in Fig. 13-6*a* for estimating the amount of secondary compression as

$$C\alpha = \frac{\Delta H_s}{\log(t_2/t_1)} \tag{13-10}$$

where ΔH_s = compression along the secondary branch from times t_1 to t_2, in or mm

t_1, t_2 = starting and ending times for ΔH_s, min

As before, you should extend the slope across one log cycle. Note that C_α is approximately 0 in Fig. 13-6*a* for both tests (an inorganic blue clay-fine sand mixture).

D: CONSOLIDATION TEST SETTLEMENT PARAMETERS

Settlement parameters are obtained from plotting certain consolidation test data. Normally one plots either

Void ratio vs. pressure, or
Strain vs. pressure

These plots will be considered separately:

1. *Plot of void ratio vs. pressure.* This plot may be made using either log pressure (Fig. 13-9) or as an arithmetic plot (Fig. 13-8). The semilog plot is more commonly used.

   From Fig. 13-4 it is evident that while the soil skeleton deforms there is a loss of voids, but the total solids remain constant. Thus, the initial void ratio e_o is

$$e_o = \frac{H_v A}{H_s A} = \frac{H_v}{H_s}$$

At the end of any load increment, the change in voids height is

$$\Delta H = D_{fi} - \Delta H_e \qquad (13\text{-}11)$$

where H_e = equipment deformation at this load increment, in or mm.

D_{fi} = one of the following:

a. D_{100}, or
b. Dial reading at some specified time that is the same for each load increment, or
c. Dial reading at the end of that load increment.

ASTM (and AASHTO) allows defining D_{fi} by either "a" or "b" above. The U.S. Army Corps of Engineers requires "b." Since there is generally little difference between "b" and "c," the latter can also be used. The choice of "a," "b," or "c" produces slight differences but they are usually less than other test limitations. ASTM allows either "a" (Method **B**) or "b" (Method **A**) with a 24 h time.

From the height of soil solids H_s, initial voids height H_v and change in sample voids height, the current void ratio is

$$e_i = \frac{(H_v - \Delta H)A}{H_s A} = \frac{H_v - \Delta H}{H_s} \qquad (13\text{-}12)$$

Since the initial void ratio is H_v/H_s it is usually more convenient to compute the current void ratio (c.f. the computer program with this manual) as

$$e_i = e_o - \frac{\Delta H}{H_s} \qquad (13\text{-}12a)$$

From an arithmetic plot of e vs. pressure one can obtain the *coefficient of compressibility* a_v (Fig. 13-8) defined as:

$$a_v = \frac{\Delta e}{\Delta p} \tag{13-13}$$

with the negative sign disregarded. Also one can obtain the *coefficient of volume compressibility* m_v as

$$m_v = \frac{a_v}{1 + e_o} \tag{13-14}$$

Note that m_v is the reciprocal of the stress-strain modulus ($1/E_s$).

From a semilog plot (most common method of presenting data) of void ratio vs. log pressure (see Fig. 13-9), we obtain, from the straight-line part past the preconsolidation pressure, the *compression index*, C_c,

$$C_c = \frac{\Delta e}{\log(p_2/p_1)} \tag{13-15}$$

The negative sign should be ignored. If there is an unload branch we can obtain the *swell index* C_s as

$$C_s = \frac{\Delta e_s}{\log(p_2/p_1)} \tag{13-16}$$

Again ignore the negative sign. The initial branch of the usual consolidation test on an "undisturbed" field soil has a characteristic slope of the first two plotted points (25 and 50 kPa) of Fig. 13-9. If you let the sample rebound from the 1600 kPa point back to 100 kPa and then recompress to 1600 kPa, you also obtain the characteristic slope of an "undisturbed" soil. The best-fit slope of this curve part is called the *recompression index* C_r and is computed as

$$C_r = \frac{\Delta e_r}{\log(p_2/p_1)} \tag{13-17}$$

Again neglect the negative sign. If you have several recompression branches you should use an average value of C_r from all of the branches.

In Equations (13-15), (13-16), and (13-17) you should extend the line slopes across one log cycle so that $\log(p_2/p_1) = \log 10 = 1$. Project both cycle ends to the ordinate to get the corresponding change in void ratio Δe.

2. *Plot of strain ϵ vs. log pressure.* This method of presenting consolidation test parameters is relatively recent but is becoming a preferred method. It has the advantage of not requiring so many computations, and plotting progresses with the test. The strain at the end of any load increment is simply

$$\epsilon = \frac{\Delta H}{H_i} \tag{13-18}$$

where ΔH is from Eq. (13-11) and H_i = initial sample height. These values are plotted as in Fig. 13-10. The slope of the curve portion past the preconsolidation pressure is called the *compression ratio* C'_c, which is computed as

$$C'_c = \frac{\Delta \epsilon}{\log(p_2/p_1)} \tag{13-19}$$

Neglect the negative sign. By analogy one can obtain the recompression ratio C'_r and the swell ratio C'_s.

The following relationship exists between the compression index C_c and the compression ratio:

$$C_c = C'_c(1 + e_o) \tag{13-20}$$

F: PRECONSOLIDATION PRESSURE

When the void ratio vs. log pressure or strain vs. log pressure curve has been drawn, it will be found that if the test has been performed on an undisturbed sample from the field, it will have a characteristic sharp[8] curve bend such as point 0 of Fig. 13-10. This characteristic shape is attributed to the sample being unloaded of its overburden pressure when it was recovered from the field. This conclusion has been drawn from observing a similar curve shape obtained by loading and unloading consolidation-test samples in the laboratory as illustrated by the single load cycle in Fig. 13-9.

Special Note: It has been found that the characteristic shape is not clearly defined if the sample is permitted to swell during the reload branch. If it appears the sample may swell, the saturation ring should not be filled until the second or third load increment has been applied. You can ascertain if this is a problem by observing the dial gauge after the first load when you fill the ring. If it indicates swell, remove the water in the ring with a battery bulb.

From observation of the void ratio vs. log pressure (or strain vs. log pressure) curve, Casagrande (1936) proposed that the preconsolidation pressure could be estimated as follows (see Figs. 13-9 and 13-10):

a. At the sharpest part of the curve, estimated by eye, draw a tangent.
b. Through this point of tangency, draw a horizontal line such as line $0 - C$ of Fig. 13-9.
c. Bisect the angle formed in steps *a* and *b* above.
d. Extend the straight line of the e vs. $\log p$ (or ϵ vs. $\log p$) curve until it intersects the angle-bisector line.
e. Drop a vertical line to the abscissa and read the corresponding value of pressure p.

If	Preconsolidation state
$p \le p_0$	*normally* consolidated—use p_0
$p > p_0$	*preconsolidated*—use $p_c = p_0$

In all cases p, p_o and p_c are intergranular or *effective* stresses.

f. On the e vs. $\log p$ curve, the in situ void ratio e_o can be estimated as follows:
   (1) For a normally consolidated soil extend the intersection of the p_o *point on the curve horizontally to the ordinate.*
   (2) For a preconsolidated soil extend a horizontal line from the angle bisector point B of Fig. 13-9 to the ordinate.

Carefully note that the initial void ratio of any undisturbed field specimen will be somewhat larger than in situ from sample expansion. This expansion, due to loss of overburden pressure, always occurs. Some empirical procedure is necessary to estimate the in situ value.

If the e vs. $\log p$ or ϵ vs. $\log p$ curve appears curved throughout without a clearly defined sharp curvature, one must either estimate the preconsolidation pressure or use

[8]Since this is a remolded sample, the "sharp" break characteristic of an undisturbed soil is not well identified. In this type of situation you must identify the preconsolidation point as best you can.

some other method. A method proposed by Becker et al. (1987) computes the work per unit volume as

$$W_i = \frac{(\sigma_i + \sigma_{i-1})}{2}(\epsilon_i - \epsilon_{i-1}) \qquad (13\text{-}21)$$

where i = current increment and
$i - 1$ = previous increment of stress σ and strain ϵ. The W_i values are plotted against the corresponding σ_i stresses to an arithmetic scale. Draw best-fit straight lines (1) through the initial three or four points and (2) through the end several points. The intersection of these two straight lines is a best estimate for the preconsolidation pressure p_c.

Procedure

This is suggested as a lab section project.

Special Note: The following procedure assumes use of a floating-ring device. Adjust instructions if using a fixed-ring consolidometer.

1. Prepare the equipment. Obtain the two porous stones, check their fit into the sample ring, and put them in a beaker of warm water to deaerate and saturate. Get your dial indicator and loading block. Identify which consolidation load device you will use.
2. Carefully trim a sample to fit the consolidation ring (or use the sample provided by the instructor).[9]

   Optional: Collect a representative sample of the trimmings for a water content sample. Also use some of the trimmings to determine the specific gravity G_s.

   Trim the sample to the nominal height of 20 mm or as produced by the spacer disk and record this value on your data sheet. Also determine the initial mass of the sample + ring and record on your data sheet. Next put a piece of filter paper on each face of the sample (to minimize particle penetration into the porous stones). Cover with a damp towel to avoid sample dehydration.
3. Calibrate your equipment by putting the loading block onto your two porous stones with two pieces of filter paper between. Fix the dial gauge to the cross-arm assembly and apply the load sequence to determine the equipment compression for each load increment. Record these ΔH_e values on your Data Sheet 13*c*.
4. Carefully place the soil sample in the consolidometer with one of the saturated porous stones on each face, and be sure the stones partly penetrate into the sample ring. Place the consolidometer into the loading device and attach the dial gauge; remember to allow for possible sample compression of 4 to 12 mm.
5. Apply a seating load of say 5 (for soft soil) to 10 (for stiff soil) kPa. Again check that the porous stones do not hang on the ring. Within 5 min of load application, zero the deformation dial gauge (leave the seating load on the soil).
6. At a convenient starting time, apply the first load increment (sufficient additional load to develop first desired load increment) and simultaneously take deformation readings at the specified elapsed times as given on Data Sheet 13*b* + others, perhaps 2, 4, 8, 12, 16 and 24 h (change loads at 24 h). Modify the time sequence if you are doing the complete test in one lab period. After the 15 or 16 min reading fill the saturation ring with enough water that it will not evaporate below the top porous stone and dehydrate the top of the sample. Check whether the sample is swelling and, if so, quickly remove

[9]The author suggests preparing a sample of fine, silty sand that will consolidate in about 30 min per load so the test can be performed in one extended lab period. The sample will not have a preconsolidation pressure (unless previously made and consolidated) but the other aspects of the test can be observed. Advanced classes should perform the test using an undisturbed soil and change the load every 24 h.

the saturation water with a battery bulb. In this case put damp towels, wet cotton or other material around the sample to keep it from dehydrating.

7. After 24 h or as directed, or when ΔH between two successive readings is sufficiently small, change the load to the next value and again take elapsed-time interval readings as in Step 6 above. If your saturation ring was emptied in Step 6, fill it after the 15 min reading. No swell should occur under a 50 kPa pressure.

   If using a "quick" laboratory test with a remolded soil, on the third or fourth load increment, run the test sufficiently long to establish a valid secondary consolidation slope for use with all the other load increments.
8. Continue changing loads and taking elapsed time vs. deformation dial readings through the load range of the consolidometer (or until arbitrarily terminated). At least one load increment past the preconsolidation pressure p_c should be taken. The final load increment should be at least $4p_c$, subject to equipment limitations.
9. Weigh a clean, dry evaporating dish for a water content container. Now carefully remove all the water from the saturation ring using a battery bulb, sponges, paper towels, and so on. Next remove the sample load and *quickly* remove the sample ring from the consolidometer. Push the wet soil cake out of the ring into the evaporating dish and rapidly weigh the dish + wet soil. Put these data on a water content sheet from this lab manual. Now put the sample in the oven to dry for 12 to 16 h. Also, if not previously done, determine the mass of the empty sample ring. Clean the porous stones. Boiling is suggested to get any clay particles out of the pores before the clay dries and becomes very difficult to remove. In lieu of boiling, brush the stones with a metal brush while washing until they appear visibly clean.
10. On the next day return to the lab and determine the mass of the oven dry soil cake + dish. Record on the data sheet and compute the mass of dry soil solids M_s and record. Also compute the final water content and the volume of water in the soil cake at the end of the test V_{wf}. Refer to Fig. 13-5*a* and/or use a block diagram to make this computation.
11. Now compute the height of solids H_s, final voids height H_{vf} and initial voids height $(H_i - H_s)$. Use these data to compute the initial and final void ratios.
12. For each load increment, plot curves of dial reading vs. log time using 5-cycle semilog paper from your data sheet section at the end of this manual. Find D_{100}, D_0, D_{50}. Use D_{50} and obtain t_{50} as previously described. Show these data on your curve and record necessary data on your data sheet. You may plot more than one load increment on a sheet but do not excessively bunch the plots—use additional sheets. Refer to Figs. 13-5 to 13-10 for computations you will have to make and methods of displaying data.
13. *Optional:* Plot dial readings vs. $\sqrt{\text{time}}$ for any two load increments, find D_0 and D_{90}, and compute D_{50} and t_{50}. Now compare D_{50} and t_{50} with those obtained in Step 12 above. Comment in your report on possible reasons if there is a large discrepancy.
14. Using your time vs. dial reading sheets and your computations for e_o, H_s, ΔH, and average sample height H (corrected for equipment deformation as previously outlined) complete Data Sheet 13*c* titled "Computation sheet for e, ϵ and c_v" (refer to Fig. 13-7).
15. Plot either *a* or *b* (or both) following as directed by your instructor:
    a. Plot a curve of ϵ vs. log p using a sheet of 3-cycle semilog paper from the back of this manual. Compute the strain ϵ using Eq. (13-18). Also use $D_{fi} = D_{100}$ unless specified otherwise by the instructor.

       From the plot obtain the preconsolidation pressure p_c (if applicable) and compute C_c' by extending the curve slope across one log cycle. Also compute C_c using Eq. (13-20).
    b. Plot a curve of e vs. log p using 3-cycle semilog paper from the back of this manual. Compute the void ratio e_i using Eq. (13-11). In this equation use $D_{fi} = D_{100}$ unless specified otherwise by the instructor.

       From this plot obtain the preconsolidation pressure p_c (if applicable) and the compression index C_c by extending the curve slope across one log cycle. Show the computations on the graph sheet.

16. Plot on the graph of Step 15 the curve of c_v vs. log p with the ordinate marked on the right-hand scale as shown in Fig. 13-9. Indicate your recommendation for what to use for c_v if this test is for a design project.
17. Prove that

$$C'_c = \frac{C_c}{1 + e_o}$$

18. Comment on the effect of including the equipment calibration ΔH_e in the computations for e, ϵ, and the H used for c_v. Does its inclusion improve the results or simply imply a precision that is not realistic for this test?

It is recommended that you check the computations (Data Sheet 16) using the enclosed computer program. If you use only the computer program, you must show at least two lines of sample computations.

At this point you should appreciate that this is one of the most complicated laboratory tests in geotechnical engineering and is one of the most expensive to perform in a commercial laboratory. For these reasons there is sometimes a hesitancy—clients are reluctant to pay—to do enough tests to define the soil parameters. Unless you do at least two tests on any soil, there is no assurance that a data reduction error has not been made or that the soil sample is representative of the project.

Sample Computations (For Figs. 13-5 and 13-7)

For Fig. 13-5 most of the data are self-explanatory.

Compute height of soil solids H_s using Eq. (13-4):

Final dial reading	= 364 units at 0.01 mm/unit	= 3.640 mm
Equipment calibration correction		= 0.064 mm
	Net total change	= 3.576 mm

with M_w = 16.75 g → 16.75 cm³ the height of solids is

$$H_s = H_f - V_{wf}/\text{Area of Ring}$$

$$= (2.00 - .3576) - \frac{16.75}{30.33} = 1.0901 \text{ cm}$$

$$= 10.901 \text{ mm}$$ (You may verify this with a block diagram.)

Since $H_s + H_{vi}$ = 20 mm, obtain H_{vi} = 20.000 − 10.901 = 9.099 mm

The initial void ratio $e_o = H_{vi}/H_s$ = 9.099/10.901 = 0.835

DATA REDUCTION ON FIG. 13-7a:

First plot all the curves for DR vs. log time and obtain D_{50}, D_{100} and t_{50} and enter on the sheet. Note format for showing D_{50} and D_{100} separated by slashes (also use this format for displaying void ratio e and strain ϵ). Enter the several values of ΔH_e in the column provided.

We start with initial e = 0.835 from Fig. 13-5a and initial strain ϵ = 0. At the end of the first load increment D_{50} = 0.155 mm (from time curve), D_{100} = 0.31 mm and t_{50} = 10 min.

VOID RATIO COMPUTATIONS (ASTM METHOD B AND D_{100})

$$e = e_o - (D_{100} + \Delta H_e)/H_s$$

for 25 kPa : $e = 0.835 - \dfrac{(.31 - 0.000)}{10.90} = 0.807$

for 50 kPa : $e = 0.835 - \dfrac{(.58 - 0.002)}{10.90} = 0.782$

for 100 kPa : $e = 0.835 - \dfrac{(.96 - 0.004)}{10.90} = 0.747$

for 1600 kPa : $e = 0.835 - \dfrac{(3.60 - 0.064)}{10.90} = 0.511$

Note: $e = 0.511$ differs from final computed value of $e_f = 0.506$. The discrepancy comes from using D_{100} therefore some secondary compression is not included.

STRAIN COMPUTATIONS

$$\epsilon = \frac{(D_{100} - \Delta H_e)}{H_i} \quad \text{where } H_i = 20 \text{ mm}$$

for 25 kPa : $\epsilon = \dfrac{(0.31 - 0.000)}{20} = 0.016$

for 50 kPa : $\epsilon = \dfrac{(0.58 - 0.002)}{20} = 0.029$

for 100 kPa : $\epsilon = \dfrac{(0.96 - 0.004)}{20} = 0.048$

for 1600 kPa : $\epsilon = \dfrac{(3.60 - 0.064)}{20} = 0.177$

AVERAGE SAMPLE HEIGHTS AND H FOR c_v

$H = H_i - D_{50}$ For c_v, $H = H/2$

for 25 kPa : $H = 20.000 - 0.155 = 19.845$ mm

for c_v : $H = \dfrac{1.9845}{2} = 0.992$ cm

for 50 kPa : $H = 20.000 - 0.450 + 0.002 = 19.552$ mm

for c_v : $H = \dfrac{1.9552}{2} = 0.978$ cm

for 100 kPa : $H = 20.000 - 0.805 + 0.004 = 19.199$ mm

for c_v : $H = \dfrac{1.92}{2} = 0.96$ cm

for 1600 kPa : $H = 20.000 - 3.25 + 0.064 = 16.814$ mm

for c_v : $H = \dfrac{1.6814}{2} = 0.84$ cm

Note rounding of values to a reasonable number of digits.

COMPUTE EXAMPLE c_v

$$c_v = \frac{0.197H^2}{t_{50}}$$

At 25 kPa: $$c_v = \frac{0.197(.992^2)}{10} = 0.019 \text{ cm}^2/\text{min}$$

At 50 kPa: $$c_v = \frac{0.197(.978^2)}{22} = 0.009 \text{ cm}^2/\text{min}$$

At 1600 kPa: $$c_v = \frac{0.197(.840^2)}{7.4} = 0.019 \text{ cm}^2/\text{min}$$

COMPUTE EXAMPLE WORK FOR p_c

You should complete and plot using Eq. (13-21).

At 25 kPa: $$W_{25} = 0.5(25 + 0)(0.016 - 0.) = 0.2 \text{ kPa}$$

At 50 kPa: $$W_{50} = 0.5(50 + 25)(0.029 - 0.016) = 0.5$$

At 100 kPa: $$W_{100} = 0.5(100 + 50)(0.049 - 0.029) = 1.5$$

At 200 kPa: $$W_{200} = 0.5(200 + 100)(0.076 - 0.049) = 4.1$$

At 1600 kPa: $$W_{1600} = 0.5(1600 + 800)(0.177 - 0.141) = 43.2 \text{ kPa}$$

From a large-scale plot (with load increments σ_i on the *abscissa*) obtain p_c between 110 and 140 kPa. Note you can also plot the average load increment vs. W_i with the same approximate results.

CONSOLIDATION TEST (Sample data) Data Sheet 13*a*

Project Lab Manual Data Job No. —

Location Soil Laboratory Boring No. — Sample No. —

Description of Soil Blue Clay with Very Fine Sand Depth of Sample —

Tested By K. M. & M. C. Date of Testing Nov 23 – 29, 19– –

Ring: Diam. = 62.14 mm; Area, A_r = 30.33 cm²; Ht. = 24.00 mm

Sample Ht., H_i = 20.00 mm Soil, G_s = (2.71)

Mass of ring + wet sample at start of test = 292.02 g

Mass of ring = 174.90 g

Mass of wet soil = 117.12 g

Initial water content[1], W_i = — %

Mass of dry soil computed from init. water content M_s = — g

Equations for H_s

$$H_s = H_i - \Delta H - H_{vf}$$

also

$$H_s = M_s/(G_s \rho_w A_r)$$

Final Water Content Determination — End of Test

Mass dish + ring + wet soil cake = 706.45 g

Mass dish + ring + dry soil cake = 689.70 g

Mass dish + ring = 600.00 g

Mass of oven dry soil cake M_s = 89.70 g

Mass water in soil cake M_{wf} = 16.75 g

Final water content $W_f = M_{wf}/M_s$ = $\frac{16.75 \times 100}{89.70}$ = 18.7 %

see below

Compute: Ht. of soil solids H_s = $H_i - \Delta H - H_{vf} = \frac{20}{10} - 0.358 - 0.552 = 1.090$ cm

Initial ht. of voids, $H_{vi} = H_i - H_s$ = 2.00 − 1.090 = 0.910 cm

Initial void ratio $e_o = H_{vi}/H_s$ = 0.910/1.090 = 0.835 cm

Final Test Data

sht 13c col 4

Dial readings: Initial D_1 = 0.00 Final D_2 = 364 × 0.01 = 3.64 mm

Change in sample ht., $\Delta H = D_2 - D_1 - \Delta H_e$ = 3.64 − 0.00 − 0.064 = 3.576 → 0.358 cm

Final ht. voids, $H_{vf} = M_{wf}/A_r$ = 16.75/30.33 = 0.552 cm

Final void ratio, $e_f = H_{vf}/H_s$ = 0.552/1.090 = 0.506

[1]Use water content sheet from data sheet set (Data Sheet 1).

Figure 13-5*a*
Consolidation test sample data.

CONSOLIDATION TEST (Time-compression data)

Project Lab Manual Data — Job No. — (Typical Set)

Location of Project Soil Laboratory — Boring No. — Sample No. —

Tested By K.M. & M.C. — Date of Testing Nov. 23-29, 19--

Loading Test Data

Load 100 kPa, ~~ksf~~ — Load 200 kPa, ~~ksf~~

Date applied 11/25 — Date applied 11/26

Applied by K.M. — Applied by M.C.

Clock time and date	Elapsed time, min	*Dial readings × 10^{-2} mm/div Original	Adjusted**	Clock time and date	Elapsed time, min	*Dial readings × 10^{-2} Original	Adjusted**
11-25 8:27	0	59	0.59 mm	11-26 8:04	0	98	
	0.1	66			0.1	109.5	1.095 mm
	0.25	67.5			0.25	112	
	0.5	69.5			0.5	113.5	
	1	70			1	116	
	2	72			2	117.5	
	4	73			4	120	
	8	75			8	123.5	
	15	78			15	127	
	30	80			30	132.5	
	60	83.5			60	138	
10:27	120	88.5		10:04	120	143.5	
12:43	256	94		12:00	256	148	
3:27	420	97		5:43	579	151	
11-26 8:04	1417	98	0.98 mm	11-27 7:34	1410	152	1.52 mm

*Insert gauge subdivisions 0.01 mm/div, etc.; if LVDT insert deformation, in or mm.
**Adjust if reset dial gauge for large compression.

Figure 13-6
Dial reading vs. time curves.

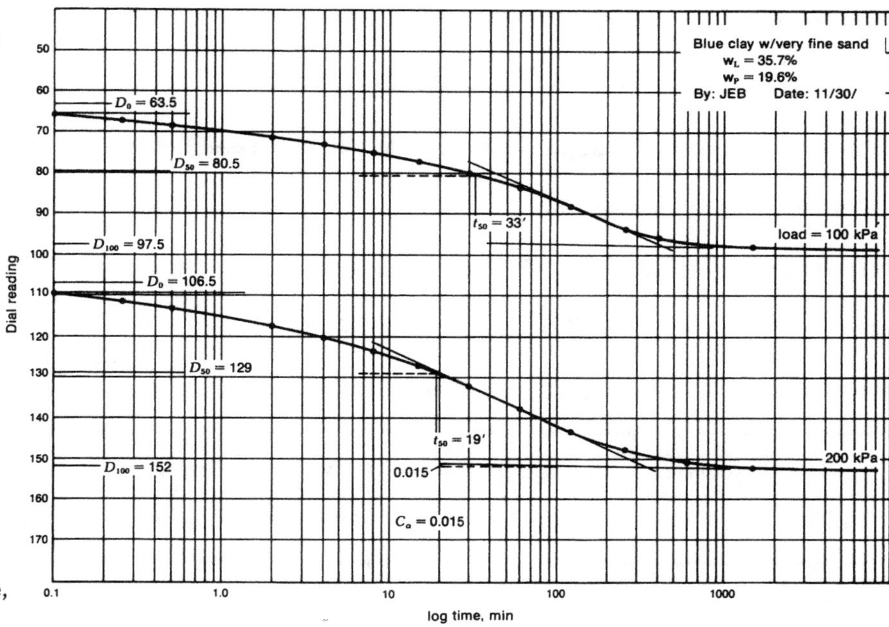

(*a*) Dial reading vs. log time, min.

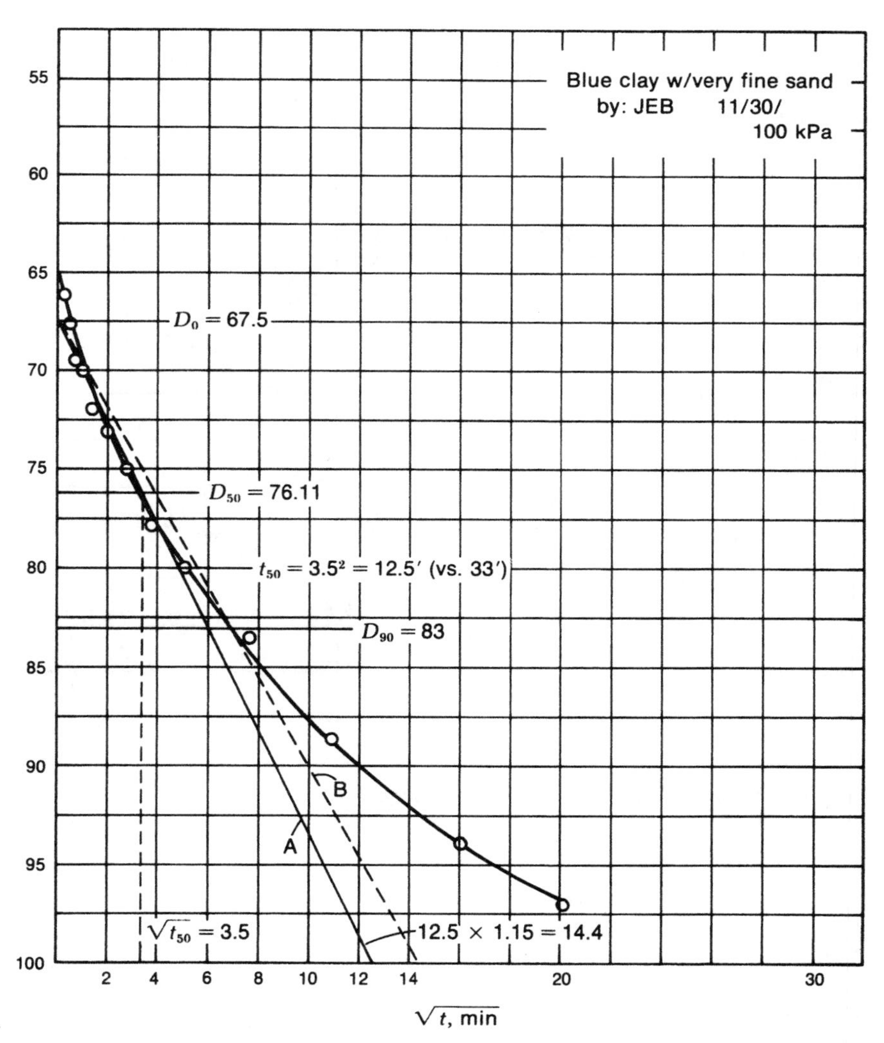

(*b*) Dial reading vs. $\sqrt{\text{time, min}}$.

CONSOLIDATION TEST (Computation sheet for e, ϵ and c_v) Data Sheet 13*c*

Project Lab Manual Data Job No. ——

Computed by JEB Date 11/30/19-- Test Date 11/29/19--

Sample Data (fixed ring, floating-ring) Ring diam. = 62.14 mm

Sample heights: H_i = 20.00 mm H_v = 9.10 mm H_s = 10.90 mm

Sample void ratios: e_o = 0.835 e_f = 0.506

Method for D_{100}: (use D_{100}) ✓ (D at end of load) ______ (D at specified time) ______ (check one)

Load increment kPa (1)	Def dial reading at end of load 0.01 /div (2)	D_{50}/D_{100} mm (3)	Equipment deform ΔH_e mm (4)	VR = e Strain = ϵ VR/Strain (5)	Average sample ht H mm (6)	H used for c_v cm (7)	t_{50} from DR vs time curves, min (8)	c_v cm²/min (9)
0	0			0.835 / 0				
		0.155 / 0.310	—		19.845	0.992	10	0.0194
25	32			0.804 / 0.016				
		0.450 / 0.580	0.002		19.552	0.978	22	0.009
50	59			0.782 / 0.029				
		0.805 / 0.975	0.004		19.199	0.960	33	0.006
100	98			0.746 / 0.049				
		1.290 / 1.520	0.008		18.718	0.936	19	0.009
200	152			0.696 / 0.076				
		1.860 / 2.200	0.016		18.156	0.908	15	0.011
400	223			0.635 / 0.109				
		2.530 / 2.860	0.032		17.502	0.875	14	0.011
800	290			0.576 / 0.141				
		3.250 / 3.600	0.064		16.814	0.841	7.4	0.019
1600	361			0.511 / 0.177				
				obtain before test begins				
	Data plotted on Figs. 13-8, 13-9 & 13-10							

Note: Insert units in column headings as necessary.

[a]Average height = $H_i - D_{50} + \Delta H_e$

$e = e_o - \Delta H/H_s$; $\epsilon = \Delta H/H_i$; $\Delta H = D_{100} - \Delta H_e$

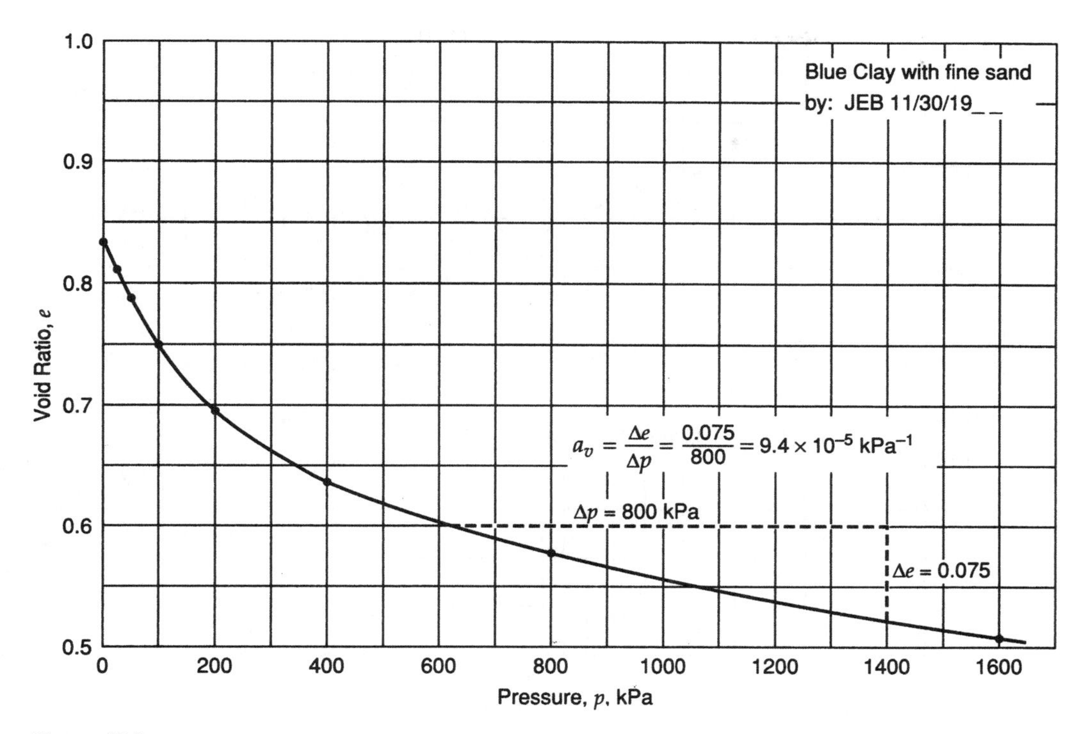

Figure 13-8
Plot of void ratio e vs. p (pressure in kPa) using data from Fig. 13-7.

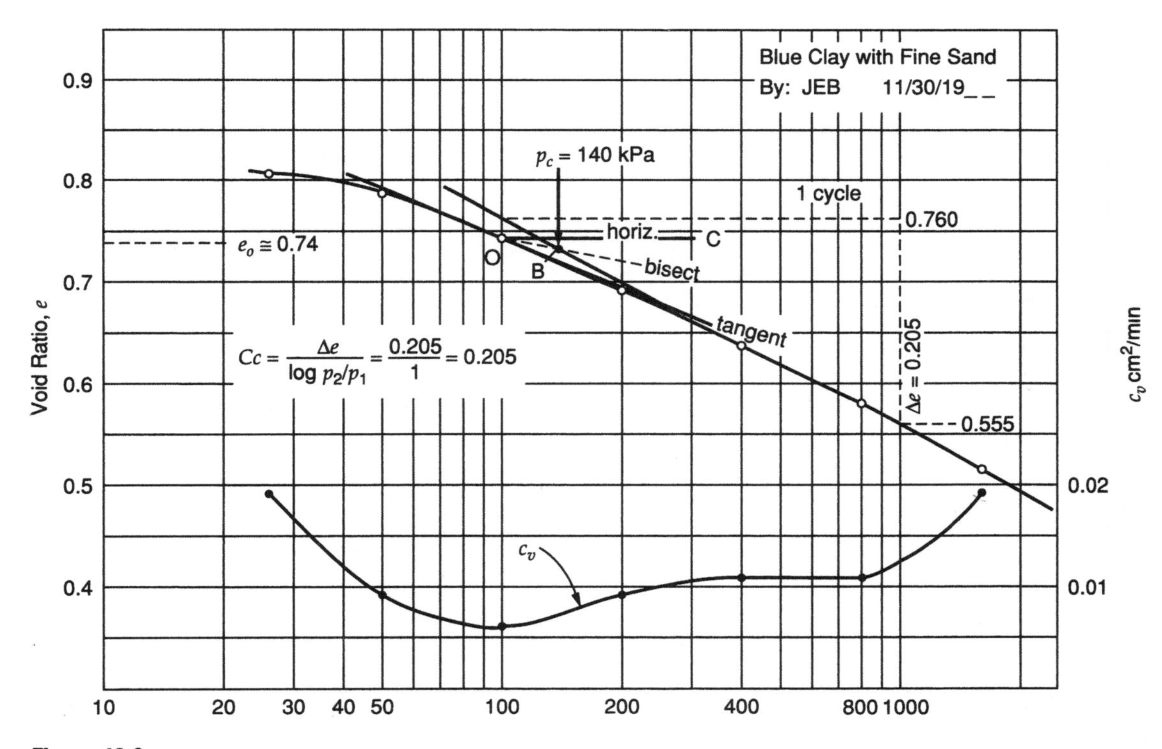

Figure 13-9
Plot of e vs. log p and c_v vs. log p using data from Fig. 13-7. Some authorities suggest plotting c_v vs. *average* log p but this is not recommended—an average H is correct for c_v but the load for this thickness is the full increment and not an average.

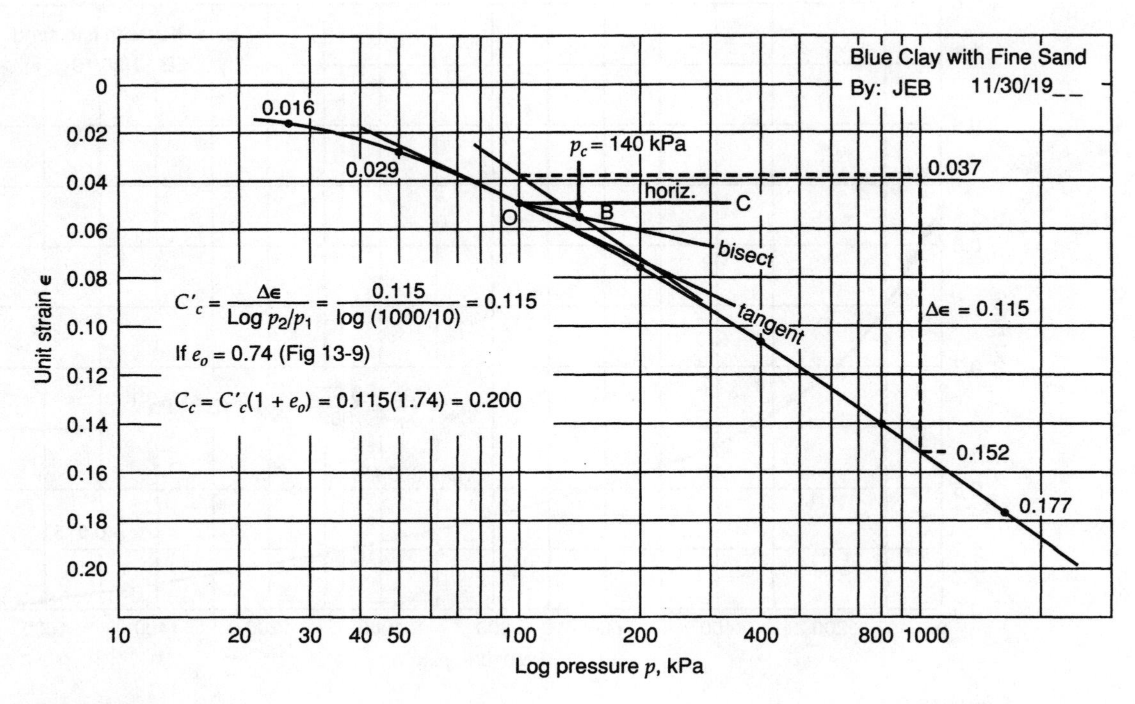

Figure 13-10
Plot of strain ϵ vs. log p also using data from Fig. 13-7. Note there is reasonable agreement between the preconsolidation pressure p_c and C_c by both methods.

14: UNCONFINED COMPRESSION TESTING

References

ASTM D 2166-85
AASHTO T 208-90

Objective

To introduce the reader to a procedure for measuring the approximate shear strength of a cohesive soil.

Equipment

Unconfined compression testing machine [any type of compression machine that has a load-reading device (load ring or a load cell) of a low enough range to obtain accurate load readings; see Fig. 14-1[1]]
Dial gauge (reading to 0.01 mm/div or to 0.001 in/div) or an LVDT of the same sensitivity.
Device to measure length and diameter (calipers or similar) of test sample
Sample extruder, if using tube samples
Trimming tools, if using tube samples
Harvard Miniature Compaction equipment, if making remolded test samples
Water content cans or evaporating dishes
Balances reading to 0.1 g and to 0.01 g

FIGURE 14-1
One type of unconfined compression test machine. Machine may be hand operated or motorized as here. The CBR system of Fig. 19-3 or similar can also be used for an unconfined compression test.

[1]This is a basic hand-operated machine with the hand crank removed. The variable speed motor is from the motorized direct shear machine of Fig. 17-1*a*. It has an alternative shop-constructed plate mount and chain drive cover as shown. Put the drive sprocket from the direct shear machine on the crank shaft and use a shorter drive chain (from a machine shop supplier). A suitable strain rate is found by varying the motor speed and measuring platen movement with a dial gauge. Mark the motor speed on the motor speed dial for future use. It takes about 10 min to switch the motor from one machine to the other so both are motorized.

General Discussion

When the method of testing tube-recovered cohesive soil samples in simple compression was first introduced, it was widely accepted as a means of rapidly measuring the shear strength of a soil. This test can only be done if a soil sample is intact (has no cracks or fissures) and has cohesion; an unconfined, cohesionless sample would simply fall apart.

With only a vertical load on the sample the major principal stress σ_1 is vertical and the minor (horizontal or lateral) stress is $\sigma_3 = 0$. From a Mohr's circle construction of this stress state we obtain the *undrained shear strength*—in this case also the cohesion (symbol c)—as

$$s_u = c = \frac{q_u}{2} \tag{14-1}$$

where q_u is always used as the symbol for the unconfined compressive strength of the soil. Be very careful with this distinction: q_u = unconfined compression strength or the peak strength as measured in a compression test; s_u = undrained shear strength (sometimes called the "cohesion") and is $q_u/2$ in this test.

The Mohr's circle construction for this test assumes (1) that σ_3, the minor principal stress, is zero or atmospheric (any surface tension confinement is ignored) and (2) the angle of internal friction ϕ of the soil is zero. This $\phi = 0$ condition is the same as is obtained in the unconsolidated, undrained compression test of Test 15, following, on a saturated soil; thus, to give the unconfined compression test more dignity, it is often called an "unconsolidated undrained" or UU (or simply a U) test.

As more knowledge concerning soil behavior became available through research and field observations, it is now accepted that the unconfined compression test only approximates the actual soil shear strength for at least four reasons:

1. Sampling disturbance. Samples are recovered by pushing thin-walled tubes with a sharpened and swaged cutting edge. The swaging produces a sample diameter slightly less than the tube ID to reduce friction between sample and tube walls during recovery. These are called "undisturbed" samples. Samples are also recovered and field tested from the standard split spoon used in Standard Penetration Testing. These samples are usually quite disturbed since the cutting edge bevels from about 8 mm down to 1.5 mm.
2. When the sample is removed from the ground the lateral restraint provided by the surrounding soil mass is lost. There is, however, some opinion that internal sample moisture provides a surface tension (or confining) effect so that the sample is somewhat "confined" for a compression test. This effect will depend on degree of saturation S and humidity in the testing area so that a quantitative estimate is impossible.
3. The internal soil conditions (the degree of saturation, the pore water pressure under stress deformation, and the effects of altering the degree of saturation) cannot be controlled.
4. Sample quality. In all cases the sample diameter should be larger than 30 mm (ASTM standard). If the sample consists of alternating thin layers of sand and clay the resulting s_u is of questionable value. Samples with gravel have the following restrictions:

Diam	Max gravel size
30 to 72 mm	Diam/10
> 72 mm	Diam/6

If after testing it is found that larger gravel particles are present make a note on the data sheet to this effect. The larger particles tend to reduce s_u.

Samples with cracks or fissures will give an s_u that is too low. These samples should be tested in the triaxial cell (see Test 15 following).

Errors from the first three factors cited cannot be eliminated at the current state-of-art. There may be some compensating effects, but their analysis is beyond the scope of this lab manual.

There are published claims that the undrained shear strength s_u from the unconfined compression test ranges anywhere between 40 and 85 percent of the true in-situ shear strength. In spite of such a wide range of claimed error this is a widely used test of shear strength. Its popularity stems from its simplicity—and from the fact that most of the time the measured strength actually is on the order of 80 to 85 percent.

In practice a small hand-operated compression device is taken to the site and samples recovered by boring are routinely tested for q_u. There is also a hand penetrometer with which one measures q_u by pushing a spring-loaded 6-mm diameter plunger a calibrated distance into the soil and directly reading from a scale the unconfined compression strength. The particular value of the latter device is that, where a sample size adequate for a compression test yields a single value, the penetrometer may measure four or five values for an average of q_u.

One may plot a curve of stress versus strain and measure the initial slope to obtain a modulus of elasticity[2] E_s. The loss of confining pressure nearly always gives a value of E_s that is too low for most geotechnical work. This will be shown to be true in Test 15 following.

The unconfined compression test may be either *strain-controlled or stress-controlled.* The strain-controlled test is almost universally used, as it is simply a matter of attaching the proper gear ratio to a motor to control the rate of advance of a loading head (no timing is needed for the test). The test has been found to be somewhat sensitive to the rate of strain, but a strain rate between ½ and 2 percent/min (e.g., a 70-mm specimen at a 1 percent strain rate would be compressed at the rate of 0.70 mm/min; 10 min = 7 mm) appears to yield satisfactory results. One should use the higher strain rate with softer soils (which may require failure definition at a strain of 15 percent) and the lower (½ percent/min) strain rate for dry or stiff soil.

Since the unconfined-compression-test specimens are exposed to the usually dry laboratory air (low humidity), they should reach failure within about 10 min; otherwise, the change in water content may increase the unconfined compressive strength—primarily from an increase in surface tension on the sample perimeter caused by surface drying, producing an apparent confining pressure.

A stress-controlled test requires incremental changing of loads and may result in erratic strain response and/or the ultimate strength falling between two stress increments. The loads are applied through a dead-load apparatus yoke. Some persons suggest developing the load by adding water to a container attached to the yoke. In most cases, however, the dead load is simply a hanger to which masses are periodically added. Both methods produce "shock" loading of the sample and are difficult to apply. For these several reasons, stress-controlled tests are seldom used in this or in any similar soil test.

THE UNCONFINED COMPRESSION TEST

The unconfined compression test consists of trimming a sample of some diameter to an adequate L/d ratio, placing it in a compression device, and measuring load and corresponding displacement at selected time or displacement intervals. The line details of this test are shown in Fig. 14-2.

The length/diameter ratio of the test specimens should be large enough to avoid interference of potential 45° failure planes of Fig. 14-3 and small enough not to obtain a "column" failure. The ASTM-specified L/d ratio to satisfy these criteria is

$$2 < \frac{L}{d} < 2.5$$

[2]More correctly, this is a *stress-strain modulus* since soil is not an elastic material—particularly in the range of strains associated with most compression tests.

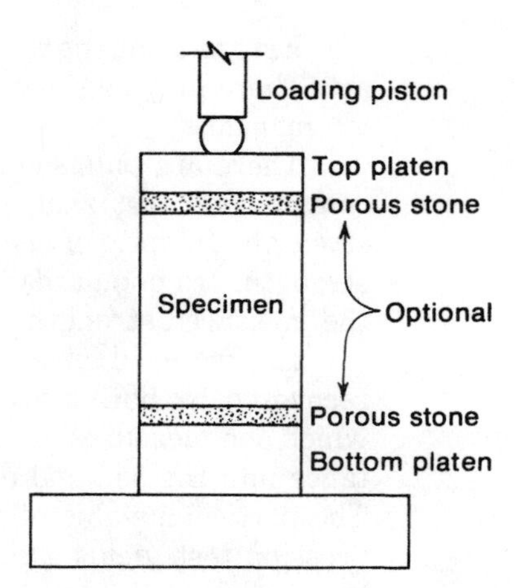

FIGURE 14-2
Line details of the unconfined compression test.

The unconfined compression test continues until the load on the sample peaks and falls slightly. Where a defined load peak and fall-off cannot be clearly identified as with many normally consolidated clays and most remolded soils the test proceeds until at least 15 percent strain[3] [e.g., a sample of $L = 100$ mm shortens by $100(.15) = 15$ mm]. Whether a remolded soil fails with a peak stress or at a specified strain depends on the molding water content. A "dry" soil tends to have a peak stress and a fairly low strain.

From the load-deformation data recorded, calculations are made of axial strain ϵ and stress σ_1 so that a stress-strain curve can be drawn to obtain the maximum stress (unless the stress at 15 percent or other designated "failure" strain occurs first). This is taken as the unconfined compressive strength q_u of the soil. The laboratory-obtained stress-strain curve is drawn to obtain an "averaged" value of q_u rather than merely taking the largest value of stress from the computation sheet. Field-obtained values are not usually plotted and in this case q_u is calculated from an observation of maximum load on the sample and the measured sample deformation using Eq. (14-3) below.

The unit strain ϵ is computed from mechanics of materials as

$$\epsilon = \frac{\Delta L}{L_o} \quad \text{(mm/mm, m/m, in/in, etc.)} \tag{14-2}$$

where ΔL = total sample deformation (axial), mm (or other)
L_o = original or initial sample length, mm (or other)

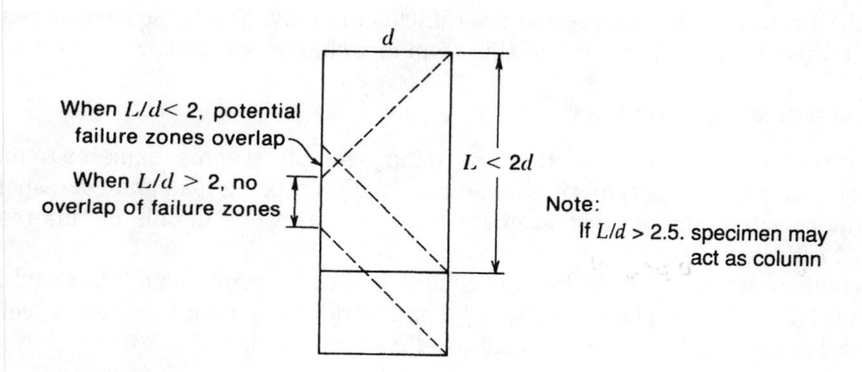

FIGURE 14-3
L/d ratios for any soil compression test (unconfined, triaxial, etc).

[3]ASTM now specifies 15% strain. A value of 20% is also used. Any other reasonable value may be specified by the designated authority. One should show on the stress-strain plot what percent strain is used to define failure if a definite stress peak is not obtained.

The instantaneous test stress σ on the sample is computed as

$$\sigma = \frac{P}{A'} \quad \text{(kPa, ksf, etc.)} \tag{14-3}$$

where P = load on sample at any instant for a corresponding value of ΔL, kN (or kips, lbs, etc.)

A' = cross-sectional area of specimen for the corresponding load P, m^2 (or ft^2, in^2, etc.)

In geotechnical work it is standard practice to correct the area on which the load P is acting. This is not done when testing metal specimens in tension or compression. One of the reasons for this area correction is to make some allowance for the way the soil is actually being loaded in the field. Applying a correction to the original sample area is also somewhat conservative since the ultimate strength computed in this way will be less than that computed using the original area. The original area A_o is corrected by considering that the total volume of the soil is unchanged as the sample shortens. The initial total soil sample volume is

$$V_T = A_0 L_0 \tag{14-4}$$

but after some change in specimen length of ΔL we have

$$V_T = A'(L_0 - \Delta L) \tag{14-5}$$

Equating Eqs. (14-4) and (14-5), canceling terms, and solving for the corrected area A' to use in Eq. (14-3) we obtain

$$A' = \frac{A_0}{1 - \epsilon} \tag{14-6}$$

Field-tested samples are directly obtained from the borings and tested to failure with the maximum load and the deformation ΔL being recorded. The deformation is used to compute the corrected area A' and Eq. (14-3) is used to compute the ultimate stress or q_u. The q_u obtained in this manner is not likely to differ significantly from the more precise laboratory method outlined here. Sample fragments may be tested for q_u using the pocket penetrometer described earlier.

The enclosed computer program can be used in the data reduction of this test and for Tests 15 and 16. This program does not plot the stress-strain curve. Inspection of Fig. 14-5 indicates that the initial tangent modulus may be very different from that shown if one uses a least squares regression analysis subroutine for curve fitting.

Procedure

This is suggested as a group project.

1. Each group will prepare two tube samples with an L/d ratio between 2 and 2.5 (or use the laboratory samples furnished by the instructor).[4] If using remolded samples prepared using the Harvard Miniature equipment use the dimensions of the mold given in Test 9 and go to Step "1c" below. If you use tube samples remove them from the

[4]Suitable test samples can be made using the Harvard Miniature compaction apparatus (see Fig. 9-3). By using the same soil but varying the water content and/or layers and tamps per layer the lab section can study the effect of shear strength s_u vs. water content and/or density.

tube by using a sample extruder. If this appears to cause too much sample degradation, split the tube lengthwise and take the samples out.[5] Note that sawing the tube heats it and may also degrade the sample. Now:

a. Carefully cut two or more samples of adequate L/d ratio and with the ends as nearly perpendicular to the longitudinal axis as possible. If a pebble is pulled from the sample during end-trimming, fill the hole with trimmings to about the density of the sample. Measure the sample length at three points 120° apart and average the values as L_0.
b. Determine the average sample diameter as follows: At the top, mid-height and bottom take three diameter measurements at 120° apart. Average these nine values as the sample diameter and record on Data Sheet 14.
c. Determine the mass of each test sample and record.

2. Place the samples in moist containers (you can use a 1000 mL beaker turned upside down over the two samples with a damp paper towel inside or leave them in the moisture room to prevent their drying while awaiting a turn at the compression machine. Compute the deformation corresponding to 15 percent strain for your samples while waiting so that you will know about when to terminate the test if the sample load has not peaked before this amount of strain occurs.

   Compute the volume and density ρ of your samples and weigh two moisture cups (or dishes) so that you can determine the water content of your samples after testing them.

3. Carefully align the specimen in the compression machine (be sure it is centered on the lower platen). If the ends are not exactly perpendicular to the specimen axis, the early part of the stress-strain curve will be flat (until the full sample area contributes to stress resistance, strains will be too large for the nominal stresses computed). Now adjust by hand or motor so that the sample is just making contact with both platens (or load plates). "Just making contact" means about 0.5 kg (1 load dial division or equivalent from a load cell). This small load is insignificant for the test but ensures that the deformations will be correct from test start.

   Now zero *both* the load device (either load dial for load ring or Digital Voltmeter (DVM) if an electronic load cell is being used) and deformation dial (or LVDT). At this time the very small load should still be on the sample but is recorded as zero.

4. After a final check for equipment clearance, start the test and and take load and deformation dial readings as follows (for a 0.01 mm/div dial gauge):

   10, 25, 50, 100, and every 50 to 100 dial divisions thereafter, until one of the following occurs:

   a. Load decreases on sample significantly
   b. Load holds constant for four readings
   c. Deformation is significantly past 15 percent strain, say to about 20 to 25 percent

   Use the data sheets provided for this test—do not recopy the data from scratch paper unless you are allowed to use the computer program (refer to Fig. 14-4). If you are allowed to use the computer program, submit the output pages instead of Data Sheets No. 14. Use as many data sheets as necessary for each test.

   Remove the sample from the compression machine and either weigh the entire sample immediately for a water content or use a portion. If you use a portion put it in the water content cup and put on the lid so it does not dry out while the next sample is tested.

5. When all samples are tested, weigh the water content cups or dishes and record the data. Use a water content sheet from the data set for this. Remove the lids from the

[5]It is not recommended in introductory laboratories to have students trim a sample diameter from a larger tube sample. Trimming ends perpendicular to the long axis is difficult, reducing the diameter of a tube sample is overly time-consuming, and is seldom done in commercial laboratories.

cups (if used) and put the cups in the oven. Return the next day, weigh the oven-dry samples, compute the water content and dry density ρ_d of each sample, and record on your data sheet.

6. Compute the unit strain, the corrected area, and the unit stress for enough of the readings (say 8 to 10 well-spaced points) to define the stress-strain curve adequately for each sample tested. Plot the results on a sheet of graph paper from your data sheet section (all curves on same sheet); show q_u as the peak stress (or at 15% strain) of each test, and show the average value of q_u for all tests. Be sure to plot strain as the abscissa.
7. a. Draw a Mohr's circle using the average value of q_u from the curves of Step 6 and show the *undrained shear strength* s_u (or cohesion). Be sure to use a compass to draw Mohr's circle (no free-hand work).
   b. Compute the cohesion of the soil using the Mohr's circle of (*a*) and also assuming the soil has an angle of internal friction, $\phi = 10°$. Discuss if s_u is conservative or "unconservative" under these conditions.
   c. Plot a dashed curve to represent an "average" curve for your test series. From this average curve compute the secant modulus of elasticity of the soil using one end of the secant at the origin and the other cutting the curve at 0.25, 0.50, and $0.75q_u$; also compute the initial tangent modulus of this average curve. Tabulate these results in your "Conclusions" and discuss their significance. Which value of stress-strain modulus do you recommend for this soil? Why is the initial tangent stress-strain modulus commonly used?
   d. Comment on the density and water-content effects vs. s_u if applicable.

Sample Computations (refer to data sheet Fig. 14-4)

Compute sample area $A_0 = 0.7854(45.5/10)^2 = 16.26\ \text{cm}^2$.

For Line 2:

$$\text{Def dial} = 50 \qquad \text{Load dial} = 8$$

$$\text{Total deformation } \epsilon = 50(0.01) = 0.5\ \text{mm}$$

$$\text{Load P} = 8(0.34) = 2.72\ \text{kg}$$

$$\text{Strain } \epsilon = \frac{e}{L_o} = \frac{0.5}{108} = 0.0046\ \text{mm/mm} \rightarrow 0.0050$$

$$\text{Corrected } A' = \frac{16.26}{(1 - 0.0050)} = 16.34\ \text{cm}^2\ \text{[Eq.(14-6)]}$$

Now use Eq. (14-3) and 98.07 to convert kg to kN

$$\text{Sample stress } \sigma = P/A = \frac{2.72(98.07)}{16.34} = 16.3\ \text{kPa}$$

For Line 10:

$$\text{Def dial} = 1200 \qquad \text{Load dial} = 49$$

$$\text{Total deformation e} = 1200(0.01) = 12.0\ \text{mm}$$

$$\text{Load P} = 49(0.34) = 16.66 \text{ kg}$$

$$\text{Strain } \epsilon = \frac{12}{108} = 0.1111 \text{ mm/mm}$$

$$\text{Corrected } A' = \frac{16.26}{(1 - 0.1111)} = 18.29 \text{ cm}^2$$

$$\text{Sample stress } \sigma = \frac{16.66(98.07)}{18.29} = 89.3 \text{ kPa}$$

These and the rest of the data are plotted on Fig. 14-5 to obtain S_u.

UNCONFINED COMPRESSION TEST Type of Sample (~~Undisturbed~~, Remolded) Data Sheet 14

Project Unconfined Comp. Test #1 Job No. —

Location of Project Soil Laboratory

Description of Soil Brown Silty Clay w_L = 37.3 % w_P = 19.8 %

Tested By JEB & K.M. Date of Testing 12/7/19--

Sample Data

Strain rate

Diam. 45.5 mm Area A_o 16.26 cm² Ht., L_o 108 mm

Vol. 175.61 cm³ Mass 379.5 g Wet density ρ_w = 2.16 g/cm³

Water content, w% 14.5 % Dry density, ρ_d 1.89 g/cm³ LRC 0.34 kg/div

Deformation dial[1] reading (0.01 mm)	Load dial[1] (units)	Sample deformation ΔL, (mm)	Unit strain, ϵ $\Delta L/L_o$ $\times 10^{-2}$	Area CF $1-\epsilon$	Corrected area A'. (cm)²	Total load on sample[2] (col. 2 × LRC) kg.	Sample stress, σ_1 kPa
1	2	3	4	5	6	7	8
Note:	Data	somewhat	edited &	cols 4 & 8	plotted	Fig. 14-5	
0	0	—	—	—	16.26	0	0
50	8	0.5	0.46	0.9954	16.34	2.72	16.3
100	15	1.0	0.93	0.9907	16.41	5.10	30.5
200	20	2.0	1.85	0.9815	16.57	6.80	40.2
400	27	4.0	3.70	0.9630	16.88	9.18	53.3
500	30	5.0	4.63	0.9537	17.05	10.20	58.7
600	33	6.0	5.56	0.9444	17.22	11.22	63.9
750	38	7.5	6.94	0.9306	17.47	12.92	72.5
1000	44	10.0	9.26	0.9074	17.92	14.96	81.9
1200	49	12.0	11.11	0.8889	18.29	16.66	89.3
1400	54	14.0	12.96	0.8704	18.68	18.36	96.4
1500	56	15.0	13.89	0.8611	18.88	19.04	98.9
1600	58	16.0	14.81	0.8519	19.09	19.72	101.3
1700	57	17.0	15.74	0.8426	19.30	19.38	98.5

Note: Insert units in column headings as necessary.
[1]omit if using LVDT or load cell
[2]insert direct reading of load if using load cell.

$\sigma_1 = \frac{2.72(98.07)}{16.34} = 16.3$ kPa

Av. q_u = 91.6 (2 tests); Av. s_u = 91.6/2 = 45.8

Unconfined compressive strength q_u = 101.5 Cohesion $s_u = q_u/2$ = 50.8 (45.8 = Av.)

FIGURE 14-4
Data from an unconfined compression test together with necessary computations for plotting the stress-strain curve of Fig. 14-5.

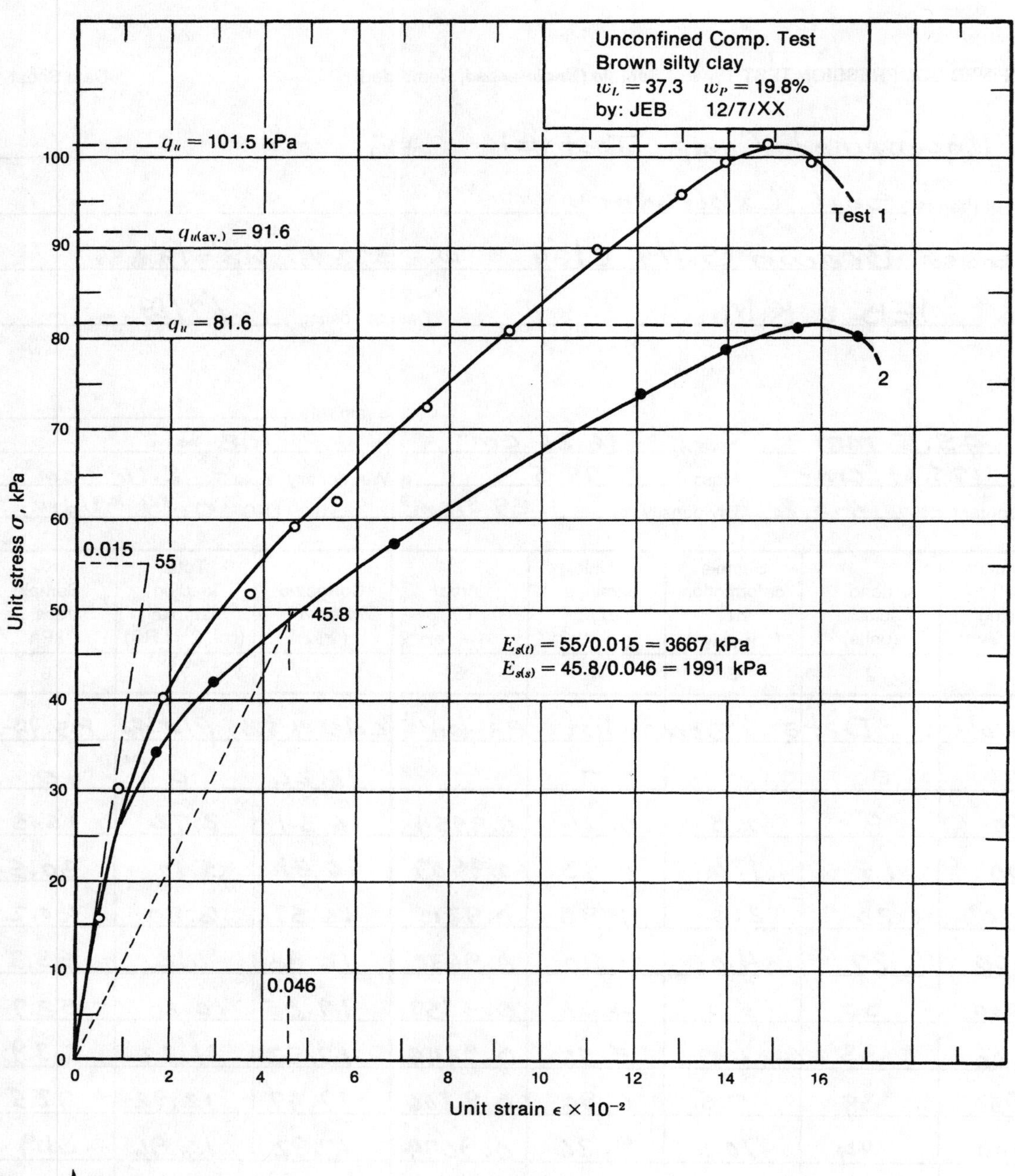

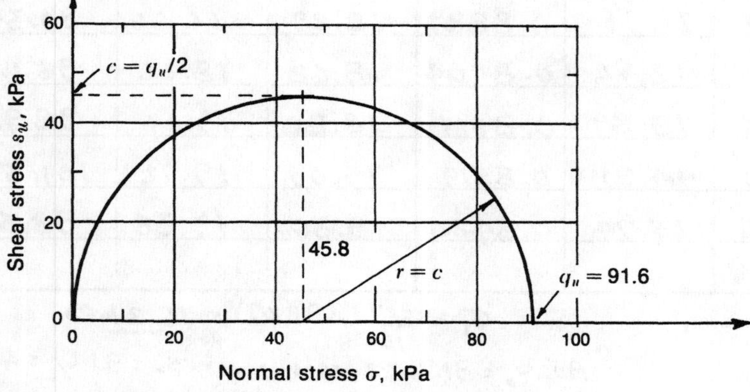

FIGURE 14-5
Plot of unconfined compression test data to obtain the stress-strain modulus E_s and a best value of the unconfined compression strength q_u. A Mohr's circle can be plotted using the average q_u from one or more tests to obtain the undrained shear strength s_u as shown; however, it is usually obtained as $s_u = q_u/2$ without plotting.

15: TRIAXIAL TEST—WITHOUT PORE-PRESSURE MEASUREMENTS

References

ASTM D 2850-87 (Unconsolidated-Undrained)
ASTM D 4767-88 (Consolidated-Undrained)
AASHTO T 234-85 (1990) (Unconsolidated-Undrained)
ASCE (1960), Research Conference on Cohesive Soils, Boulder, Colorado, *Proceedings.*
ASTM (1988), *Advanced Triaxial Testing of Soil and Rock*, ASTM Special Technical Publication STP 977, 896 pages. A series of papers including state-of-art summaries through 1988.
Berre, T. (1982), "Triaxial Testing at the Norwegian Geotechnical Institute," *Geotechnical Testing Journal, ASTM*, Vol. 5, No. 1/2, pp. 3–17.
Bowles, J. E. (1984), *Physical and Geotechnical Properties of Soils*, McGraw-Hill Book Co., 578 pages (in particular Chap. 10). (Or a similar text).
Hurtado, J. E., and E. T. Selig (1981), "Survey of Laboratory Devices for Measuring Soil Volume Change," *Geotechnical Testing Journal, ASTM*, Vol. 4, No. 1, March, pp. 11–18.
Khosla, V. K., and R. D. Singh (1983),"Apparatus for Cyclic Stress Path Testing," *Geotechnical Testing Journal, ASTM*, Vol. 6, No. 4, December, pp. 165–172.
Lade, P. V. (1978), "Cubical Triaxial Apparatus for Soil Testing," *Geotechnical Testing Journal, ASTM*, Vol. 1, No. 2, June, pp. 93–101.
Lambe, T. W. (1964), "Methods of Estimating Settlement," *Jour. Soil Mechanics and Found. Div., ASCE*, Vol. 90, SM 5, September, pp. 43–68.
Ramana, K. V., and V. S. Raju (1981), "Constant-Volume Triaxial Tests to Study the Effects of Membrane Penetration," *Geotechnical Testing Journal, ASTM*, Vol. 4, No. 3, September, pp. 117–122.
Testing Forum (1988), "Correction of Strength for Membrane Effects in the Triaxial Test," *Geotechnical Testing Journal, ASTM*, Vol. 11, No. 1, March, pp. 78–82. (A survey of practice).

Objective

To introduce the reader to the basic procedure for determining the "undrained" soil shear strength parameters ϕ and c and the "elastic" parameters of stress-strain modulus E_s and Poisson's ratio μ of a soil. The methods for both cohesionless and cohesive soils are given.

Equipment

Compression machine (strain-controlled)
Triaxial cell (refer to Figs. 15-1 to 15-3)
Specimen mold, rubber membrane, membrane stretcher, rubber binding strips, and porous stones
Vacuum pump and air-pressure source with necessary pressure indicators.[1]
Calipers
Specimen trimmer for cohesive tube samples (optional)

[1]Pressure gauges are usually used. Also it is usual to use atmospheric pressure as the reference. Absolute pressure is seldom used in soil testing.

Figure 15-1
Triaxial cells and peripheral equipment. Two types of cells are shown: quick-clamp type (separated), which is preferred for most work for rapid set-up. On left are two sizes of split sample sleeves used to square ends of cohesive samples. On right are two membrane stretchers used for cohesive or cohesionless samples. Miscellaneous equipment shown includes spatulas, sample compactor, and trimming saw.

General Discussion[2]

Read the "General Discussion" of Test 14 and note that some of the equations given there are also used in this test. Also it will be useful to read the "General Discussion" of Test 16 but this can be optional.

One of the primary purposes of this test is to determine the shear strength parameters of the soil. These parameters are defined by Coulomb's shear strength equation

$$s = c + \sigma_n \tan \phi \tag{15-1}$$

where s = shear strength, kPa, ksf, psi, etc.

c = soil cohesion or interparticle adhesion, kPa, ksf, etc.

σ_n = intergranular pressure—may be either an *effective* or *total stress* value, kPa, ksf, etc.

ϕ = angle of internal friction, degrees

The cohesion term is self-explanatory as being similar to glue sticking particles together. The $\sigma_n \tan \phi$ term is the usual description of friction between two particles, used in physics as μN, where $\mu = \tan \phi$ and $N = \sigma_n$, the normal stress between the two particles.

If we apply a normal pressure σ_n on a volume of saturated soil the pressure is carried by both interparticle stresses and an increase Δu in the pore water (this is the same Δu of Fig. 13-3). This σ_n is the total pressure. If the shear stress and cohesion are determined in some manner we can solve Eq. (15-1) for ϕ, which is termed a "total stress" parameter. If we reformat Eq. (15-1) to remove the increase in pore pressure we obtain

$$s = c' + (\sigma_n - \Delta u) \tan \phi' \tag{15-1a}$$

[2]There is an entire book on the triaxial test: Bishop and Henkel (1962), *The Measurement of Soil Properties in the Triaxial Test*, second ed., considered a classic. The discussion here is intended to give some background. To keep this text to a readable size for a laboratory manual, select sources have been referenced for additional background and for details too advanced for routine tests.

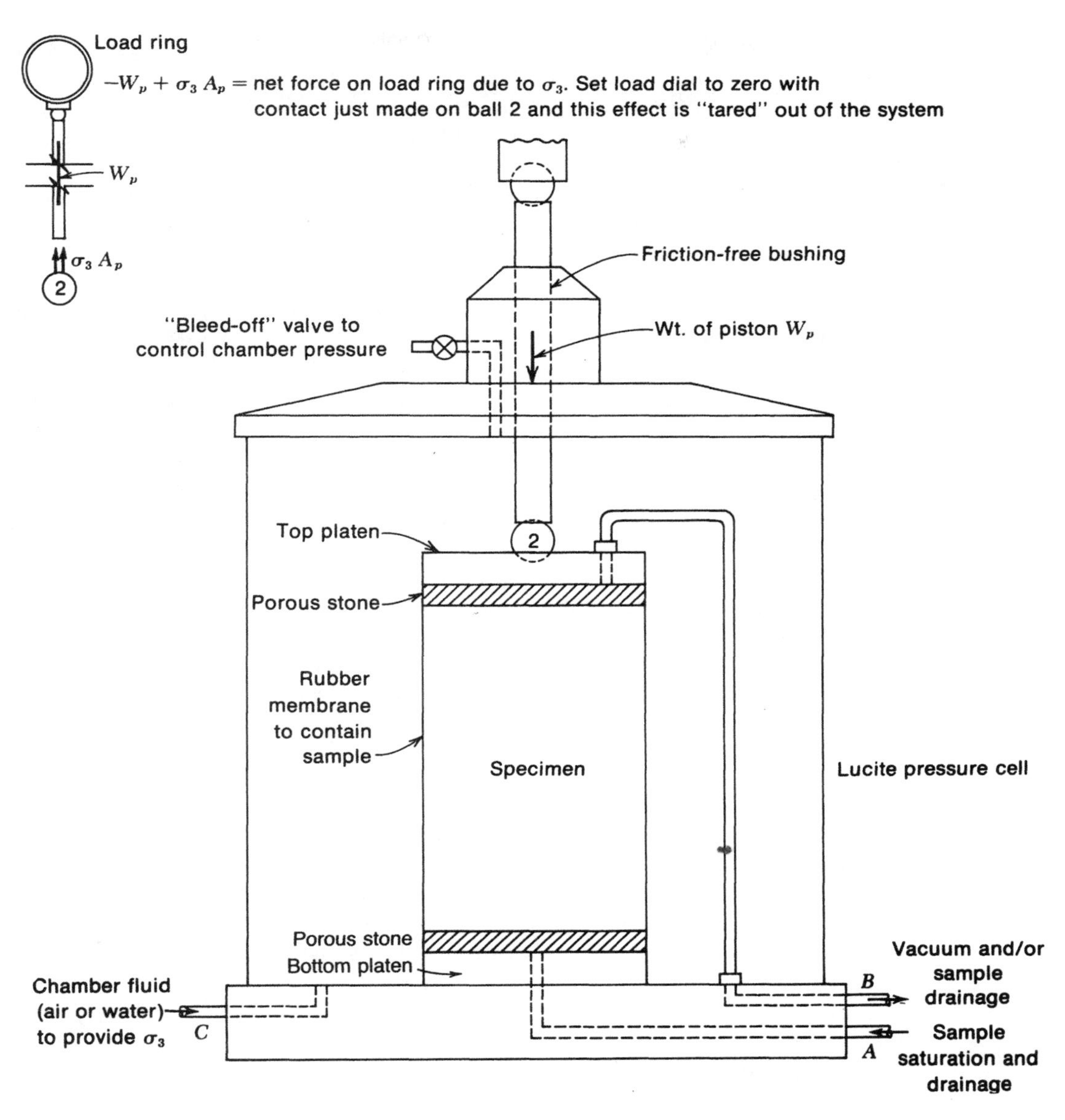

Figure 15-2
Line details of the triaxial cell. Shutoff valves are located on tubes A, B, and C. If an electronic pore-pressure transducer is used, it would be on a T connection on tube A and would measure pore pressure at the base of the sample. Substitute "load cell" for "load ring" if appropriate.

where the values of c' and ϕ' are termed *effective* stress parameters. Even if a soil has an angle of internal friction ϕ, a test (as well as field loadings) can give an apparent friction component of zero, or

$$s_u = c \tag{15-1b}$$

We used this form of Eq. (15-1) in Test 14 and it is obtained by having a case where $\sigma_n - \Delta u = 0$ in Eq. (15-1*a*). This gives us the "undrained shear strength" s_u as in the unconfined compression test. The triaxial test allows us to use a range of test states that can produce shear strength parameters that range from total stress to effective stress values. Effective stress values are generally accepted as the "true" soil shear strength parameters.

From an inspection of the triaxial apparatus (Figs. 15-1 and the simplified line details of Fig. 15-2), the reader should be able to conclude that any soil pore-fluid state can be obtained with this equipment from a negative (or vacuum) to a fully saturated condition and even to an excess pressure in the pore-fluid (if desired). Manipulation of the pressure in the cell fluid surrounding the membrane-covered sample can replicate field conditions. Specially constructed cells can allow the vertical pressure on the sample to be different from the lateral pressure.

Either *drained* or *undrained* states can be investigated. For a *drained* test, as the load is applied to the soil specimen the sample pore fluid is allowed to drain by opening the appropriate valve. An *undrained* test is obtained by closing the soil system to the atmosphere so that no pore fluid can escape during the test.

To most persons, the term *drained* test is used if the proper cell valve is opened so that the soil pore fluid can escape during sample compression. Strictly speaking, however, a drained test is one in which the proper pore-fluid drain valve is opened and the rate of stress application is so slow that only a minimal *excess pore pressure* Δu *develops within the sample* during the test. It is impossible to have a state of no excess pore pressure during any shear[3] test since during compression there is strain. A strain is necessary to produce an excess pore pressure and an excess pore pressure is necessary to cause drainage of the pore fluid so the excess pore pressure can dissipate. What is necessary is to limit the magnitude of the excess pore pressure by testing at a very low strain rate.

A drained test can take up to a week. For a test of this duration, membrane leaks, soil set-up and equipment corrosion can be significant. As a consequence, very few drained tests are performed in practice. Other means, such as Test 16 following, are used to obtain drained shear strength parameters.

To distinguish among the different types of tests that can be performed by opening or closing the sample pore-fluid exit valve, the three triaxial tests most commonly used will be briefly described:

1. *Unconsolidated-undrained test*—also called a U or UU test. This test is performed by placing the sample in the membrane and into the cell while valves on lines A and B of Fig. 15-2 are closed. At this time to be strictly correct the porous stones and lines A and B should be completely filled with water. Some cell pressure is then applied and the test begun with load and deformation readings taken exactly as for the unconfined compression test (Test 14). The measured soil parameters of ϕ and c depend heavily on degree of saturation of the sample (i.e., whether $S = 100\%$) and whether the sample is cohesive. This test cannot be used to test a fully saturated ($S = 100\%$) cohesionless soil.

   Note that if the sample is not fully saturated some consolidation takes place when the cell pressure is applied, and regardless of whether the drain valves are open or closed. Some additional consolidation may occur during the test as well.

   The unconfined compression test is a special case of this test where the triaxial cell is not used and the confining pressure is atmospheric (or zero).
2. *Consolidated-undrained test*—also termed a consolidated-quick or CU test. This test takes place after the sample has consolidated from the cell pressure with the drain valve open (but lines A and B of Fig. 15-2 saturated) under some confining cell pressure. Some type of sample-monitoring device is needed to determine when the sample volume change has ceased (or the pore fluid ceases to drain if $S = 100\%$), indicating that consolidation is complete. Monitoring may be by either observation of sample drainage or using a liquid confining cell fluid. As the sample decreases in volume from consolidation one must add cell fluid. When no more cell fluid can be added (after a modest time elapse) it may be concluded that the sample has "consolidated."

   Because consolidation takes a considerable time for clay samples, special techniques are used to speed drainage such as placing a sand core in the specimen or using filter paper strips or cotton threads spaced around the sample perimeter adjacent to the

[3] Even though this is a "compression" test the sample actually fails in shear, thus the soil strength parameters define the shear strength of the soil.

confining rubber membrane. Strictly some consolidation always takes place even in a UU test since the sample compresses some amount when any cell pressure is applied. Note this consolidation is somewhat three-dimensional (all around the sample + vertical compression).

The drain valves are closed when consolidation is complete (but drain lines fully saturated) and then the sample is tested with load and deformation readings taken exactly as in both the unconfined compression test (Test 14) and UU test previously described. Excess pore pressure of varying intensity depending on the degree of saturation develops on the shear surface.

There are two basic types of consolidation in compression tests: *isotropic* and *anisotropic*. Isotropic consolidation is more common and is produced by a constant all-around pressure. This is typified by simply putting the sample in the triaxial cell and applying some cell pressure of perhaps 2, 5 or more N/m^2. This pressure acts both on the sides and top of the sample. Anisotropic consolidation occurs when the vertical pressure on top of the sample is not the same as the side (or lateral) pressure. This requires a specially constructed cell. Samples recovered from in situ are usually anisotropically consolidated from the overburden and lateral confining pressure being different. As a laboratory convenience they are usually isotropically consolidated if a CU test is used.

If the sample is isotropically consolidated (equal all-around pressure) the test may be called a CIUC (Consolidated Isotropically Undrained Compression) test. The last C for compression is to distinguish between whether the test was compression or extension since both types of tests can be done (but require cell modification). The test procedure of this manual is limited to compression or CIUC testing.

If the sample is anisotropically consolidated (lateral pressure not equal to vertical), the test may be termed a CKUC test. If the ratio $K = \sigma_h/\sigma_v$ is that existing in situ (σ_h being the lateral cell pressure on the sample) K is subscripted as K_0 and the test is a CK_0UC type. In most cases one would perform CK_0UC tests if anisotropy is to be taken into account—but this is not a trivial task. Anisotropic sample consolidation requires specially constructed triaxial cells, but they can also be used for extension tests.

3. *Consolidated-drained test*—also called CD or consolidated-slow test. In this test the sample drain valves are opened, the cell pressure is applied, and the sample is monitored for complete consolidation as for the consolidated undrained test. At the end of consolidation and with the drain valve still open, the sample is tested similar to the unconfined compression test but at a much slower strain rate.

   The slower strain rate tends to minimize the amount of excess pore pressure $\Delta\mu$ that develops on the shear surface so that the shear strength parameters are not much affected.

   Because close monitoring and a considerable length of time are required for this test, it can be economically justified only for large projects. A CD test is claimed to produce drained (or true) shear strength soil parameters ϕ and c. Fortunately, on most projects the apparent parameters obtained using either UU or CU tests are more realistic—depending on how rapidly the in situ loading, for example, a fill or footing load, is applied.

Test Results

The following several comments concerning the shear strength parameters ϕ and c and the elastic parameters of E_s and μ will give some indication of reliability of the test.

SHEAR STRENGTH PARAMETERS ϕ AND c

1. For any dry soil (both cohesionless and cohesive), about the same soil parameters would be obtained from any of the three tests (UU, CU, or CD).

2. For a saturated or partly saturated cohesionless soil, the CD test will yield about the same ϕ angle as for a dry soil unless the material is very fine-grained (low coefficient of permeability) and/or the test is performed at an extremely rapid rate of strain.
3. Any saturated cohesive soil will yield $\phi = 0$ in a UU test to possibly a small ϕ-angle in a CU test.
4. For any *saturated, cohesive* soil, the results are highly dependent on which of the three tests is used. That is, the soil parameters will range from

   $\phi \approx 0$ and $c > 0$ in the UU test

   to

   $\phi =$ true value and $c \approx 0$ in the CD test

   for the same soil. Results will depend also on whether the soil is *normally* consolidated, *overconsolidated*, or a remolded sample. The data presentation should indicate these factors and the type of test used to obtain the results. Note that an overconsolidated soil can give both c and ϕ for normal stresses less than the preconsolidation pressure. Tests that produce larger normal stresses produce a shear envelope[4] which, projected back to the origin, gives $c = 0$ [refer to Bowles (1984, pp. 450–451) for an extended discussion].
5. For any partially saturated cohesive soil, the results depend on both the degree of saturation S and the type of test performed. The shear strength parameters can range from

   $\phi = 0$ for $S = 100\%$ in a UU

   $0 < \phi < \phi$ true for $S < 100\%$ in either UU or CU

   c varies both with type of test; S; and whether soil is normally consolidated or overconsolidated

ELASTIC PARAMETERS E_s AND μ

The stress-strain modulus E_s is *generally too low.*

1. In cohesionless soil one can only test remolded samples. While recovery methods such as injection or freezing and recovering a core are possible these samples are not in an "undisturbed" state when they are tested. Even a sample built to the same void ratio as in situ is remolded. Any grain contact cementing from aging or mineral precipitation is lost on recovery and while for individual particles this may be almost nil the statistical accumulation can be large. Water surface tension is not large, but the cumulative effect can allow a vertical cut in a sand mass.
2. It is not possible to recover tube samples of cohesive soil, trim them to size and put them in the triaxial cell without considerable remolding. Loss of overburden pressure always occurs. Further it is extremely difficult to replicate field in situ stress conditions since the only true reconsolidation is to use the CK_o and it is nearly impossible to produce the K_o state in a triaxial cell—we do not know what K_o to use.
3. Cohesive tube samples are usually some variable in composition since they are at different levels in the tube (or stratum). This produces an unknown effect on results. Some organizations claim to avoid the depth variation by taking large diameter samples [say about 125 mm (5 in)], removing a length, dividing it into four quadrants and trimming three samples of diam 35 to 38 mm for the test—one quadrant is saved for a check if required. There is clearly much optimism that this substantial amount of trimming together with the membrane insertion produces a sample of sufficient quality to warrant the effort.

[4]A shear envelope is produced as the best fit line tangent to a series of Mohr's circles from triaxial testing.

4. Poisson's ratio μ is defined in theory of elasticity as the ratio of lateral to vertical strain, or

$$\mu = \frac{\epsilon_3}{\epsilon_1}$$

Strictly, this ratio has a sign, but when ϵ_1 is compressive and when ϵ_3 produces "bulging"—as in most soil tests—the sign is (+). It is not practical (or easy) to measure Poisson's ratio in any current soil test, so it is usually estimated. Commonly used values are 0.25 to 0.4 for all cohesionless materials and dry cohesive soil. Values of 0.35 to 0.5 are often estimated for wet-to-saturated cohesive soils.

Poisson's ratio has a theoretical range of −1 to 0.5 for elastic materials; since soil is not an elastic continuum, however, the theory of elasticity is used for computational convenience. For soil it appears that μ may have a range from 0 to 1.0, with values greater than 0.5 for very loose soil in compression.

These several factors put severe restrictions on the quality of the triaxial test data. In spite of these limitations these test data are considered more reliable than those obtained from the unconfined compression test. On the other hand, because of these several limitations and additional test effort the triaxial test is not considered a "routine" test as is the ordinary unconfined compression test. Since the triaxial test is about four to five times as costly as the unconfined compression test and with no quantifiable means to assess the data improvement, most commercial laboratories perform the triaxial test only if specifically requested by the client.

Data Obtained

At least two cohesionless samples and at least three cohesive samples are tested at a different cell pressure σ_3 for each test. Load and deformation readings are taken for each test so that stress-strain curves can be plotted to obtain the maximum stress (or stress at 15 percent strain). This stress is the deviator stress $\Delta\sigma_1$ and should increase with increasing cell pressure σ_3—if it does not this data set should be rejected and another test run. A sketch of each sample at failure should be made. This may be compared to the other tests to see if the failure mode is different to ascertain if the sample had a crack (or fissure) or some large gravel pieces which induced an early failure.

The axial load (obtained exactly the same as for an unconfined compression test) at some deformation is also called the *deviator* load. As illustrated in Fig. 15-7*a*, the data are reduced similar to the unconfined compression test using the measured sample load P at some deformation. The deformation e is reduced to strain $\epsilon = e/L_o$. Equation (14-6) is used to correct the sample area and Eq. (14-3) to compute the deviator stress at this strain level.

A curve of stress versus strain can be plotted and the peak value obtained. This peak value (or at 15 percent strain) is the deviator stress (symbol $\Delta\sigma_1$) of interest.

The two or three curves should be plotted on the same graph sheet. The initial tangent modulus is taken as E_s and Poisson's ratio μ is usually estimated. The peak or "failure" deviator stress from each curve is used next to obtain the shear strength parameters from a Mohr's circle plot.

We may do the following computations using Hooke's general stress-strain law (see Bowles, Chap. 10) for which the vertical *principal* strain is

$$\epsilon_y = \frac{1}{E_s}[\Delta\sigma_1 - \mu(\Delta\sigma_2 + \Delta\sigma_3)] \tag{15-2}$$

where $\Delta\sigma_1$ = stress changes; usually $\Delta\sigma_2 = \Delta\sigma_3$ = cell pressure in the triaxial test. The in-situ analogy is that as one goes from the initial steady-state condition to an increase in vertical pressure $\Delta\sigma_1$ there is an increase in confining pressure $= K\Delta\sigma_1$. There are principal strains along the other two orthogonal axes (x and z) but you are referred to Bowles (1984). In Eq. (15-2) there are two unknown values, namely, E_s and μ. We may make an estimate of these by taking two points along the stress-strain curve of constant $\Delta\sigma_3$, read the strain and deviator stress $\Delta\sigma_1$ and solve the following equation [rearranged from Eq. (15-2) and with $\Delta\sigma_3 = \Delta\sigma_2$]:

$$\epsilon_y = \frac{1}{E_s}(\Delta\sigma_1 - 2\mu\sigma_3)$$

With two strain and deviator stress values we can solve this equation for the two unknown values of E_s and μ. Strictly, this equation should not be used except when

$$-1 < \mu < 0.5$$

While it would appear that revised Eq. (15-2) could be used across two curves (i.e., constant $\Delta\sigma_1$ and $\Delta\sigma_{3(1)}$ and $\Delta\sigma_{3(2)}$), there is evidence that the stress-strain modulus is dependent on the confining pressure $\Delta\sigma_3$ and would be different for the two values of cell pressure subscripted as (1) and (2).

Note that Eq. (15-2) clearly shows that the unconfined compression test strain is larger for the same E_s and $\Delta\sigma_1$ than from a confined compression test. Thus an unconfined compression test will always give too low a value for E_s because with the loss of confining pressure $\Delta\sigma_3$ Eq. (15-2) reduces to

$$E_s = \frac{\Delta\sigma_1}{\epsilon}$$

The triaxial test data are used to plot Mohr's stress circles using major σ_1 and minor σ_3 principal stresses as from any mechanics of materials text. For most routine testing the cell pressure is the minor (σ_3) principal stress and the major principal stress (σ_1) is the minor stress + the stress from the additional axial load obtained from the stress-strain curve previously plotted for E_s. The principal difference in plotting Mohr's circles from mechanics textbooks is that even though the stresses are compression they are plotted in the first quadrant. The two stresses for Mohr's circle are then

$$\sigma_3 = \text{cell pressure (also } \Delta\sigma_3)$$

$$\sigma_1 = \sigma_3 + \Delta\sigma_1 \qquad \text{(or cell + deviator stress)}$$

From this it is obvious that the deviator stress $\Delta\sigma_1$ = diameter of the resulting Mohr's circle.

With two shear strength parameters c and ϕ to obtain it will take at least two tests to solve Eq. (15-1) using simultaneous equations. Alternatively, it is preferable to perform at least three tests using a different value of cell pressure σ_3 (or $\Delta\sigma_3$) for each test but samples of the same state (S, ρ, soil type) otherwise and plot the three resulting Mohr's circles.

By drawing a best-fit tangent to these three (or more) circles we can obtain a graphical solution for Eq. (15-1). While two tests are sufficient, with only ϕ and c the unknowns, it is preferable to use at least three tests. If the curve envelope does not fit reasonably well to all three Mohr's circles either one or more tests are bad. This gives you a chance to check the data and, if necessary, run one or more additional tests. The intercept of the curve envelope with the ordinate is the cohesion intercept c and its best fit slope is the angle of internal friction ϕ.

For cohesionless soils, the cohesion intercept should be small unless the soil is so damp that there is appreciable surface tension which becomes measured as cohesion. There is generally a small intercept (usually termed *apparent cohesion*) with cohesionless soils due to the rubber membrane used to surround the sample and to test errors, but values of 7 to 14 kPa are generally neglected. If the apparent cohesion is larger than this the test data should be checked as something is not correct.

While the preceding paragraphs indicated that at least two tests are required to solve Eq. (15-1) for shear strength parameters, in fact, since $c = 0$ for cohesionless soils, only one test is required (but at least two for a check). For $c = 0$ the angle of internal friction can be computed from the geometry of a Mohr's circle as

$$\sin \phi = \frac{\sigma_1 - \sigma_3}{\sigma_1 + \sigma_3} \tag{15-3}$$

STRESS PATHS

Triaxial data are sometimes presented using stress paths. The general methodology was given by Simons [see ASCE (1960)] and later by Lambe (1964). Instead of plotting Mohr's circles using σ_1 and σ_3 Simons plotted a single circle point defined by

$$p = \frac{\sigma_1 + \sigma_3}{2} \qquad q = \frac{\sigma_1 - \sigma_3}{2}$$

The plot coordinates p, q may be from either total or effective stress values of σ_1 and σ_3. The best-fit line through the locus of points obtained from a test series is called the K_f- line. This line if projected back to the p (horizontal) axis cuts the q axis with an intercept a; the slope of the K_f-line is scaled for angle α. From the plot geometry you can obtain

$$\sin \phi = \tan \alpha \qquad c = \frac{a}{\cos \phi}$$

To use this equation first compute tan α from which ϕ can be computed. Using the a-intercept directly, compute the cohesion c. A principal advantage of this procedure is a graphical presentation of the possible path traced by an increase in stress. Another advantage is this plot is simpler than a Mohr's circle construction, where circles may overlap and which require using a compass.

General Test Details

The deviator load is applied to the sample inside the cell through a load piston. This device is a carefully machined rod fitted into a bushing in the cell cover so that fluid leakage due to cell pressure is a minimum. One end is usually machined for a ball-to-sample connection both to attempt sample alignment and to produce as nearly an axial load as possible. The other end connects to a load ring or load cell for measuring the instantaneous load on the sample during the test.

When placing the sample in the cell and installing the cover, care must be taken not to damage the sample. Once the cell is attached with a tight seal (usually has rubber O-rings at base) the sample is loaded with the cap and piston. The mass of the cap and porous stone is usually ignored. When the cell pressure is applied the piston usually rises from the upward pressure on its base and a load appears on the load ring or load cell readout.

When this occurs you must carefully raise the cell until the piston again contacts the sample (or ball on top of the upper cap). You ascertain this by raising the sample slowly and observing the load ring/cell. When it shows about 0.5 kg of load you have remade contact. Now check the cell pressure, restabilize as necessary (use the bleed-off valve in the cell cap), and remake piston contact. When everything is stable you should then zero your load device.

If a calibrated buret is attached to the sample drain outlet and if the sample is saturated, changes in sample volume can be observed under the different deviator loads. Completion of consolidation for consolidated tests can also be determined using the buret connection, since no volume change or change in water level in the buret should occur after the sample consolidation is complete and prior to application of a deviator stress. If the sample is not saturated, volume changes occurring in the sample may not be indicated accurately by water-level changes in the buret, since the water may not drain but may instead remain in the sample, resulting in a slight change in water content distribution.

Special attachments are available, or can be constructed, to measure the pore pressure in the sample and changes in pore pressure under load. This method of triaxial testing is considered in Test 16 following.

Test-specimen diameters are standardized so that different-sized load platens and porous stones required for a range of sample diameters can be installed in a triaxial cell. A single cell being able to accommodate several sample diameters reduces test costs considerably. Common *nominal* sample diameters that can be tested are

33, 36, 47, 50, 63.5, 71, 76, and 100 mm

Rubber membranes are required to confine the specimen, and these are available in the above sizes.[5] Do not use a commercial membrane whose diameter is less than 90% that of the sample as the stretch will produce an excessive confinement stress.

It is necessary to use standard sample diameters so that trimming devices for the standard thin-wall sample tubes required to produce ends perpendicular to the longitudinal axis can be provided. Also this allows commercial availability of devices to build cohesionless samples. Membrane stretchers can be provided in order to enclose the sample in the rubber membrane.

Where the rubber membrane may contribute a significant apparent shear strength, ASTM gives the following equation

$$\Delta\sigma_c = \frac{4E_m t\epsilon}{D'} \qquad (15\text{-}4)$$

where E_m = Modulus of elasticity of membrane material, use 1400 kPa for latex;

t = average membrane thickness, mm, and ranges from about 0.15 to 0.25 mm;

ϵ = strain at this load increment and D' is computed as

$$D' = \sqrt{\frac{4A'}{\pi}} \qquad (15\text{-}5)$$

with A' as previously defined by Eq. (14-6).

Note that condom membranes do not require a membrane correction. A correction for latex membranes may be required when testing very soft soils where the membrane strength may be a significant part of the test strength. A correction may be required where the sample bulges considerably and "failure" is defined by 15 percent strain. You can always check and compute the membrane correction for the peak or failure load and see what percent of that load is carried by the membrane. If it is over about 5 percent you probably should make the correction.

As for the unconfined compression test, the length/diameter ratio should be

$$2 < \frac{L}{d} < 2.5$$

[5]Condoms can be used for sample diameters from 30 to 50 mm and are usually cheaper than commercial membranes. Take a pair of scissors and snip off the end, put over the lower platen and porous stone; put the sample on the porous stone and simply roll the condom on. This does not require a membrane stretcher and the sample can be prepared in about half the time needed using a membrane stretcher.

The strain rate for UU and CU tests can be taken as about that of the unconfined compression test, namely in the range of 1/2 to 2% per minute with the larger strain rate for softer soils.

Cell Pressure σ_3 and Normalization of Data

The cell pressure σ_3 (or $\Delta\sigma_3$) should have some relationship to pressure in situ. Usually three or more triaxial tests are performed, and σ_3 should include a value reflecting the current stress state and other values based on the expected increase in vertical stresses from project loading.

Some authorities suggest using isotropic cell pressure values of

$$\sigma_3 = \frac{\sigma_v + 2\sigma_h}{3} \tag{15-6}$$

where σ_v = vertical stress (either current in situ or current + some expected load increment $\Delta\sigma_v$) and σ_h = corresponding lateral stress value as $\sigma_h = K\sigma_v$.

For *normally* consolidated soils, the lateral pressure might be related to the vertical as follows:

$$\sigma_h = (\gamma h + \Delta\sigma_v)(1 - \sin\phi) \tag{15-7}$$

where γ = *unit weight* of in situ soil; h = depth to sample from ground surface; ϕ = effective angle of internal friction. Before testing ϕ will have to be estimated since the test is to determine its value. $\Delta\sigma_v$ was defined in Eq. (15-6).

It has been found that triaxial stress-strain data can be *normalized* with respect to σ_3 (or $\Delta\sigma_3$) to provide a more compact data presentation. To normalize the data we simply divide the deviator stress by the cell pressure to obtain the normalized pressure NP as

$$\text{NP} = \frac{\Delta\sigma_1}{\sigma_3} \tag{15-8}$$

A plot is made of NP vs. ϵ instead of $\Delta\sigma_1$ vs. ϵ. All the normalized data from a test series are plotted on a single graph, and an average curve is drawn through the locus of points thus obtained. We are justified in doing this only because substantial research work using data presented in this manner has been found to be adequate.

The enclosed computer program used in Test No. 14 can also be used here and will normalize your data on request so long as $\sigma_3 > 0.0$ (to avoid a divide-by-zero error).

RESILIENT MODULUS

Sometimes a triaxial test is taken up to some stress level, say, 0.25 to 0.50 of the estimated ultimate value; then the deviator load is reduced to zero and then reapplied. This load cycling is repeated three to five times; on the last cycle the deviator load is then continued to sample failure (see Fig. 15-7*a*). There is some body of opinion that the *initial tangent modulus* (which is somewhat larger than on the initial cycles, with the increase considered due to "strain hardening") is a better estimate of the stress-strain modulus (or modulus of elasticity) than otherwise obtained. This initial tangent modulus is termed the *resilient modulus* to distinguish it from one obtained from the usual triaxial compression test. This modulus is sometimes used in pavement design.

In performing this test, care is necessary in unloading the sample of the deviator stress to obtain the new strain at a deviator stress of zero, since the sample undergoes permanent (plastic) deformation and very little elastic recovery is obtained as shown in the actual data of Fig. 15-6. This can be accomplished by:

1. Unloading the sample (deviator stress = 0) immediately since the unload branch of the stress-strain curve is of academic interest only.
2. Carefully reapplying a very small load increment and then reading the deformation. The difference between the zero reading and this reading will be the plastic deformation of the sample at the end of that load cycle.

Procedure

This is suggested as a group or lab section project.

A. TESTING A COHESIONLESS SOIL—LABORATORY WORK

1. Fasten the base platen to the base of the cell.[6] Attach a rubber membrane of the proper diameter to the base platen (refer to Figs. 15-2 and 15-3), using rubber strips or bands. To provide a more impervious joint, the base platen may be lightly coated with silicone grease prior to attaching the membrane; this will increase the seal between the membrane and the platen.

   Some persons measure the wall thickness of the rubber membrane so that an area adjustment can be made for the initial area of the sample, but for ordinary computations it is the author's opinion that this introduces a fictitious precision into the results.

   Place a porous stone on the base of the platen. If the sample is damp or is to be saturated, the porous stone should be saturated[7] prior to use.
2. Weigh a container of dry sand so that the sample density can be obtained and approximately duplicated for succeeding tests.
3. Place a specimen mold around the rubber membrane and fold the top portion of the membrane down over the mold. Do not fold, however, if the mold is so large that the membrane will be torn in the process. If they are available, membrane stretchers may be used for forming the sample. When a split-barrel type mold that can be attached to a vacuum is used, it may be necessary to apply silicone grease along the split to produce a more efficient vacuum seal.

   For the split stretcher and other types of stretchers using a vacuum, it may also be necessary to use threads or small strips of filter paper between the membrane and stretcher wall so that the vacuum will be effective the full height of the membrane stretcher (you can tell if this needs to be done as there will be loose flaps). If the vacuum is fully effective the membrane is everywhere tight to the inside wall of the mold.
4. Carefully place the sand in the membrane, using a tamper to maintain the sample shape and density. When producing a sample of required density it may be desirable to divide the height into equal increments, compute the quantity of (wet) soil for each height, and place that portion in each sample segment.

   This procedure is of aid in producing a sample with the density evenly distributed through the height. To obtain a specified sample density is not an easy process, and the sample may have to be done over several times before the desired density is obtained.

   If it is necessary to produce a sample of a given density you will have to saturate the sample since it is nearly impossible to compact dry sand. Saturate the sample by opening the base inlet valve in line A of Fig. 15-2 and letting water into the partially filled mold to above the soil level; close and tamp the sample, open again while pressing your tamper against the soil in the mold and again submerge the soil, tamp, and so on.
5. Place a porous stone (damp for wet soil) and then the top platen on the sample. It may also be necessary to coat the outer rim of the platen with silicone grease to obtain a

[6]For sand the platen should be at least 63 mm in diameter. If gravel is present the diameter should be six times the largest gravel particle size.

[7]See Step 1 of Procedure A, Test 16.

more leakproof seal. Roll the membrane off the mold onto the top platen and seal with rubber strips or bands. Take a small level and level the top platen.

6. Attach the tube from the top platen to the vacuum (outlet B of Fig. 15-2) Close the valve of outlet A in the base of the cell and apply a vacuum of 200 to 250 mm of mercury to the sample.
7. Now remove the specimen mold and examine the membrane for holes and obvious leaks. If any are found, the sample must be remade using a new membrane.
8. Obtain three height measurements approximately 120° apart, and use the average value for the initial specimen height L_o. Using a pair of calipers (or other measuring device), take three diameter readings 120° apart at the top, at midheight, and at the base. Measure to the nearest 1 mm. Compute the average diameter of the specimen at each height and then compute a final average specimen diameter as

$$d_{av} = \frac{d_t + 2d_{mid} + d_b}{4} \tag{15-9}$$

where d_t = average top diameter and other d_i are similarly defined. Compute the corresponding value of initial sample area A_o using the average diameter just computed.

9. Place the Lucite cylinder on the cell base, being sure the base is free of soil grains so that an airtight seal can be obtained (there should be a rubber O-ring on the cell base as an aid in producing a good seal). Place the cell in the compression machine, and just barely make load contact of the loading piston and the loading bar (or crosshead) of the compression machine. At "just in contact" the sample should have a load of not over about 0.5 kg (or about 5 N).
10. Using compressed air[8] apply a predetermined lateral pressure σ_3 to the cell (preferably in even multiples of N/m^2 for computational ease) and simultaneously reduce the vacuum on the interior of the sample to zero.

    Some persons prefer to use a liquid chamber fluid (often water with a rust inhibitor or glycerin added) rather than compressed air. A liquid cell fluid has the advantages of providing a more uniform pressure and of being more viscous and hence not so sensitive to membrane leaks. For student laboratories it has the disadvantages of requiring more laboratory time (for siphoning out the cell at the end of a test) and of being more messy—especially if any leakage occurs.

    Now refer to Fig. 15-2 and with the vacuum shut off at B, open outlet A and check if any air is coming out under pressure. This would indicate a sample leak and the necessity to start over.

    When the chamber pressure is applied, observe that the load dial (or load cell DVM) records an upward load that is the *difference* between the mass of the load piston and the upward chamber pressure on the piston base area. Carefully raise the cell until the load (or DVM) dial just records a load of about 0.5 kg (about 5 N), indicating that the piston is in contact with the soil sample. Now zero the load dial (or DVM) to commence the test. By doing this the corrected deviator load is recorded for the corresponding sample deformation and the computations are simplified.
11. If it is desired to saturate the sample, open the appropriate valves of lines A and B and observe the sample until it is saturated.[9] A slight vacuum on the sample can be used to speed up the saturation process.

    Alternatively, a positive pressure (termed *back pressure*) of about $\sigma_3/2$ can be applied to the sample saturation reservoir. If you use a back pressure that is larger than σ_3 the sample will visibly expand inside the cell and self-destruct. Be sure during this process that the load piston remains in a "just-contact" position with the loading ball on top of the sample.

[8]It is also possible to produce a lateral pressure on dry samples using a vacuum. If vacuum is used it is not necessary to use a cell cover. Depending on efficiency of vacuum pump you may be able to obtain an effective σ_3 from about 90 to 95 kPa.

[9]See Test 16 for efficient sample saturation using back pressure.

Figure 15-3
Details of a cohesionless triaxial sample construction and test.

(*a*) Equipment needed includes cell, two porous stones, membrane stretcher, membrane, and top and bottom platens of appropriate size. Also needed is a vacuum source.

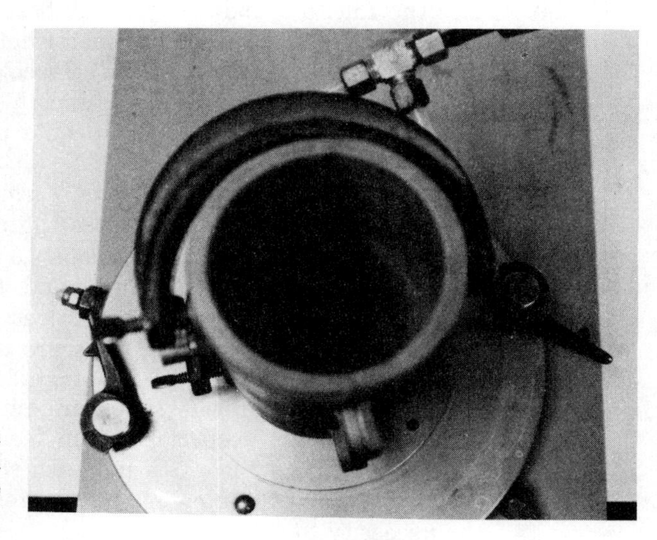

(*b*) Membrane stretcher with membrane rolled over both ends (as for top end shown) and vacuum applied. System has been placed over the bottom platen with porous stone so soil can be poured into membrane.

(*c*) Sample has been nearly constructed of damp sand (here) and leveled to top of membrane stretcher (which also functions as a sample mold). Porous stone is positioned on top, top platen is then placed and membrane will be rolled off stretcher onto platen. Later, vacuum will be applied to sample interior so membrane stretcher can be removed.

(*d*) Sample now built with top platen positioned; vacuum on interior maintains sample as shown. A load ball is placed in recess of top platen for load piston to apply an axial load. The next step is to attach cell cover with piston aligned with load ball.

(*e*) Cell in position in one type of compression machine. Upper dial in load ring is for load, lower dial on top of cell is for sample deformation. Dial gauge on wall behind system is for pore pressure (not being used). Large dial gauge on top of tank at right is for monitoring cell pressure using valve on extreme right attached to table (hand shown on it is adjusting cell pressure).

12. Attach a deformation dial or LVDT that can read directly to 0.01 mm/div to the machine so that the sample deformation can be measured. Zero the deformation device. If using a dial gauge you should manually compress and release the dial plunger several times and observe the zero reading. Readjust the gauge to zero if necessary. While an LVDT may be used instead of a dial gauge it is not recommended for student laboratories.

    Make a final check of deformation gauge, load gauge, and cell pressure gauge and that the load piston is still in contact with the sample (the sample may be consolidating under the cell pressure). Rezero gauges and adjust the cell pressure.
13. Check that the compression machine is set to the desired strain rate (generally between 0.5 and 1.25 mm/min) as specified by your instructor.
14. Turn on the compression machine, and take simultaneous load and deformation readings using Data Sheet 15*b* from your data sheet section. Readings may be taken at

    5, 15, 25, 50, and every 50 to 100 dial divisions or deformation equivalent until either

a. Load peaks and then falls off, or
b. Deformation is somewhat greater than 15 percent strain, or
c. Load holds constant for three or four successive readings

Be sure to monitor the chamber pressure gauge continually and do not let the chamber pressure vary by more than 0.05 kg/cm^2 (5 kPa). It takes only slight pressure changes to alter considerably the deviator stress that defines "failure."

15. After the sample has failed, shut off and/or reverse the compression machine; release the chamber pressure, and remove the sample load. Remove the deformation equipment if necessary and remove the triaxial cell from the compression machine. Carefully disassemble and clean as necessary for the next test or for storage.
16. Prepare a new specimen to the same approximate density (within 0.2 to 0.5 kN/m^3) and repeat steps 1 through 15 for at least two additional tests (a total of three). This completes the laboratory work; refer to "C: Computations," following, to complete your report.

B. COHESIVE SOIL-LABORATORY WORK

While cohesionless samples are rapid to test for student laboratories, it is often preferable to test remolded, non-saturated cohesive samples produced from a sand-clay mixture so the test will measure both cohesion c and angle of internal friction ϕ. These samples can be conveniently made using the Harvard Miniature compaction equipment shown in Fig. 9-3. The soil should already be prepared at an adequate water content and sufficient in quantity for each lab group to make up three samples compacted at a different energy level (possibly three layers at 20 tamps/layer; 25 tamps/layer; 30 tamps/layer . . . etc., as necessary).

1. Prepare two or three tube samples of adequate L/d ratio. Alternatively, use remolded samples as described above with nominal dimensions of diam = 33.3 mm $\times L_o$ = 71.0 mm (measure your mold for actual dimensions as there are usually small variations).
2. Use your mold dimensions for the remolded samples but if you use tube samples get sample measurements as outlined in A (8) preceding.
3. Take the correct-size (exact to say, + 6 mm larger diam) membrane stretcher and membrane,[10] and fit the membrane smoothly into the stretcher, folding the ends of the membrane over the ends of the stretcher (see Fig. 15-4). Attach the tube on the membrane stretcher to a vacuum source and apply a vacuum. If there are no leaks, the membrane will form a smooth cover on the inside of the stretcher. It may be necessary to place some small threads or filter-paper strips between the membrane and the walls of the stretcher to make the vacuum effective when stretching the membrane tight prior to the next step.
4. Depending on the sample, either lightly lubricate the membrane with petroleum jelly, powder it with Teflon powder (an inert substance), or, as a last resort, moisten it with water for ease of insertion of the sample into the membrane.
5. Insert the sample into the membrane and attach the lower platen and porous stone by rolling the bottom part of the membrane down over the lower platen. Now use rubber bands or strips to seal the membrane to the platen. A more airtight seal may be obtained if the platen perimeter is lightly greased with silicone grease. Attach the upper platen with its porous stone at this time if possible—if not, simply roll the membrane off the upper part of the stretcher and close the vacuum. Be sure to use damp or saturated porous stones on the platens, depending on the sample condition; a dry stone will absorb water from the ends of the sample.
6. Remove the sample from the membrane stretcher and attach the lower platen to the base of the triaxial cell. Also attach the upper platen if this has not already been done. Use extreme care not to damage the soil specimen.

[10]Omit this step if you use remolded samples and condoms and go to Step 6.

Figure 15-4
Triaxial test using cohesive soil.

(*a*) Membrane has been placed in a membrane stretcher and vacuum applied.

(*b*) Membrane stretcher now placed on bottom load platen covered with a wet porous stone. With the stretcher slightly larger than the sample it can be inserted without much difficulty (sometimes requires wetting, or greasing rubber for better insertion).

(*c*) Attach top porous stone and platen, roll membrane up and remove membrane stretcher. Attach cell cover, put in machine, apply cell pressure and test similar as for cohesionless soil.

7. Connect the top-platen tube to the vacuum line, *but do not at this time apply a vacuum*. If a consolidated-undrained test (on a saturated sample) is to be run, flood the lines into and out of (i.e., the vacuum line) the sample. Then close the valve of the line into the sample, leaving the vacuum line open for drainage. Some ingenuity is required to ensure that the lines, top porous stone, and top platen orifice are saturated. For unsaturated samples, line flooding is not necessary and may actually be harmful.
8. Place the Lucite cover on the cell, and place the cell in the compression machine. Bring the load bar into contact with the load piston until a load just flickers on the load dial.
9. Apply a predetermined chamber pressure (preferably compressed air) for the lateral pressure σ_3. For the consolidated-undrained test on a saturated sample, observe the flow of fluid out of the drain line and, when the flow is zero, proceed with Step 11 following.[11]

   The drain line may be connected to a calibrated buret so that when the water level remains constant the consolidation can be assumed to be complete. This type of connection will also yield a volume change during consolidation. If the sample is not saturated, it becomes an exercise in judgment[12] as to when the consolidation is complete.
10. If it is desired (on initially unsaturated samples) to saturate the sample, open the vacuum line to the atmosphere and open the inlet valve to the sample from the saturation reservoir. Do not apply a vacuum to the sample to speed the saturation process, as this will remove water already in the sample on the vacuum side faster than water will flow into the sample from the source, because of the low permeability of the sample. It will take a long time to saturate even small cohesive samples. It may be possible to put a positive pressure into the saturation reservoir to speed the process as long as the pressure is well below the chamber pressure.
11. If the sample is already saturated (and/or consolidated) and it is desired to measure volume changes, open the appropriate valves to fill the lines on both sides of the sample and then close the drain outlet. Open the line to the volume-change indicator (the buret). To speed the consolidation process of saturated samples when performing consolidated tests, one may place a series of small threads or strips of filter paper between the soil sample and the membrane in Step 5.

From this point on, refer to Steps 12 through 16 of Procedure A (cohesionless soil) since the compression test data collection is identical.

C: COMPUTATIONS

Compute the corresponding stress (kPa) and strain for sufficient readings to define the stress-strain curve (about 8 to 12 points). If you use the computer program it is simpler to compute every reading taken.

1. Compute the unit strain from the deformation and load reading for that load using Eq. (14-2) and fill in the appropriate column of Data Sheet 15*b*. Also compute the adjusted area A' using Eq. (14-6) and place this in the column 7 of the data sheet (see Figs. 15-5 and 15-6).
2. Compute the *deviator* load P using the load-dial (or cell) readings. If a load ring is used, the deviator load is simply the load ring constant multiplied by the number of dial divisions. Most load cells are calibrated to read the instant load directly and in this case column 3 of your data sheet gives the actual instant sample load P.

[11]Fill the cell exactly full if you use a liquid (use drain petcock as an aid). Continue adding cell fluid as sample consolidates. When it is not possible to add more fluid the sample should be consolidated.

[12]Refer to Test 16; alternatively, attach a deformation dial to the system and monitor when vertical movement has halted as an indication of consolidation (or U) = 100 percent.

3. Use the deviator load P and corrected sample area A' and compute the instant deviator stress, and if necessary $\Delta\sigma_c$ using Eq. (15-4), as

$$\Delta\sigma_1 = \frac{P}{A'} \qquad \text{or with membrane correction: } \Delta\sigma_1 = \frac{P}{A'} - \Delta\sigma_c$$

Put this value in column 9 of your data sheet. Then normalize the stress using the following computation:

$$NP = \frac{\Delta\sigma_1}{\sigma_3}$$

Enter the normalized stress in column 10 of your data sheet.

4. a. Plot a curve of unit deviator stress vs. unit strain (on the abscissa), and obtain the stress at the peak point unless the stress at 15 percent strain occurs first (see Fig. 15-7). Show this value of deviator stress on the graph. Plot all three tests on the same graph sheet, but be sure the curves are identified for the corresponding values of confining pressure σ_3.

   b. Also plot NP vs. σ_3 for the three tests but as a single curve as shown in Fig. 15-7*c*. Draw a best-fit curve through the locus of points and note that your curve may not be as smooth or points as well in agreement as shown on this plot.

5. a. With the maximum deviator stress from Step 4*a*, compute the major principal stress for each test as

$$\sigma_1 = \sigma_3 + \Delta\sigma_1$$

   Using σ_1 from each test and its corresponding cell pressure σ_3, plot the three resulting Mohr's circles on a sheet of graph paper to a scale large enough to read stress values to the nearest integer. Draw a best-fit tangent to the three circles and measure the ordinate intercept for cohesion c and the slope for angle of internal friction ϕ. Be sure to show the numerical values on the plot.

   If you used a cohesionless soil and there is a large cohesion value, indicate in your report what might have caused this.

   b. Make an alternative p, q plot of the data, locate the K_f-line and compute the shear strength parameters and compare to those from 5*a*. Be sure to show the measured angle α and any value of a.

6. If you used a cohesionless soil, compute ϕ for each test using Eq. (15-3). Average these values and compare with the value obtained from the Mohr's circle plots of Step 5 above. Also in this step derive Eq. (15-3) on the sample computation sheet.

7. Using the stress-strain curves, compute the initial tangent modulus for each of the three tests and the secant modulus using the origin and $\Delta\sigma_1/2$. Also compute the normalized initial tangent modulus and convert this to the same units as used for the tangent modulus from the stress-strain curves. Comment on any differences.

   Using the rearranged Eq. (15-2) and for the stress-strain curve using the largest σ_3, compute E_s and μ using $\Delta\sigma_1$ at 0.1, 0.2 and 0.3 of peak value. Comment on the computation if μ is out of the allowable range (between -1 and 0.5).

8. Using Eq. (15-4) compute the membrane correction for each test using the strain at the "failure" $\Delta\sigma_1$ stress and an average membrane thickness of 0.15 mm. Comment if the correction is large enough (in your opinion) that the data should be corrected. Note that the computer program allows this with minimal input.

9. If cohesive soils have been tested, be sure to comment in the report on which type of test was done (UU or CU). Is the soil saturated?

TRIAXIAL COMPRESSION TEST(Cohesive, ~~Cohesionless~~) **Data Sheet 15a**

Project Resilient Modulus Test Job No. —

Location of Project Soil Laboratory Boring No. — Sample No. —

Description of Soil Blue Clay (G_s = 2.72) Depth of Sample —

Tested By JEB & R.G.L. Date of Testing 1/12/19--

Fill in the blanks with data and appropriate units.

Sample Data

Sample dimensions: Diam. D_0 = 33.35 mm Area A_0 = 8.735 cm² L_0 = 70.9 mm

Vol. V_0 = 62.02 cm³ Water content w = 13.7 % Degree of saturation S = 61.0 %

Mass M_0 = 119.08 g

For Cohesionless soils

Initial mass of dish + sand = —

Final mass of dish + sand = —

Mass of sand used in specimen, M_o = —

Specific gravity of sand G_s = —

Vol. of soil solids in test specimen V_s = —

Vol. of voids in test specimen (initial) V_v = —

Initial void ratio of test specimen e_i = —

Void ratio of sand at minimum density e_{max} = —

Void ratio of sand at maximum density e_{min} = —

Relative density of test specimen D_r = —

Density of test specimen (cohesive, ~~cohesionless~~) $\rho = M_0/V_0$ = 1.92 g/cm³ $\frac{119.08}{62.02}$

Machine Data

Rate of loading 17.7 mm/min (insert mm or in)

The following data may not be applicable if machine can be adjusted to tare these effects out of the load readings.

Cross section area of loading piston, A_p = — cm²

Upward load on piston = $A_p \sigma_3$ = —

Mass of loading piston = —

Computed value of initial sample load = —

Figure 15-5

Preliminary sample data for a cohesive triaxial test. Not all data spaces above are required for any given test.

Sample Computations

Refer to Data Sheet 15*b*.
Fourth Cycle at DR = 540 (load dial set to 0)—no membrane correction

$$\Delta L = 540(0.01) = 5.40 \text{ mm}$$

$$\epsilon = \frac{\Delta L}{L_o} = \frac{5.40}{70.9} = 0.076$$

At DR = 550

$$\Delta L = 550(0.01) = 5.50 \text{ mm}$$

$$\epsilon = 5.50/70.9 = 0.078$$

$$1 - \epsilon = 1 - 0.078 = 0.922$$

$$A' = \frac{A_o}{(1-\epsilon)} = \frac{8.735}{0.922} = 9.474 \text{ cm}^2$$

$$\text{sample load } P = \text{Load dial Read} \times \text{LRC}$$

$$= 15(0.579) = 8.685 \text{ kg}$$

$$\Delta\sigma_1 = \frac{P}{A'} = \frac{8.685(98.07)}{9.474} = 89.9 \text{ kPa}$$

Normalizing:

$$NP = \frac{\Delta\sigma_1}{\sigma_3} = \frac{89.9}{98} = 0.92$$

Other points are computed similarly and except for the normalized values are identical to computations for Test 14.

These data are plotted on Fig. 15-7*a*. Two other tests were run but not cycled for resilient modulus and one set is plotted on Fig. 15-7*b*. Note the steeper slope for Fig. 15-7*a* for cycle 4. The initial slope (cycle 1) is quite similar to Fig. 15-7*b*.

The normalized NP vs. ϵ data are plotted in Fig. 15-7*c*. Only the last two tests are used since the cycled test is not the same.

Mohr's circles are plotted in Fig. 15-8 using all three tests. The shear strength envelope defined by Eq. (15-1) did not require much "fairing" in. The slope for ϕ is measured and the c intercept is read from the plot.

Also shown on the Mohr's circles is the p, q plot with the K_f line through the locus of points. Reading a = about 32 and $\alpha = 28°$, one can solve

$$\sin\phi = \tan\,\alpha \rightarrow \phi = \sin^{-1}(\tan 28) = 32.1°$$

$$\text{cohesion } c = \frac{a}{\cos}\phi = \frac{32}{\cos}32.1 = 38 \text{ kPa}$$

These computations are consistent with scale of drawing and check the two methods rather well. **REMEMBER,** however, these were tests on remolded samples using the Harvard Miniature apparatus and are more homogeneous than field tube samples.

TRIAXIAL COMPRESSION TEST Data Sheet 15*b*

Project Resil. Modulus Test Job No. —

Tested By JEB & R.G.L. Date of Testing 1/12/19--

Sample Data: Area A_o = 8.735 cm² Length L_o = 70.9 mm

Machine Data: Load rate = 12.7 mm/min. Load ring constant LRC = 5.68 N/div.

(Cohesive soil) Cell pressure σ_3 98 kPa

	Deform. dial 0.01 (units)	ΔL mm	Load dial units	Sample load P kN	ϵ $\Delta L/L_o$	$1-\epsilon$	Corr Area A^1 cm²	Deviator stress $\Delta\sigma_1 = P/A^1$ kPa	Normal stress $\Delta\sigma_1/\sigma_3$
1	2	3	4	5	6	7	8	9	10
1st Cycle	0	0	0			1.000	8.74	0	0
	25	0.25	3	17.04	0.0035	0.9965	8.77	19.4	
	50	0.50	9	51.12	0.0071	0.9929	8.80	58.1	
	75	0.75	16	90.9	0.0106	0.9894	8.83	102.9	
	85	0.85	17	96.6	0.0120	0.9880	8.84	109.3	
2nd Cycle	78	0.78	0	0	0.0110	0.9890	8.83	0	0
	100	1.00	17	96.6	0.0141	0.9859	8.86	109.0	
	125	1.25	26	147.7	0.0176	0.9824	8.89	166.1	
	150	1.50	29	164.7	0.0212	0.9788	8.92	184.6	
	175	1.75	32	181.8	0.0247	0.9753	8.96	202.9	This data not plotted on Fig 15-7a
	200	2.00	34	193.1	0.0282	0.9718	8.99	214.8	
3rd Cycle	171	1.71	0	0	0.0241	0.9759	8.95	0	0
	180	1.80	10	56.8	0.0254	0.9746	8.96	63.4	
	~					~			
4th Cycle	540	5.40	0	0	0.0762	0.9238	9.46	0	0
	550	5.50	15	85.2	0.0775	0.9224	9.47	90.0	0.92
	650	6.50	46	261.3	0.0917	0.9083	9.62	271.6	2.8
	750	7.50	55	312.4	0.1058	0.8942	9.77	319.8	3.26
	1200	12.00	66	374.9	0.1693	0.8307	10.51	356.7	3.64
	1300	13.00	67	380.6	0.1834	0.8166	10.70	356.4	3.64
20% =	1418	—							
		See	Plot of data on Fig 15-7a						

Note: Insert units in column headings as necessary.

[a]The Deviator stress computation shown is based on taring the loading system so that the load reading is the deviator load value.

Computed Data

Maximum deviator stress (from stress-strain curve) $\Delta\sigma_1$ = 357.5 kPa (Fig. 15-7a)

Maximum value of vertical stress $\sigma_1 = \sigma_3 + \Delta\sigma_1$ = 455.5 kPa

Figure 15-6

Stress-strain data for a resilient modulus triaxial test using cell pressure shown. Data are plotted in Fig. 15-7*a*.

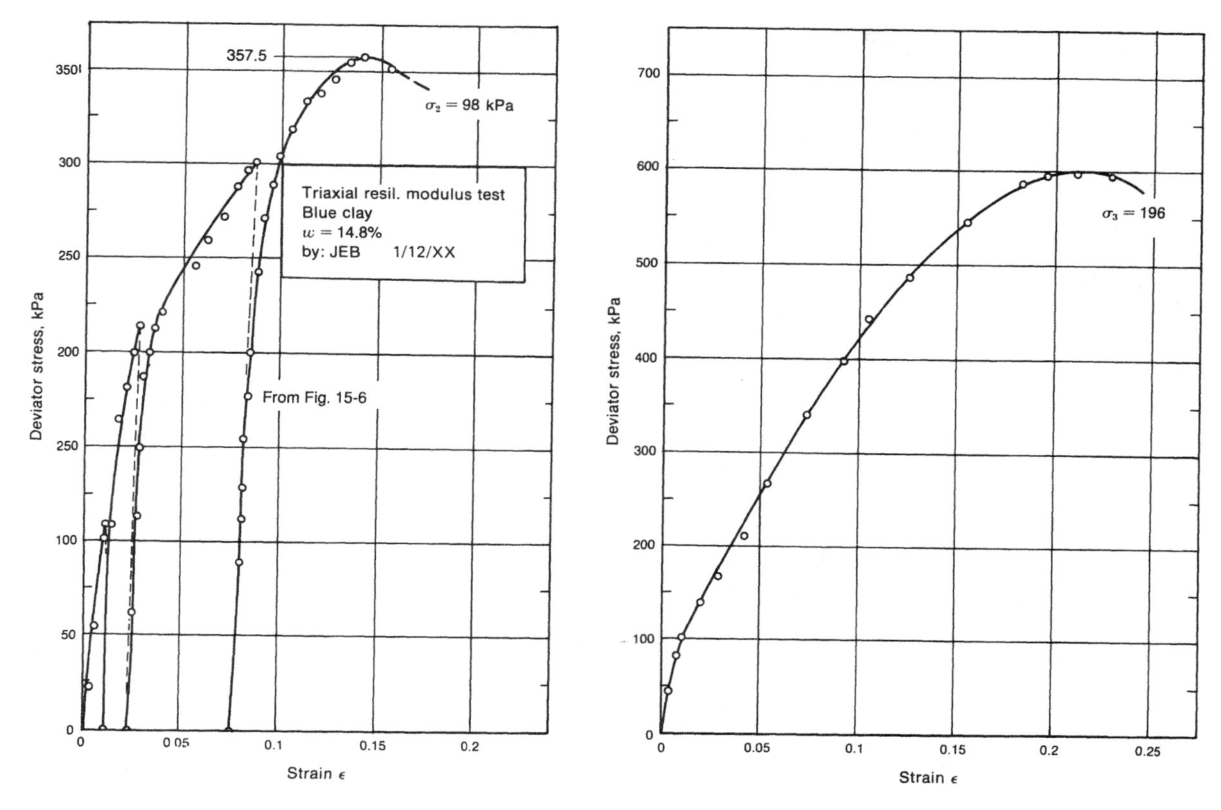

(*a*) Resilient modulus test from edited stress-strain data of Fig. 15-6.

(*b*) A second test using σ_3 = 196 kPa.

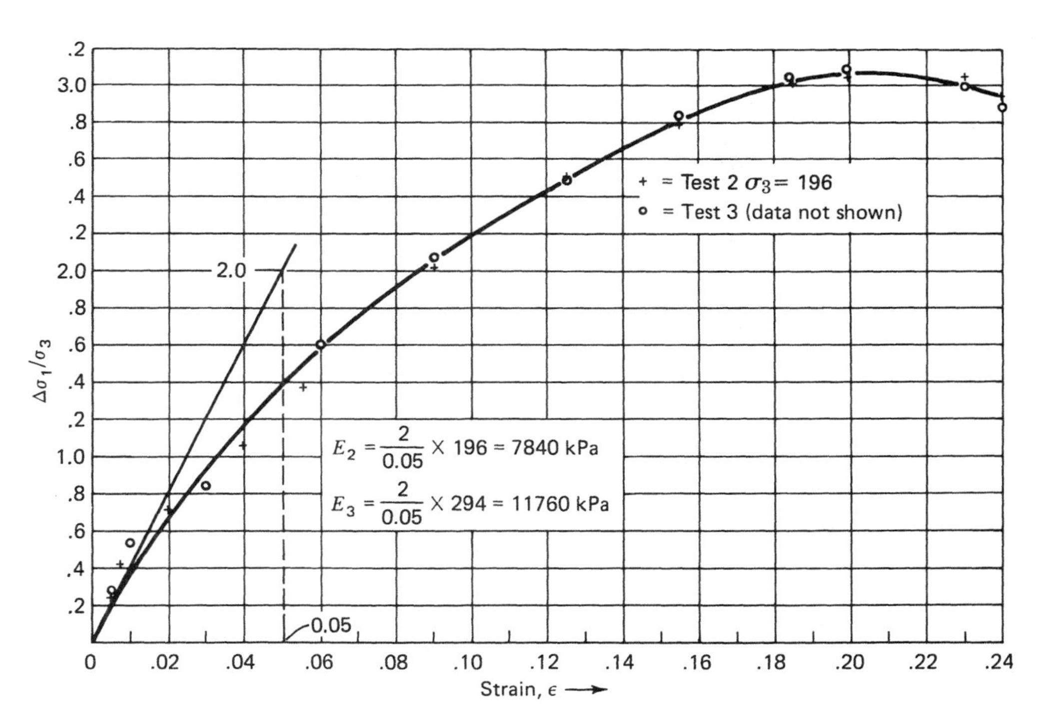

(*c*) Plot of normalized stress-strain data. *Question:* Why are data from Test 1 not included?

Figure 15-7
Stress-strain plots.

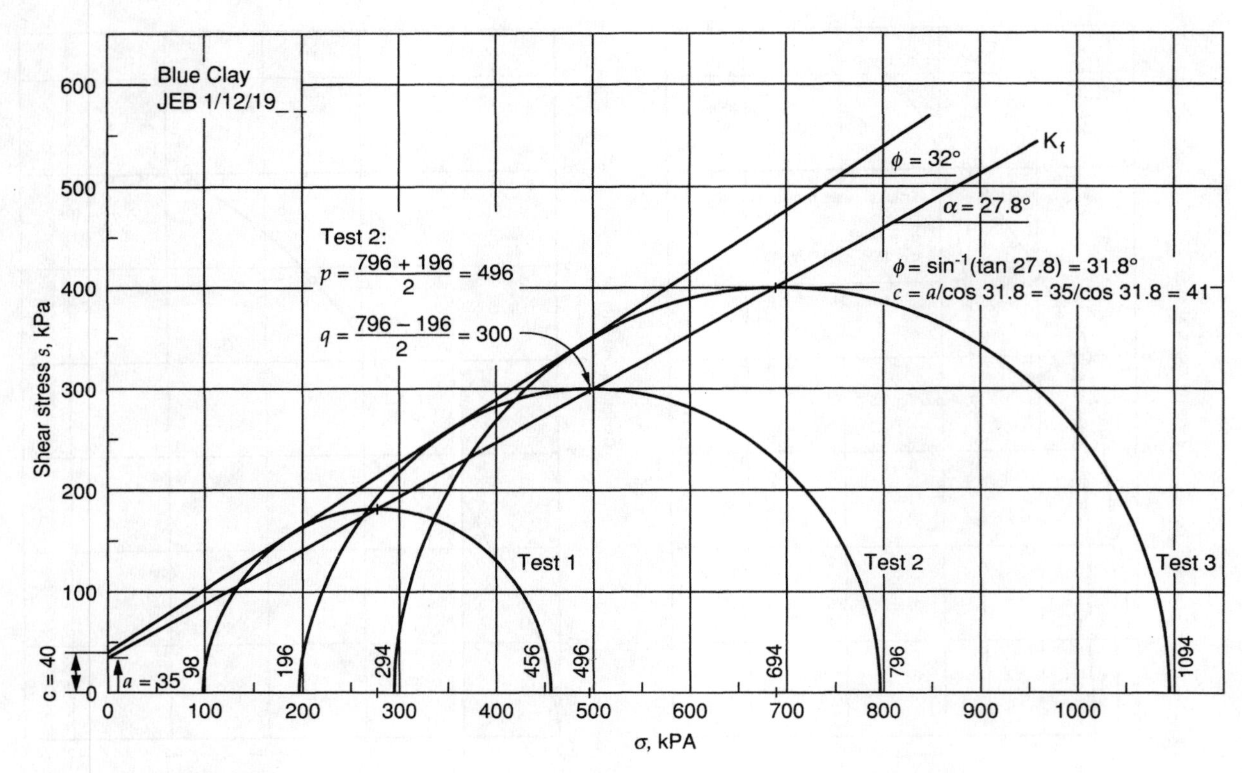

Figure 15-8
Mohr's circles and p, q plot using data from Fig. 15-7 and additional test data not shown because of space limitations.

16: TRIAXIAL TEST—WITH PORE-PRESSURE MEASUREMENTS

References

See Test No. 15 (This procedure is not standardized by either ASTM or AASHTO).

ASTM (1988), *Advanced Triaxial Testing of Soil and Rock*, ASTM Special Technical Publication STP 977, 896 pages.

Bishop, A. W., and D. J. Henkel (1962), *The Measurement of Soil Properties in the Triaxial Test*, 2d ed., Edward Arnold Ltd., London, 228 pages.

Black, D. K. , and K. L. Lee (1973), "Saturating Laboratory Samples by Back Pressure," *Jour. Soil Mech. and Found. Div., ASCE*, Vol. 99, SM 1, January, pp. 75–94.

Chaney, R. C., E. Stevens, and N. Sheth (1979), "Suggested Test Method for Determination of Degree of Saturation of Soil Samples by *B* Value Measurement," *Geotechnical Testing Journal, ASTM*, Vol. 2, No. 3, September, pp. 158–162.

Saada, A. S., and F. C. Townsend (1981), *State-of-Art: Laboratory Strength Testing of Soils*, ASTM, STP No. 740, pp. 7–77.

Skempton, A. W. (1954), "The Pore Pressure Coefficients *A* and *B*," *Geotechnique*, London, Vol. 4, No. 4, December, pp. 143–147.

Wissa, A. E. (1969), "Pore Pressure Measurement in Saturated Stiff Soils," *Jour. Soil Mech. and Found. Div., ASCE*, Vol. 95, SM 4, July, pp. 1063–1074.

Objectives

To present procedures for obtaining pore-water pressure and/or volume changes during a consolidated-drained triaxial shear test. To use excess pore pressure to obtain the "effective" stress parameters from the measured total stress shear strength parameters c and ϕ.

Equipment

Compression machine (preferably strain-controlled)
Triaxial cell (see Fig. 15-1)
Pore-pressure apparatus (Fig. 16-1 is one type that was locally fabricated) or pressure transducers and electronic readout equipment
Specimen mold, rubber membranes, rubber binding strips, and porous stones
Calipers (or other sample-measuring equipment)
Sample-trimming equipment as necessary

General Discussion

Read the "General Discussion" of both Tests 14 and 15.

Laboratory and field observations in the early 1930s recognized that Coulomb's shear strength equation as given by Eq. (15-1) was a total stress case. A more correct formulation to include the pore pressure term and "effective" stress parameters is given by Eq. (15-1a) and repeated here as Eq. (16-1) is

$$s = c' + (\sigma_n - \Delta u)\tan\phi' \qquad (16\text{-}1)$$

The effective stress parameters require that the normal stress σ_n on the shear plane be reduced by any excess pore pressure Δu that develops on that plane during shear. Actually,

Figure 16-1
Pore pressure panel (noncommercial). Valve on lower left connects to 3-mm diam nylon tubing from soil sample inside cell. Plastic device at left is a null indicator. The U-tube on right can measure negative pore pressures. Buret at extreme right measures volume change during sample consolidation. Large disc at bottom is the handle of a manufactured hydraulic metering device used to add water to system to maintain the null reading so that the pore pressure can be developed and measured on the large dial gauge in center. All tubing must be saturated (absolutely no air bubbles), and all critical valves must be non-displacement type. Use plastic tube connections where mercury contamination is possible. Nylon plastic tubing is used because it allows for visual inspection of air bubbles and because it is rigid enough to undergo little expansion at reasonably high internal pressures. A somewhat similar pore pressure panel is available from at least one laboratory supply house. This type panel is particularly useful in any laboratory where pore pressures are measured, as the user can see what is happening.

Eq. (16-1) is the "general" Coulomb shear strength case since varying the pore pressure term produces anything from the undrained case to the actual effective stress parameters.

Equation (16-1) appears simple enough, however, its solution is anything but simple. There are two major problems:

1. Defining (or locating) the shear plane on which σ_n acts, and
2. Measuring the excess pore pressure Δu that develops during shear.

It is usually easier to measure the excess pore pressure Δu that develops during shear if the sample being tested is fully saturated. Theoretically the excess pore pressure in this case is evenly distributed through the sample. This allows one to attach a pressure-measuring device to the end of the sample. If the sample is not saturated the excess pore pressure Δu on the shear plane may not be the same as on the ends of the sample. This requires that pressure sensors be inserted into the sample across the estimated shear plane location. This is not a very desirable situation since the pressure sensor (usually a small perforated tube) may influence both the location of the shear plane and the measured shear strength.

Normal stresses, the σ_n of Eq. (16-1), are carried by both the soil particles and the pore fluid. However, as given in Test 15, only the "effective" component, carried by the soil particles as intergranular normal pressure, develops the friction resistance defined as μN or here as $(\sigma_n - \Delta u) \tan \phi'$. It is a convenience to rewrite Eq. (16-1) as

$$s = c' + \sigma' \tan \phi' \qquad (16\text{-}2)$$

where $\sigma' = \sigma_n - \Delta u$ and other terms are as previously defined.

One may obtain the effective stress σ' in two ways:

1. Perform the shear test so that the excess pore pressure $\Delta u \to 0$. The pore pressure due to structure reorientation can never be zero, but if the reorientation is at a slow enough rate the effect on the soil parameters is not significant.

2. Perform the shear test using equipment that allows measurement of the excess pore pressure Δu.

The *consolidated-drained* test described in Test 15 is an attempt to satisfy the condition of

$$\Delta u \rightarrow 0$$

The disadvantage of this test is that the time-duration may be in the range of a week or more. It has the advantage, however, of being the more precise—particularly if the degree of saturation $S < 100$ percent.

If the soil is saturated ($S = 100\%$) direct pore-pressure measurements during a consolidated-undrained test provide the most rapid means to obtain the "effective" stress parameters and with careful attention to detail are probably as accurate as any method. Certainly we avoid the time penalty and accompanying labor costs (a critical item in commercial laboratories).

In CU tests the major problem is to ascertain if the soil sample is saturated. In order to be sure that the excess pore pressure measured at sample ends also exists on the shear plane it is essential to use saturated samples.

Skempton (1954) suggested that the excess pore pressure in either saturated or partially saturated soil under some applied stresses $\Delta\sigma_i$ can be described as

$$\Delta u = B[\Delta\sigma_3 + A(\Delta\sigma_1 - \Delta\sigma_3)] \qquad (16\text{-}3)$$

where Δu = change in pore pressure due to any incremental increase in confining pressure $\Delta\sigma_3$ or in the *deviator stress* $(\Delta\sigma_1 - \Delta\sigma_3)$. Note that $\Delta\sigma_i$ and σ_i are often used interchangeably and you have to look at the context of usage to apply the equation correctly.

A, B = Skempton's pore-pressure coefficients (or parameters)

From Eq. (16-3), one can determine the B coefficient by applying a cell pressure of σ_3 to a sample in the triaxial cell. After consolidation is complete the excess pore pressure should read some value (possibly the cell pressure σ_3). Now apply an increment (say from 5 N/m^2 to 8 N/m^2) and measure the change in pore pressure Δu; do this about three times. Now compute the pore pressure parameter B. Note that with only $\Delta\sigma_3$ changed, the deviator stress is zero and the A parameter vanishes. Take an average of the three computations ($B = \Delta u/\Delta\sigma_3$) for B. If B is in the range of about 0.95 to 1.0 the sample can be assumed to be saturated. Strictly, for $S = 100\%$ the B-parameter should be 1.0; but soil variability, equipment limitations and procedural errors may lead to a computation of $B < 1.0$ for a saturated sample. After computing the B parameter, applying two or three increments of deviator stress will allow the A parameter to be estimated if for some reason it is needed. The principal use of Eq. (16-3) is to estimate when (or if) the test sample is fully saturated ($S = 100\%$).

In theory Eq. (16-3) can be used to estimate pore pressure increases in the field for, say, embankment construction for dams, levees, roads, and so on. In practice this equation has had little success. However, as in the case of permeability tests, a prediction based on a theoretical basis is better than making a guess—even if the guess has some probability of success.

Pore pressure measurements can be made using two types of equipment:

1. By connecting a pressure transducer to the saturation line with the output connected to some type of readout device (digital voltmeter or other). Preferably the readout is adjusted to read the pressure directly rather than as a change in voltage. The pressure

transducer must be one requiring a very small volume displacement to be fully activated. Recall that any pore drainage will reduce the pressure and some drainage must occur to activate the transducer.

2. By using some type of null-pressure indicator device (typically as shown in Fig. 16-1 and schematic of Fig. 16-2). These devices are commercially available, but if you make your own you will know (a) how it works, and (b) what its limitations are. A device similar to that shown in Fig. 16-2 can be made at a nominal cost (depending on who pays the machinist or does the machining required). The following is a list of essential parts:

   8 valves (open-shut which do not displace any volume when opening/closing)
   1 pressure gauge (20-cm diam × 5 to 7 kg/cm^2 range at 0.05 kg/cm^2/div)
   1 hydraulic cylinder with screw control (fabricate in shop), 100-mL capacity
   1 glass U-tube for a manometer to measure negative pore pressures
   1 buret (100-mL) to measure volume changes
   1 null indicator (build in shop); refer to Fig. 16-2 for line details
   2 pieces of 5 × 5 × 5 cm plastic block; refer to Fig. 16-2 for drilling and tapping
   1 piece of glass capillary tube 1 mm ID × 6 to 10 mm OD and 20 mm long
   2 No. 20 hypodermic needles (stainless steel) 38 mm in length
   1 threaded plastic rod (use convenient threads) about 10 cm long with threads on about 6 cm of rod
   1 metric scale to observe and calibrate the mercury column
   $\frac{1}{2}$ kg mercury metal for null indicator and manometer (U-tube)
   Nylon tubing 3 mm × 3 m (use nylon to withstand high pressures without expansion)
   Plastic (polyflo) tubing 6 mm × 3 m for locations where expansion is not critical
   Sufficient fittings for both tubing sizes (3- and 6-mm) using standard pipe threads. Be sure to use nylon fittings where there is any possibility of mercury contamination since mercury will destroy brass fittings in a very short time.

   To operate the null indicator system (refer to Fig. 16-2):

1. Deaerate and saturate the system, then stabilize the mercury in the null indicator to some reference mark.
2. Connect the 3-mm nylon tubing (precharged with water) to the sample outlet.
3. Close all valves. Apply the cell pressure σ_3. Now open valves numbered 1, 2, 3, 4, and 5 on the line drawing. If the sample is saturated, the mercury will immediately drop in the null indicator due to the cell pressure.
4. Add water to the system from the hydraulic cylinder to bring the mercury column in the null indicator back to its original location. The pressure required to do this can be read on the dial of the pressure gauge and should be very nearly equal to cell pressure σ_3.
5. Apply the deviator load and, using the hydraulic cylinder, visually control the mercury column at the initial mark. Record the pressure gauge readings along with the deviator load and deformation readings on the data sheet provided.

Inspection of the tubing diagrams of Figs. 16-1 and 16-2 indicates that volume changes can be taken either to determine the end-of-consolidation for consolidated tests or (if pore pressures are not taken) to measure volume change of the test sample under load (with valves 1, 2, and 7 open).

If it is known in advance that the soil structure will collapse during application of the deviator load, one may connect the sample outlet to the mercury manometer (U-tube with valves 1, 2, 3, and 6 open) to measure negative pore pressures. For other cases of sample collapse, the operator will need to apply vacuum with the hydraulic cylinder (and not recorded on pressure gauge) to keep mercury from being drawn out of the null indicator into the sample side of the system.

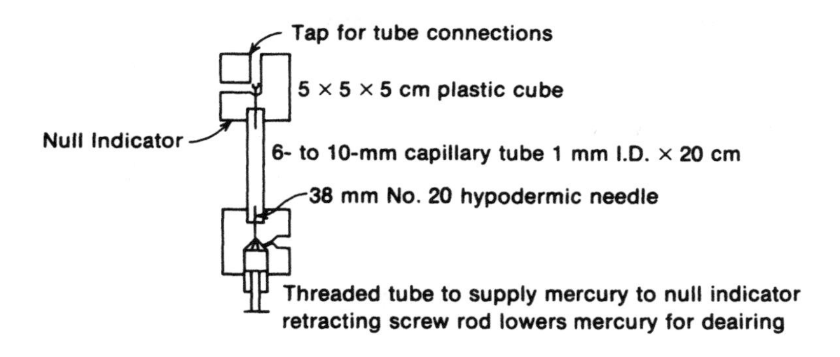

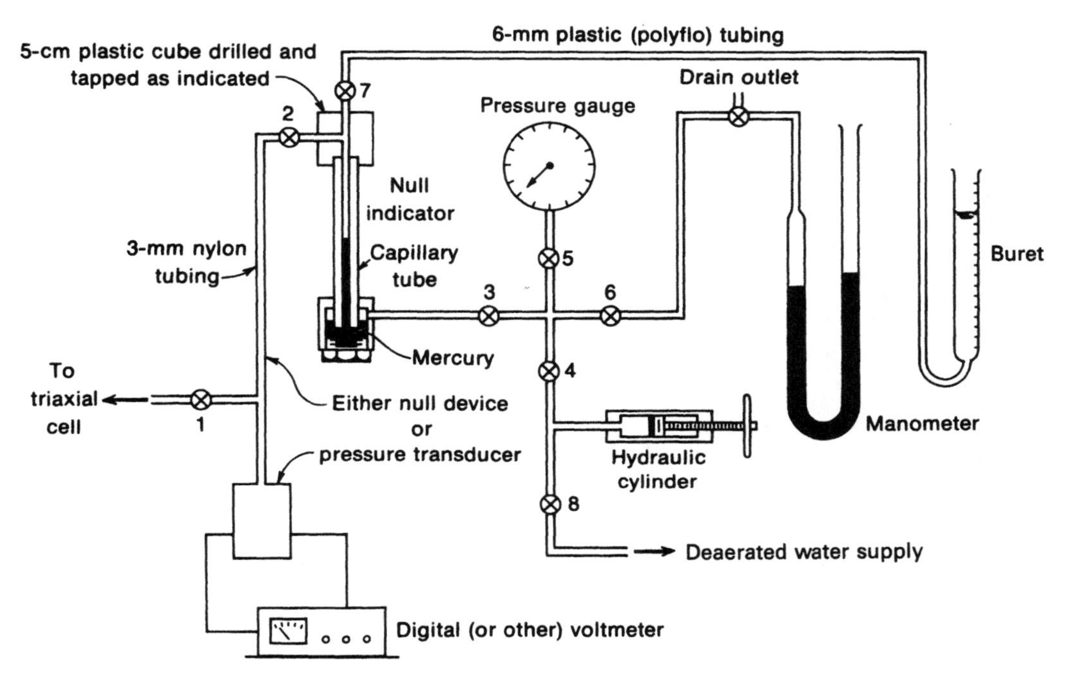

Figure 16-2
Line details of a pore-pressure measuring apparatus (of Fig. 16-1). Drill connecting holes in null device at a slight upward tilt to allow trapped air bubbles to escape.

If it is desired to measure pore pressures and volume changes simultaneously, or to measure volume changes occurring in a non-saturated sample, two test modifications are required:

1. It will be necessary to use a liquid to develop the cell pressure σ_3, as air does not have sufficient viscosity, and volume changes are difficult to measure. A glycerin-water mixture may be used. Water is sometimes used but may not have sufficient viscosity.

   If the triaxial cell containing the soil sample is completely filled with fluid (and deaerated) before the test is started, the difference of any fluid coming out of the cell is due to the load piston advance into the cell and sample volume change. Piston diameter and dial reading can be used to separate piston advance from sample volume change.
2. Use a compensating device of some type [several are available commercially (or local manufacture)] to maintain a nearly constant cell pressure. This is necessary since the fluid flow into or out of the cell will result in a change in cell pressure unless a device of this type is used. Only small cell pressure changes will produce erratic test results.

Figure 16-3 can be used to evaluate qualitatively the merits of drained and undrained testing of cohesive and cohesionless soil samples.

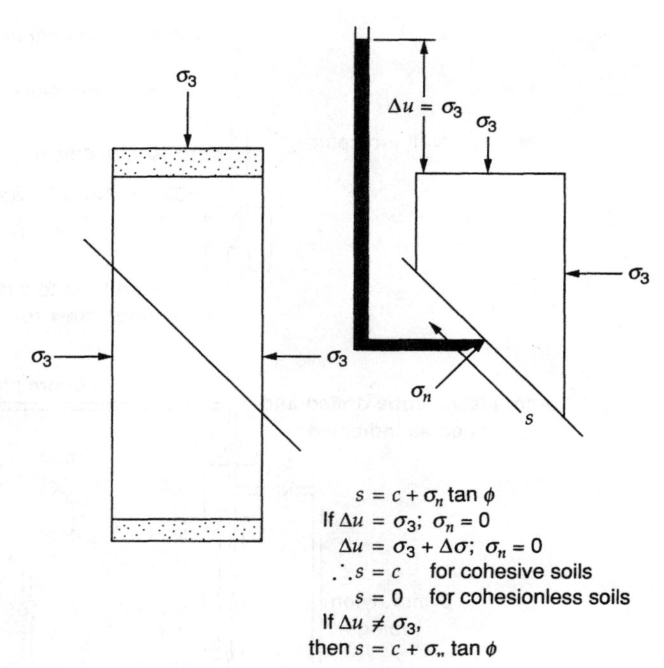

Figure 16-3
Pore-pressure effects in an undrained test. It should be evident from this figure that the undrained shear strength of a saturated cohesionless soil is the strength of the sample membrane.

Procedure

This is suggested for teams of at least two persons.

A. COHESIONLESS SOIL[1]

1. Using rubber strips or bands, attach a rubber membrane of the proper diameter (at least 50 mm for sand)[2] to the bottom platen (refer to Fig. 15-2), which has already been fastened to the base of the cell. To provide a more impervious joint, the bottom platen may be lightly coated with silicone grease to increase the seal between the membrane and platen. You may also lightly grease the platen bottom where it screws down to the cell base for a better seal and easier removal.

   Some persons measure the wall thickness of the rubber membrane prior to attaching it, so that an area adjustment can be made for the initial area of the sample, but for ordinary computations this introduces (in the author's opinion) a fictitious precision into the results.

   Place a porous stone on the top of the bottom platen. For these tests, the stone *must be saturated.* Boiling the top and bottom stones in distilled water will increase the degree of saturation and is recommended. Merely storing the stones underwater, even for a long period of time, may not deaerate them sufficiently to achieve the desired degree of saturation.
2. Weigh the container of soil being used so that the sample density can be approximately duplicated for succeeding tests.
3. Place a specimen mold around the rubber membrane and fold the top portion of the membrane down over the mold. Do not fold, however, if the mold is so large that the membrane will be torn in the process. Membrane stretchers may be used for forming the sample if the correct size is available. If a split-barrel type, which can be attached to a vacuum, is used, it may be necessary to use some silicone grease along the split to effect a vacuum seal.

[1]For student laboratories a very silty sand should be used so that measurable pore pressures can be developed and consolidation is rapid.
[2]The largest grains should be less than one-sixth the diameter of the sample.

For both the split stretchers and other types of stretchers using vacuum, it may also be necessary to use threads or small strips of filter paper between the membrane and the stretcher walls so that the vacuum will be effective the full height of the membrane stretcher.

4. a. Deaerate the water in the saturation reservoir by connecting it to a vacuum for 5 to 10 min.
   b. Open the saturation valve and allow water to flow into the bottom of the membrane for a depth of about 20 mm; then close the valve.
   c. Place sand through the water in the bottom of the membrane, using a tamper to obtain the desired density to a depth just under the water surface.
   d. Open the saturation valve and allow about 20 mm more of water but this time put the tamper on top of the existing sand and apply pressure to prevent sample expansion; then add sand as before. Repeat until the sample is just below the top of the mold. Use a pipet and remove most of the excess water.
   e. Now place a saturated porous stone on top of the sample, being sure it penetrates into the mold and is in close contact with the sample.

   Note that it is necessary to use a wet sand to produce a desired density—dry sand is very difficult to compact by tamping. The procedure here will produce a sample that is very close to (if not actually) $S = 100\%$.

   Weigh the sand container and compute the sand mass used in this sample so the density can be approximately duplicated for the next test.
5. Place the top platen on the porous stone. It may be necessary to coat the perimeter of the top platen with silicone grease to increase the seal. Roll the membrane off the mold and onto the top platen and seal it to the platen with rubber strips or bands. Level the top platen using a small level and by pushing the sample laterally or reseating the platen as necessary.
6. Attach the tube from the top platen to the vacuum outlet in the base of the cell and apply a vacuum of 150 to 200 mm of mercury. Simultaneously, open the saturation line (so you do not desaturate the sample—observe to see if more water appears to be going out of the sample than is going in and, if so, reduce the vacuum). Close the saturation line when inflow $\approx$ outflow.
7. Now remove the specimen mold and observe the membrane for holes and obvious leaks. If any are found, the sample must be remade using a new membrane.
8. Obtain three height measurements approximately 120° apart and use the average value for the initial specimen height L_o. Record this on your data sheet. Take three diameter readings 120° apart at the top, mid-height, and base using calipers (or other measuring device). Take these measurements to the nearest 1 mm. Compute the average diameter of the specimen at each height location, and then compute a final average specimen diameter using Eq. (15-9).

   Compute the corresponding value of initial sample area A_o using the average diameter just computed. Record these data on your data sheet.
9. Place the Lucite cylinder on the cell base, being sure the base is free of soil grains so that an airtight seal can be obtained. Place the cell in the compression machine and just barely make load contact of the loading piston and the loading bar (or crosshead of the compression machine).
10. Apply a predetermined lateral pressure σ_3 (preferably in multiples of 10 kPa or 0.5 kg/cm^2 for computational ease) using compressed air or other fluid (such as a water-glycerin mixture) and simultaneously reduce the vacuum on the interior of the sample to zero.

    Note that for long-duration tests, the rubber membrane will eventually leak regardless of the type of cell fluid used. Leakage can be reduced using a glycerin-water mixture and/or two membranes with silicone grease between the membranes. Two membranes will, however, require a strength correction using Eq. (15-4) for the membranes; to avoid excessive effect on the test larger diameter samples should be used.

11. If it is desired to measure volume change to detect end-of-consolidation, connect the saturation line to the volume-change buret. A plot of volume change vs. time should indicate when consolidation is nearly complete. Approximate end-of-consolidation may be obtained by monitoring the deflection dial (or LVDT), and when the sample ceases shortening, consolidation should be about complete.
12. After consolidation is complete, connect the pore-pressure measuring device[3] to the saturation-line outlet. Set the initial pressure transducer readout to zero or observe the stationary position of the null indicator.
13. When the chamber pressure σ_3 is applied you will observe the load reading device indicates an upward force that is due to the difference in weight of the piston rod and the cell pressure acting upward on the piston base. Carefully adjust the triaxial cell until piston-sample contact is reestablished, and then zero the load indicator (or load ring dial gauge). This step makes for direct reading of the deviator load and more convenient computations.

    At this point open the other exit line to the soil sample if a "drained" test is to be performed so the sample can drain from both ends. If you are going to perform a CU test with pore pressure readings be sure the sample drain valves are closed but there is direct drainage access to the null indicator or pore pressure transducer.
14. Attach either a deformation dial or displacement transducer (LVDT) that reads directly to 0.01 mm to the machine so that sample deformation can be obtained. If you use a load ring, set its dial gauge to zero; then manually compress and release the dial plunger several times, observe the zero reading, and readjust to zero if necessary.
15. Check and/or set the compression machine to the desired strain rate (between 0.5 and 1.25 mm/min, as specified by the instructor). Also
    a. Recheck the load dial for zero reading.
    b. Check the chamber-pressure gauge.
    c. Check the null indicator.

    You are now ready to commence the test.

16. Turn on the compression machine and take simultaneous load, deformation and pore-pressure readings based on equipment used. Remember (if you are not using a pressure transducer) to keep the pore-pressure gauge activated by observing the null indicator and metering water from the hydraulic cylinder to maintain the mercury column at the initial position.

    Generally, for samples of 63 to 100 mm diameter, deformation readings can be taken at each 0.5 mm of deformation (50 dial divisions of the 0.01-mm dial range). Take readings until the load holds constant and then falls off, or to somewhat beyond the estimated 15 percent strain value. If this yields a very large number of load-deformation readings, do the computations (see "Computations" following) on only enough of the readings to define adequately the stress-strain curve and its peak value.

    Be sure to monitor the chamber-pressure gauge throughout the test, and try not to let the chamber pressure vary by more than 0.05 kg/cm^2. If the chamber pressure does much varying, the stress-strain curve will be quite erratic.
17. After the sample fails, shut off and/or reverse the compression machine, take off the sample load, and gradually release the chamber pressure so that the null indicator does not lose mercury. Be sure to make a sketch of the sample failure (bulging shape or if there is a definite failure plane and if so the approximate angle measured from the horizontal).
18. Remove the Lucite cover and the tested sample. Prepare new specimens to the same approximate density (within about 0.2 to 0.4 kN/m^3) and make two additional tests

[3] If it is desired to determine whether $S = 100\%$ prior to connecting the sample to the volume-change buret, connect it directly to the null indicator (or pressure transducer) using a σ_3 less than the test value. Now increase σ_3 in increments and record the resulting measured pore pressure Δu. Use these values to compute an average B parameter; and if it is in the range of 0.95 to 1.00 you may assume sample is saturated.

at larger σ_3 valves. This completes the laboratory work for a cohesionless soil; refer to "C: Computations," following, to prepare your laboratory report.

B. COHESIVE SOIL

1. Prepare two or three tube samples of adequate L/d ratio of 2.1 to 2.5. It may be necessary to use a sample trimmer to form the samples if they are of smaller diameter than the tube samples.

   Alternatively, compact three or four samples at some constant water content and compaction effort, using the Harvard Miniature compaction apparatus (shown in Fig. 9-3). This procedure will yield samples of approximately 33-mm diameter at a suitable L/d ratio. These samples may not be saturated, however, and if you wish to test saturated samples, it may be necessary to use tube samples. Alternatively, of course, you may saturate the samples but this can take a long time even if the process is assisted by using back pressure.

   While preparing the test set-up and between tests, store these samples so they do not dehydrate. For small samples you may turn a 1000 mL beaker upside down over the samples after they have been covered with a damp paper towel. Use a controlled humidity room if available but you still may have to cover the samples so they do not get too wet on the surface.
2. Take the correct-size membrane stretcher and membrane and fit the membrane smoothly into the stretcher, folding the ends of the membrane over the ends of the stretcher.[4] Attach a vacuum tube to the membrane stretcher tube outlet and apply a vacuum. If there are no leaks, the membrane will form a smooth cover on the inside of the stretcher under the vacuum. It may be necessary to place some small threads or filter-paper strips between the membrane and the walls of the stretcher to make the vacuum effective when stretching the membrane tight (refer to Fig. 15-3*b*) prior to the next step.
3. Depending on the sample, either lightly lubricate the membrane with petroleum jelly, powder it with Teflon powder (inert substance), or as a last resort, moisten it with water for ease of insertion of the sample into the membrane.
4. Cut four strips of filter paper that are about 3-mm wide and long enough to fit under the porous stone of the lower platen and extend above the membrane stretcher when it has been slipped onto the lower platen. Fold the tops of the filter-paper strips back over the rubber membrane, and fasten them out of the way with a rubber band. Be sure the porous stone is saturated.

   Insert the sample into the membrane and seat it on the porous stone. Release the vacuum on the membrane, roll the bottom part down onto the platen, and seal it with rubber bands or strips. Silicone grease on the perimeter of the platen may increase the seal. Put the upper saturated porous stone on top of the sample and fold the filter-paper strips over it. Now place the top platen in place. Next roll the membrane from the stretcher onto the top platen and seal it with rubber bands or strips.

   Note the filter paper strips are used to speed sample consolidation and/or saturation.
5. Obtain three height measurements approximately 120° apart and use the average value for the initial specimen height L_o and record on your data sheet. Take three diameter readings 120° apart at the top, mid-height, and base using calipers (or other measuring device). Take these measurements to the nearest 1 mm. Compute the average diameter of the specimen at each height location, and then compute a final average specimen diameter using Eq. (15-9).

   Compute the corresponding value of initial sample area A_o using the average diameter just computed. Record these data on your data sheet.

[4]Omit this step if using condoms. In this case snip the end, put on the bottom platen, place the porous stone then the sample, then roll the membrane to the top. Put on top porous stone and top platen and seal with rubber bands or O-rings. Refer to Step 4, then continue to Step 5.

6. Place the Lucite cylinder on the cell base, being sure the base is free of soil grains so that an airtight seal can be obtained. Place the cell in the compression machine and just barely make load contact of the loading piston and the loading bar (or crosshead of the compression machine).
7. Apply a predetermined lateral pressure σ_3, (preferably in multiples of 10 kPa or 0.5 kg/cm^2 for computational ease) using compressed air or other fluid (such as a water-glycerin mixture). Allow the sample to consolidate using procedures previously described either in this Test or previously in Test 15. When consolidation is complete you will want to check if the sample is saturated. At this time drain lines should be open but filled with water.
8. To check sample saturation proceed as follows:
   a. Flood all the lines into and out of the sample; then close off the top-platen line. Connect the saturation line to the pore-pressure measuring device.
   b. Connect the top-platen line to the drain line and close the valve (be sure this line is saturated).
   c. Now apply three increments of cell pressure $\Delta\sigma_3$ and record the corresponding pore pressure Δu. Compute the pore pressure parameter B as previously described. If B is on the order of 0.95 to 1.00 assume the sample is saturated and continue. If it is less and you want to test a saturated sample you must do the following:
   d. Check the cell pressure. Next open the base inlet valve from the saturation container so water can flow through the sample and drain through the outlet valve so you can observe the outflow.
   e. Now apply a pressure to the saturation container *that is about 1/2 to 3/4 the cell pressure.* This is called back pressure and will speed the saturation process. When you begin to get an outflow close the inflow valve and outflow valves and repeat Step c and see if B is in the area of 0.95 to 1.00. Recycle Steps c through e as necessary.

From this point on, refer to Steps 13 to 18 of Procedure A and to all of "C: Computations" following, as the work is identical.

C. COMPUTATIONS

1. Compute the unit strain ϵ from the deformation-dial readings as

$$\epsilon = \frac{\Delta L}{L_0}$$

Also compute area correction value $1 - \epsilon$ and enter in appropriate column of the data sheet. Compute the corrected area A' as

$$A' = \frac{A_0}{1 - \epsilon}$$

and record this on the data sheet.
2. Compute the deviator load using the load-dial readings unless a direct-reading load cell was used.
3. Compute the deviator stress and use Eq. (15-4) if necessary for the membrane correction $\Delta\sigma_c$

$$\Delta\sigma_1 = \frac{P}{A'} \quad \text{or} \quad \Delta\sigma_1 = \frac{P}{A'} - \Delta\sigma_c$$

and fill in the appropriate column of the data sheet.
4. Plot a curve of unit deviator stress vs. unit strain (on the abscissa) and obtain the stress at the peak point unless the stress at 15 percent strain occurs first. Show this

value of deviator stress on the graph. Plot all the tests on the same graph, with each curve identified for the corresponding lateral pressure σ_3.

5. With the maximum deviator stress from Step 4, compute the major principal stress for each test as

$$\sigma_1 = \sigma_3 + \Delta\sigma_1$$

Also compute the pore pressure Δu corresponding to the maximum deviator stress from a plot of deviator load vs. pore pressure. Omit the plot if the values are approximately constant at the failure load and use the closest measured value. Compute the effective principal stresses σ_1 and σ_3 as

$$\sigma_1' = \sigma_1 - \Delta u \qquad \text{and} \qquad \sigma_3' = \sigma_3 - \Delta u$$

6. Plot as specified either or both:
   a. Mohr's circles for both *total* and *effective* principal stresses on the same set of axes for all the tests. Fit a tangent (stress envelope) to each set of circles, and measure the slope to obtain the apparent and "true" angles of internal friction and the two values of soil cohesion. If the testing has been done on cohesionless soils, the cohesion should be quite small. If it is not, discuss reasons for the divergence. Be sure to show all values on the graph neatly identified.
   b. A p-q diagram for both total and effective stresses (see Test 15 under heading "Stress Paths"). Draw the K_f-lines and from the stress intercepts a and slope angles α compute the parameters c and ϕ for total and c' and ϕ' for effective values.
7. Compute the apparent and "true" angle of internal friction ϕ for the cohesionless-soil tests using Eq. (15-3), and compare with the results obtained in Step 6 above.
8. Show a neat line drawing of the laboratory test setup as part of the report regardless of type of soil tested.
9. If cohesive soils have been tested, be sure to comment in the report and show on the Mohr's circle plot the type of test performed (whether a CU or CD test) and whether the soil is saturated or unsaturated.
10. Compute the initial tangent modulus and the secant modulus using the origin and a curve intercept at 50 percent of ultimate strength, and compare values. Estimate Poisson's ratio using the method of Step 7 (Computations) of Test 15.

Using the Computer Program

The computer program used for Tests No. 14 and No. 15 can also be used for this project. If you use the program it is necessary to input the several non-zero values of pore pressure measured corresponding to the input values of deformation.

SAMPLE COMPUTATIONS:

There are no sample computations for this test. You might use the computer program to run that data set which includes pore pressures and verify select columns of data.

17: DIRECT-SHEAR TEST

References

ASTM D 3080-90

AASHTO T 236-90

ASTM (1964), *Symposium on Laboratory Shear Testing of Soils*, ASTM, STP No. 361.

ASTM (1952), *Direct Shear Testing of Soils*, ASTM STP No. 131.

Kiekbusch, E., M. Kiekbusch, and B. Schuppener (1979), "A New Direct Simple Shear Device," *Geotechnical Testing Journal, ASTM*, Vol. 2, No. 4, December, pp. 190–199.

Lee, K. L. (1970), "Comparison of Plane Strain and Triaxial Tests on Sand," *Jour. Soil Mech. and Found. Div., ASCE*, SM 3, May, pp. 901–923.

Saada, A. S. , and F. C. Townsend (1981), *State-of-Art: Laboratory Strength Testing of Soils*, ASTM, STP No. 740, pp. 7–77.

Objective

To familiarize the reader with a procedure for rapidly estimating the shear strength parameters (ϕ and c) of a soil.

Equipment

Direct-shear device (see Fig. 17-1)
2 Dial gauges or LVDTs (0.01 mm or 0.001 in)
Calipers
Small level

General Discussion

The direct-shear test imposes on a soil the idealized conditions shown in Fig. 17-2. That is, the failure plane is forced to occur at a predetermined location. On this plane there are two stresses acting—a normal stress σ_n due to an applied vertical load P_v, and a shearing stress s due to the applied horizontal load P_h. These stresses are simply computed as

$$\sigma_n = \frac{P_v}{A} \tag{17-1}$$

$$s = \frac{P_h}{A} \tag{17-2}$$

where A *is the nominal area of the sample (or of the shear box)* and is ***not corrected*** *for lateral displacement under shear force* P_h. These stresses are those of Coulomb's equation given in both Tests 15 and 16 and repeated here in terms of effective stress parameters as

$$s = c' + (\sigma_n - \Delta u)\tan\phi' \tag{17-3}$$

Note that we obtain "effective" stress parameters if Δu is measured and used in Eq. (17-3). If Δu is not measured and Eq. (17-3) corrected we obtain "total" stress parameters c and ϕ.

As there are two unknown quantities (c and ϕ) in Eq. (17-3), a minimum of two tests at different values of normal stress σ_n with measured shear stress s must be made so that

Figure 17-1
Direct-shear test equipment using variable speed motor (lower right) with enclosed chain drive. Note motor and drive sprocket out of top gearbox are used to motorize the unconfined compression machine of Fig. 14-1.

(*a*) One type of direct-shear machine. All machines are sin ilar in details. Here are shown a load ring and load ring di: (could use a load cell and digital equipment). Top dial is fc vertical sample compression to monitor consolidation. Di: on left measures the horizontal shear displacement.

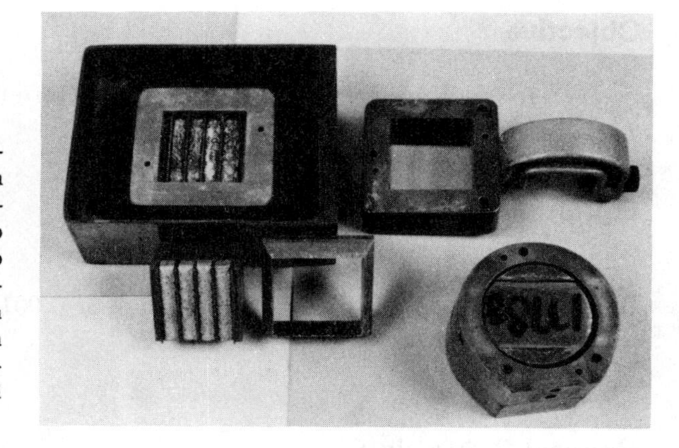

(*b*) Two shear boxes. Both round and square boxes are available. Shear-box base firmly attaches to machine, and with top half separated, relative movement to develop shear resistance is produced when a horizontal force is applied to the top half of the shear box. Serrated porous stones, to facilitate drainage, are usually used to confine the sample. The sample is usually on the order of 20 to 25 mm thick. Carefully inspect your system to determine how much of the loading block and top half of shear box are an additional sample load to be included in the total sample vertical load P_v.

the shear strength parameters c and ϕ can be computed. As before at least three tests should be done to check for test error or sample anomalies.

Since the shear stress s and normal stress σ_n have the same significance as when used in a Mohr's circle construction, rather than solving a series of simultaneous equations for c and ϕ, one may plot on a set of coordinate axes the values of shear stress s versus normal stress σ_n [or since P_h and P_v are both divided by the same sample area A in Equations (17-1) and (17-2) simply plot these]. Draw the shear envelope through this locus of points and extend the line to the shear axis to determine the cohesion c; the slope of this line is the angle of internal friction ϕ.

For cohesionless materials, the cohesion should be zero by definition and Eq. (17-3) becomes

$$s = \sigma_n \tan \phi \tag{17-4}$$

In this case there should be no difference between total and effective stress parameters. The sample is thin enough (20 to 25 mm thick) that even if it were saturated any excess pore pressure should dissipate (with two-way drainage) unless the test is done at an extremely high strain rate. Since high strain rates are not allowed, it is reasonable to assume "drained" test conditions.

Test inaccuracies and surface-tension effects of damp cohesionless materials may give a small "apparent" cohesion but this should be neglected unless it is more than 10 to 15 kPa.

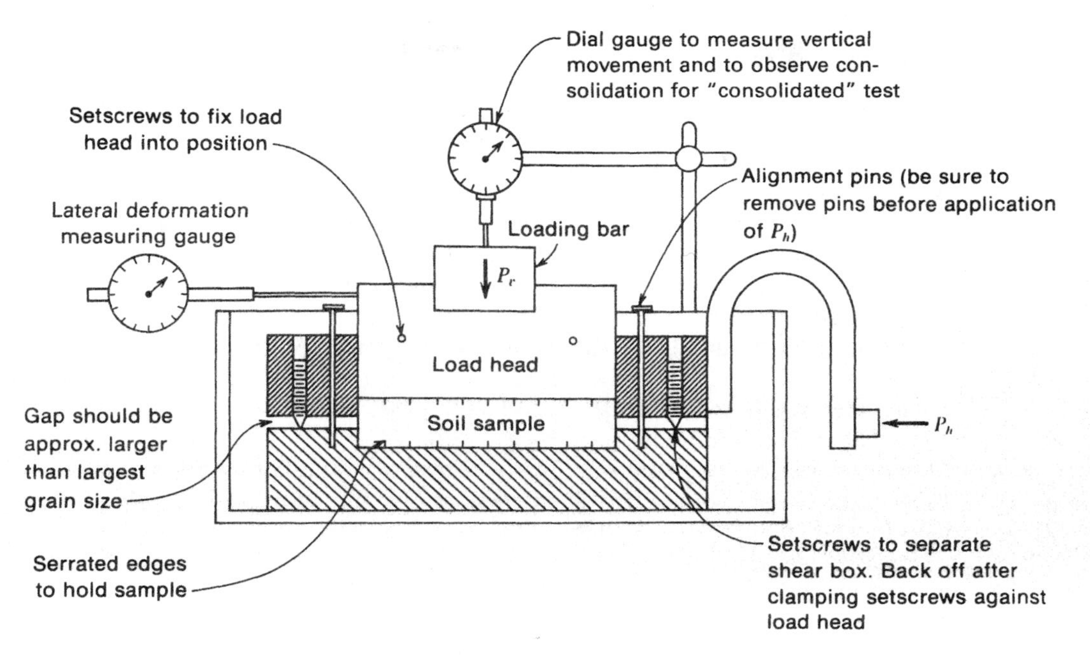

Figure 17-2
Line details of the direct-shear test.

If the cohesion intercept is large and the soil appears to be cohesionless you should investigate if the test has been incorrectly done.

The direct-shear test was formerly quite popular. Then, as the state of the art advanced, it tended to become less popular for several reasons:

1. The area of the sample changes as the test progresses but the change may not be very significant, as most samples "fail" at low deformations.
2. The actual failure surface is not plane, as is assumed or as was intended from the way the shear box was constructed, nor is the shearing stress uniformly distributed over the "failure" surface, as is also assumed.
3. The test uses a small sample, with the result that preparation errors can become relatively important.
4. The size of the sample precludes much investigation into pore-water conditions during the test.
5. Values of modulus of elasticity E_s and Poisson's ratio μ cannot be determined.
6. Triaxial test apparatus was developed.

With a further advance in the state of the art, however, the direct-shear test has regained some of its former popularity. Some of the reasons for this are as follows:

1. The triaxial test is much more difficult to perform and interpret—especially if pore pressures are measured.
2. The size of the sample makes it less time-consuming to perform consolidated-undrained and consolidated-drained tests. The time for sample consolidation under the normal load P_v is relatively short since there are two drainage faces yielding a drainage path on the order of 10 to 12 mm. Unless the coefficient of permeability k is extremely small, consolidation should occur in about 20 to 30 min—at most in 4 or 5 h.
3. Square sample boxes were introduced so that the reduction in area during the test can be easily accounted for, if desired. The use of square boxes is relatively recent, and

many older machines currently in service still use round sample boxes. In some cases it is only necessary to obtain the alternative box from the laboratory equipment supplier.

4. The direct-shear machine is quite adaptable to electronic readout equipment, so that an operator is not required to be on continuous duty for consolidated-drained tests, which may be of several days' duration if a very small horizontal shear displacement rate is used to minimize excess pore pressures during shear.
5. It has been found that the soil parameters c and ϕ obtained by the direct-shear test method are about as reliable as triaxial values[1] (probably this is more the result of an operator problem than of the equipment being equal in performance). This does not mean to imply that the triaxial test is undesirable; rather, if soil parameters are all that are desired, the direct-shear values have usually been found to be quite acceptable.

Direct-shear tests may be categorized as follows:

1. *Unconsolidated-undrained or UU tests.* Shear is begun before the sample consolidates under the normal load P_v. If the soil is cohesive and saturated, excess pore pressures may develop. This test is analogous to the UU triaxial test described in Test 15. In a practical sense the test is often begun as soon as the sample is put in the machine and the various dials initialized.
2. *Consolidated-undrained or CD tests.* The normal force is applied and the vertical dial movement is monitored until settlement stops before the shearing force is applied. This test is something between the CU and CD triaxial tests described in Test 15.
3. *Consolidated-drained tests.* The normal force is applied, and the shear force is delayed until all settlement stops; the shear force is then applied so slowly that the small pore pressure that develops in the sample can be ignored. This test is analogous to the consolidated-drained triaxial test.

For cohesionless soils, all three of the above tests give about the same results. They are almost independent of the saturation state unless the test is performed at a very high horizontal strain rate.

For cohesive materials, the soil parameters are markedly influenced by the test method, degree of saturation *and* whether the soil is *normally consolidated* or *overconsolidated.* Generally, two sets of shear-strength parameters can be obtained if the soil is overconsolidated: one set for normal loads less than the preconsolidation pressure p_c; a second set for normal loads $> p_c$. Where a preconsolidated soil is suspected, it may be necessary to perform six or more tests to ensure that the appropriate shear-strength parameters are obtained.

DIRECT SIMPLE SHEAR TEST

The test equipment and procedures described here are for a direct shear test. The *direct simple shear* (DSS) test details are described in Kiekbusch et al. (1979) and in Saada and Townsend (1981) together with limitations. The DSS uses a closed shear box fixed at the base with the top free to translate under a horizontal force. The shear box may

1. Use hinged sides
2. Use a wire-reinforced rubber membrane for the sides

The closed shear box configuration allows user control of pore pressures and also allows a reversing force to be applied to the top (with either the hinged or rubber membrane side configuration) so that cyclic stress effects can be measured. The test has found particular application in soil liquefaction studies. It is also used in lieu of direct-shear tests for soil stability/strength studies.

[1]On sands with $\phi > 35°$ the direct-shear test values may be from 1 to 4° larger than triaxial values [see Lee (1970)]; for $\phi < 35°$ the values are about the same by either test.

PROCEDURE

This is suggested as a laboratory section project.

A. COHESIONLESS SOIL—LABORATORY WORK

1. Weigh a large dish of dry sand of sufficient quantity to do at least three tests at the same density.
2. Carefully assemble the shear box (back off the sample box spacing and clamping screws as necessary) and fix into position. Obtain sufficient dimensions that the sample density can be computed. Obtain the sample cross-sectional area A. Saturate the porous stones as best you can if you use a damp or saturated sand.
3. Carefully place the sand in the shear box to about 5 mm from the top and place the load block (including porous stone) on top of the soil[2]. Take the small level and level the load block. Weigh the container of sand to determine mass of dry sand used in making up the sample. Obtain a reference thickness of the soil sample by marking the load block at several points around the perimeter with respect to the shear box.
4. Apply the desired normal load P_v and attach the vertical displacement dial gauge or LVDT (direct reading to 0.01 mm/div). Remember to include the mass of the load block and upper half of the shear box[3] as part of the sample load P_v.

   If this is a "consolidated" test observe the vertical displacement dial and commence the test after settlement has halted. For cohesionless soil this will be almost immediately after application of P_v.
5. Separate the two parts of the shear box by advancing the spacing screws in the upper half of the shear box. The space should be slightly larger (by eye) than the largest soil grains in the sample. Now set the load block by tightening the three lateral setscrews provided for that purpose in the sides of the upper box half. Next back off the spacing screws holding the upper half of the box. This half with load block and load is now carried by the soil area.
6. Attach the dial gauge or LVDT (direct reading to 0.01 mm/div) to measure the shear displacement.
7. For a saturated test, saturate the sample by filling the shear box with water, and allow a reasonable time for saturation to take place.
8. Start the horizontal (shear) loading and take readings of load dial, shear displacement, and, if required by the instructor, vertical (volume-change) displacements. If a strain-controlled test is performed take these readings at horizontal displacements of

   5, 10, and then every 10 or 20 horizontal dial displacement units.

   Use a strain rate on the order of 0.5 to not more than 2 mm/min. Do not use too fast a strain rate, or the shear load may peak between readings. The strain rate should be such that the sample "fails" in about 3 to 5 min.

   Take readings until the horizontal (or shear) load peaks then falls for two additional readings.
9. Remove the sand from the shear box and repeat steps 1 through 8 on at least two additional samples. These samples must have an initial mass within 5 to 10 g of the first sample and occupy the same volume (use the reference marks of Step 3).

   In Step 4 use a different value of P_v for each test (suggest doubling the exterior mass—say 4, 8 and 16 kg for the three tests). If you find that the shear load when using a larger P_v is less than before you must repeat the test—something has not been done correctly.

[2] If you have difficulty obtaining the desired density you may flood the box and tamp the soil. Use of damp soil may produce "bulking" from surface tension.

[3] Some shear machines allow taring this along with the load hanger (or yoke) so the added mass is the effective normal load P_v. If your machine allows this, be sure to do it to simplify your computations.

B: COHESIVE SOIL—LABORATORY WORK

1. Carefully trim three or four samples of the same size (and of the same mass or density) from the larger sample block, tube sample, or other sample source. Use a sample cutter so that the size can be accurately controlled for a good fit into the shear box and so that all samples are of the same thickness. Any sample whose mass is appreciably different (more than about 5 to 10 g) should be discarded and another sample trimmed.

   Note: You may need six samples if the soil is undisturbed and preconsolidated. Keep the samples in a controlled humidity while trimming, preparing the shear machine, and taking care of other test details.
2. Loosen the spacing and clamping setscrews in the top half of the shear box and assemble the two parts. Be sure that the porous stones are saturated unless you are testing a dry soil. Measure the shear-box dimensions in order to compute the area.
3. Carefully place the soil sample in the shear box. It should just fit into the box and be within about 5 mm from the top. Place the loading block in place, add the normal load P_v and attach the vertical dial gauge or LVDT.

   For a consolidated test, monitor the vertical displacement as for a consolidation test (Test 13) to determine when consolidation is complete.
4. Carefully separate the shear-box halves using the spacing screws to set a gap slightly larger than the largest soil grains present. Clamp the loading head in place using the setscrews for that purpose and then back off the separation screws. Be sure the normal load reflects the applied vertical load + the weight of the load block and top half of shear box[4].

   When testing soft clay be very careful in separating the shear box that material is not squeezed out between the two box halves—use of small vertical loads and/or consolidation prior to box separation may be required.
5. Attach the shear-deformation dial gauge or LVDT and set both vertical (if used) and horizontal gauges to zero. Fill the shear box with water for saturated tests and wait a reasonable time for saturation to be complete.
6. Start the horizontal (shear) loading and take readings of horizontal load, shear displacement, and vertical (volume-change) displacements. If a strain-controlled test is performed, take these readings at horizontal displacements of

   5, 10, and every 10 or 20 horizontal dial displacement units.

   Use a strain rate on the order of 0. 5 to not more than 2 mm/min. Do not use too fast a strain rate or the shear load may peak between load readings. The strain rate should be such that the sample "fails" in 5 to 10 min unless a CD test is being run.

   The strain rate for CD tests should be such that the failure occurs in a time t_f on the order of

$$t_f = 50 \cdot t_{50} \tag{17-5}$$

   where t_{50} is the time for 50 percent consolidation to occur under the normal load P_v. If t_{50} is not readily available use the formula

$$t_f = 35 \cdot t_{60} = 25 \cdot t_{70} = 12 \cdot t_{90} \tag{17-6}$$

   A plot of vertical dial reading vs. log time as for a consolidation test can be made to determine when the soil is completely consolidated. When P_v is very large, it may be necessary to apply the load in increments rather than all at once for reasons outlined in Test 13.

   Test the sample to two load values beyond the "failure" load. The failure load is the peak load obtained.

[4]See footnote 3.

7. Remove the soil and take a water-content sample. Repeat Steps 3 through 6 for two or more additional tests. If soil is preconsolidated and you use six tests, be sure to use a range of normal loads so you have three on each side of the preconsolidation pressure.

C. COMPUTATIONS

The following computations are applicable to either cohesionless or cohesive soil.

1. Compute the nominal normal stress as

$$\sigma_n = \frac{P_v}{A}$$

where A = cross-sectional area of shear-box soil sample

P_v = total normal load (and including load block and top half of shear box if it has not been tared out of the loading)

2. Plot the horizontal displacement σ_h, vs. horizontal shear force P_h to obtain the best value of ultimate shear force[5], and compute the maximum shear stress as:

$$s = \frac{P_{h(\max)}}{A}$$

Note: One may use the residual shear force (value somewhat less than the ultimate at a displacement beyond that for ultimate shear force) in this computation and make a plot to obtain the *residual* shear strength parameters.

3. Plot the value of maximum shear stress s vs. σ_n for the tests. Construct a best-fit straight line through the plotted points (see Fig. 17-5). Be sure to use the same scale for both the ordinate (stress s) and the abscissa (always σ_n). Obtain the cohesion (if any) as the intercept with the ordinate axis, and measure the slope of the line to obtain angle of internal friction ϕ.

   You can obtain the residual shear-strength parameters by plotting the residual shear stress s_r vs. σ_n.
4. On the graph of δ_h vs. P_h and using the same horizontal displacement scale, make a plot, if data taken, of vertical displacement vs. δ_h (refer to Fig. 17-4). This plot will display volume change vs. shear displacement. Make appropriate comments in your report concerning the magnitude and shape of the plot.
5. In your report make appropriate comments on the shear-strength parameters obtained. Consider whether you should have used a corrected area in computing the shear stress (and normal stress) or whether the results are conservative or unconservative without the correction for area. Comment on why it is necessary in Tests 14 through 16 to plot strain vs. stress to obtain the maximum stress when the maximum shear stress can be obtained from a plot such as Fig. 17-4 in this soil test.

SAMPLE CALCULATIONS:

Refer to Fig. 17-3 which is the first sheet of a series of three required for this test. Use one data sheet for each vertical load P_v (only the sheet for P_v = 5 kg is shown; sheets for P_v = 10 and 20 kg are omitted).

[5]Alternatively, you can plot horizontal displacement dial reading versus load dial units (or load if using a load cell) as in Fig. 17-4 to obtain the maximum shear force.

DIRECT SHEAR TEST (~~Cohesive Soil~~, Cohesionless Soil) **Data Sheet 17**

Project Direct Shear Test ~~Job No.~~ (Edited data)

Location of Project Soil Laboratory Boring No. — Sample No. —

Description of Soil Medium Coarse Sand Depth of Sample —

Tested By JEB & R.G.L. Date of Testing 1/14/19--

Sample Data: Soil state (wet, dry) Soil sample (disturbed, undisturbed)

Data to Obtain Sample Density if not an Undisturbed Sample

Initial mass container + soil = 1376.7 g
Final mass container + soil = 1236.0
Mass of soil used = 140.7 g

Shear specimen data
Sample Dimensions:
~~Dia. or~~ side = 5.08 × 5.08 cm
Ht. = 3.42 cm
Area = 25.81 cm²
Vol. = 88.26 cm³

Density: ρ_{wet} = — g/cm³ w = — % ρ_{dry} = 140.7/88.26 = 1.594 g/cm³

Normal load = 5 kg (49 N) Normal stress σ_n = 49×10/25.81 = 19 kPa

Loading rate = 0.50 mm/min Load ring constant LRC = 1.379 N/div

Vert. dial[a] reading (0.01 mm)	Vert. displace. ΔV, (mm)	Horiz. dial[a] reading (0.01 mm)	Horiz. displace. ΔH, (mm)	Corr.[b] area A'	Load dial[c] reading (N/div)	Horiz. shear force, (N)	Shear stress s, kPa
0	0	0	0	88.26	0	0	
+0.5	0.005	10	0.1		14	19.31	7.5
+2.5	0.025	20	0.2		19	26.20	10.2
+3.0	0.030	40	0.4		24	33.10	12.8
+2.0		60			25		
+2.0		75			26		
+1.5		100			29		
+1.5		150			30		
+1.5		175			31		
+1.5		200			31		
+1.5		250		↓	29.5		
+1.5		300		88.26	26		
	19.31(10)/25.81 = 7.5 kPa ... etc						
	Note: Plot on Fig 17-4 using dial readings						
	+ΔV = Sample expansion						

Note: Insert units in column headings as necessary; not all columns used with LVDT's or load cell.
[a]omit it using LVDT and enter displacements ΔV and ΔH.
[b]for square samples, may use corrected specimen area at failure as $A' = A_o - \Delta H$ to compute σ_n and s.
[c]omit if using a load cell and enter "Horiz. shear force."

Figure 17-3
Direct-shear test data for a cohesionless soil. Cohesive soil data are similar.

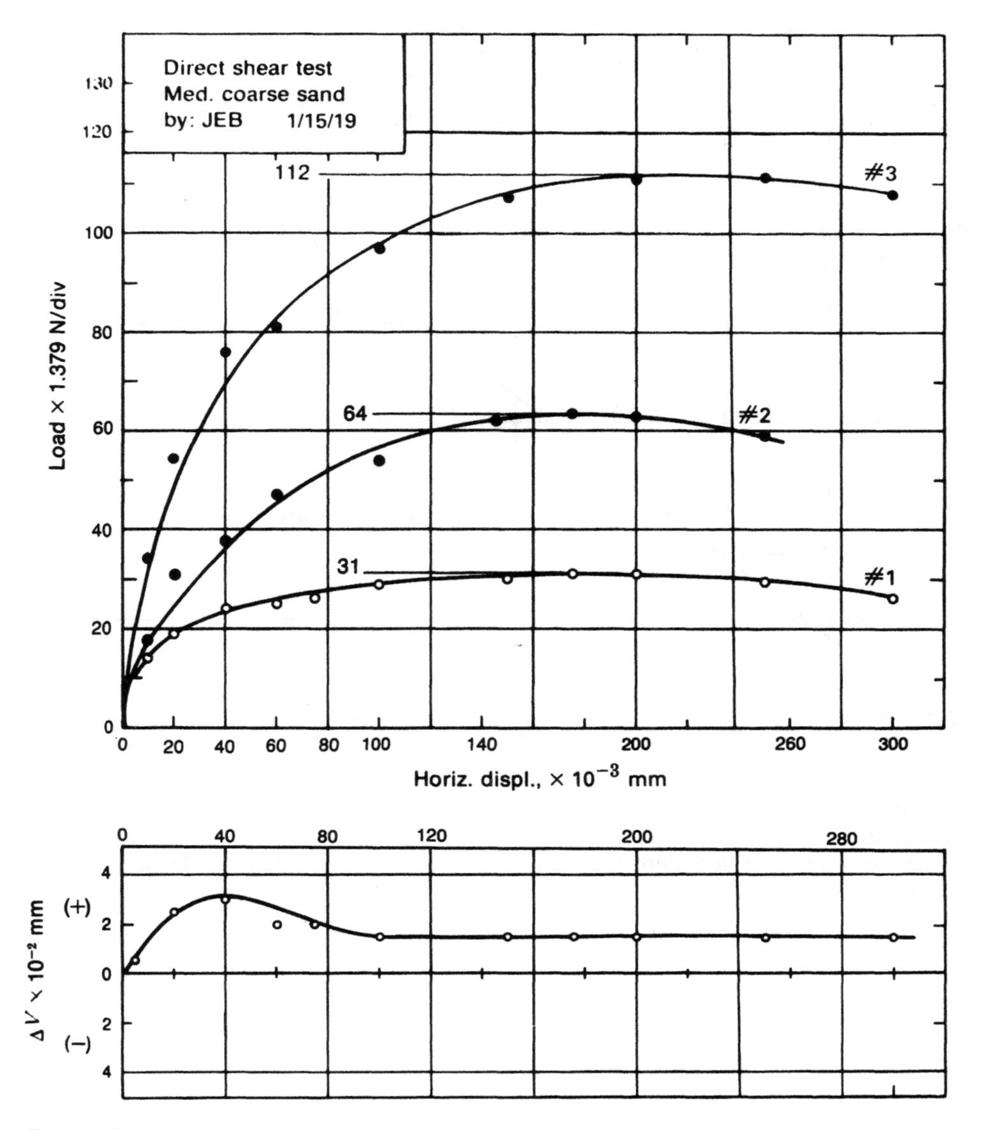

Figure 17-4
Plot of shear and volume change vs. horizontal displacement. Note use of load (not stress) and ΔV to save time since curve shape would be identical. Data shown for Test 1 is from Fig. 17-3.

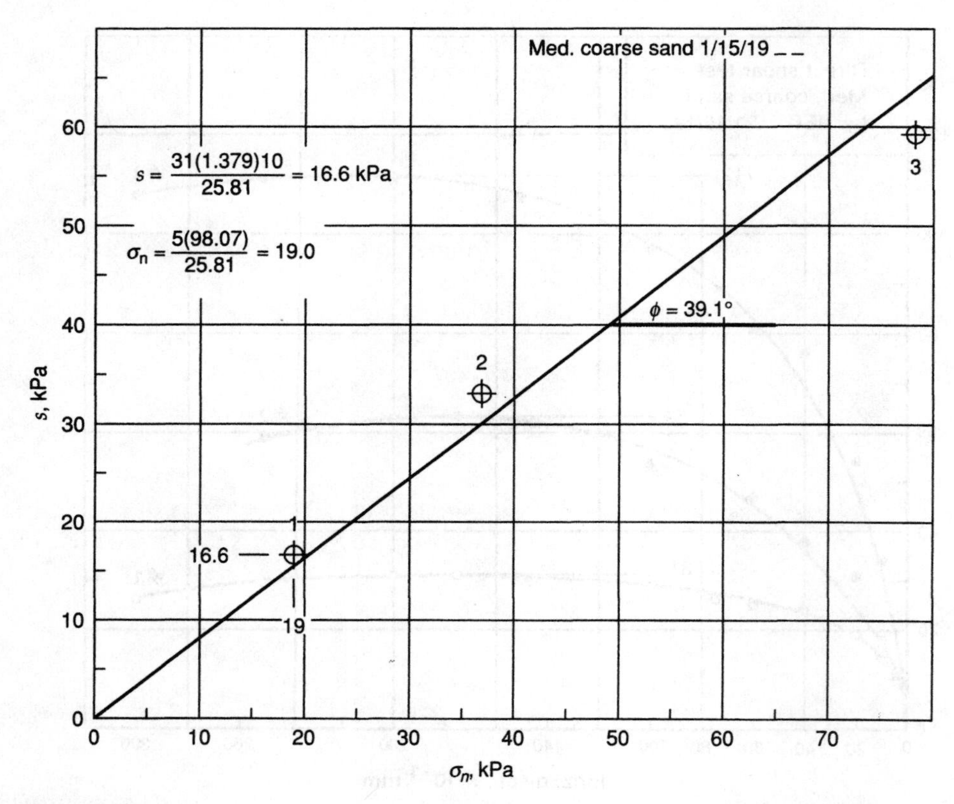

Figure 17-5
Plot of shear stress vs. normal stress to obtain soil parameter(s). The measured slope is ϕ and the *s-axis intercept is the cohesion c*. Here $c = 0$ since the soil is cohesionless. Above computations for s from Fig. 17-4 and σ_n from "normal load" recorded on Fig. 17-3.

18: RELATIVE-DENSITY DETERMINATION

References

This test does not have a current ASTM standards designation; the previous designation was D 2049-69.

ASTM D 4253-83 (Maximum Index Density)

ASTM D 4254-83 (Minimum Index Density)

ASTM (1973), *Relative Density of Cohesionless Soils,* ASTM, STP No. 523, 510 pages.

Objective

To determine the state of density of a cohesionless soil with respect to its maximum and minimum densities. The reader will also be introduced to a simple method of obtaining a compaction control density for a cohesionless soil.

Equipment

Standard compaction mold or a calibrated volume measure as used in concrete-mix design laboratories

Hand (as in Fig. 18-1) or mechanical vibration equipment

Container with a pouring spout (a glass cylinder or a large spoon); container should have a volume between 1500 and 2000 mL.

Figure 18-1
Relative density (or density) equipment suggested by the author. Equipment consists in 1000-cm^3 (or standard) compaction mold, rubber mallet to rap mold for vibration, steel straightedge to strike mold flush, and confining device constructed by welding a 10.2-cm diam plate to a section of pipe. One person confines the soil in the mold by pushing down on the pipe handle while another person sharply raps the mold with the rubber mallet.

General Discussion

The impact method for obtaining the moisture-density curve of Test 9 does not work well for cohesionless soils (either wet or dry). Some commercial laboratories obtain a control density in the laboratory for this type of material by filling a standard compaction mold in several layers, confining each layer in some manner (Fig. 18-1), and vibrating the mold by rapping it sharply on the sides with a rubber mallet. The largest density value obtained from several trials is then taken as the job control value.[1] All compaction-control (both here and in Test No. 9) is for "end results" criteria.

Some—but by no means all—authorities suggest that a better criterion for field control is the *relative density*, which has been defined[2] in terms of void ratios as

$$D_r = \frac{e_{max} - e}{e_{max} - e_{min}} \tag{18-1}$$

where e_{max} = void ratio in loosest state

e_{min} = void ratio in densest state, and

e = void ratio of the natural state

Since void ratios are difficult to obtain (they require specific gravity G_s of the soil) the equation form usually used for D_r is

$$D_r = \left(\frac{\rho_f - \rho_{min}}{\rho_{max} - \rho_{min}}\right)\left(\frac{\rho_{max}}{\rho_f}\right) \tag{18-2}$$

where ρ_{max} = maximum laboratory density (or unit weight)

ρ_{min} = minimum laboratory density (or unit weight)

ρ_f = field or in situ soil density (or unit weight)

The problems in this test arise from properly defining the soil in its densest and loosest states. For the most dense state (minimum void ratio e_{min} for ρ_{max}) ASTM specifies the use of a mechanical vibration device and mold. Under an arbitrarily selected confining pressure and length of time of vibration, the soil is defined as being in the most dense state.

While not overly scientific the author's method, suggested here, produces adequate results for the most dense state if several trials are made. Several trials (reusing the soil) can be made in relatively short time. One should always try to check any soil—here, doing several trials provides an adequate check.

The loosest material state is a maximum void ratio case. Any method that produces this state can be used. ASTM gives three methods. These follow, as well as a fourth alternative:

1. Place soil in the standard compaction mold using a container with a volume between 1500 and 2000 mL and a pouring spout. The spout should be at least 150 mm in length with an orifice on the order of 12 mm for material passing the No. 4 sieve and 25 mm for material smaller than 10 mm. This is ASTM Method A.
2. Fill a tube of diameter about 0.7 D_{mold} and a volume at least 1.25 V_{mold}. The tube is placed in the mold, filled with soil, and carefully (and slowly) withdrawn to allow the soil to flow out. This is ASTM Method B.

[1]A vibrating table available from laboratory suppliers is recommended by ASTM D 4253 for obtaining the maximum density. This table is expensive and, although having the prestigious endorsement of ASTM, does not produce significantly increased densities than those obtained using the simple equipment of Fig. 18-1.

[2]The origin of this equation is somewhat obscure. It is believed to have been first proposed by the late D. M. Burmister, "The Importance and Practical Use of Relative Density in Soil Mechanics," *Proceedings, ASTM*, Vol. 48, pp. 1249–1268 (1948).

3. Weigh 1000 g of oven-dry material and put in a 2000-mL graduated cylinder nearly filled with water. Cap the cylinder with a rubber stopper. Now turn it upside down and back and set on work table. Record the volume the sand occupies. Do this at least two more times and record the average volume. Compute the minimum density as

$$\rho_{\min} = 1000/\text{Volume}_{\text{ave}}$$

   This is ASTM Method C.
4. This fourth method is not an ASTM procedure but is suggested as an alternative. Place your weighed mold + base plate on the workbench with the collar attached. Using a large spoon, carefully spoon oven-dry material from the stock source into the mold until it extends into the collar. Carefully remove the collar with minimum mold vibration. Take a steel straightedge and strike the mold flush. Now you can handle the mold of sand with less care. Weigh the mold with sand (and the base plate) and record the mass of sand in the mold. Pour the sand back into the stock container and remix. Repeat this procedure at least two more times and reject any trials which differ by more than 10 to 15 g. Average the three best trials to obtain the minimum mass. The minimum density is

$$\rho_{\min} = \text{Mass}/V_{\text{mold}}$$

The term "relative density" is not used for cohesive soils. Instead density, unit weight and soil consistency (soft, stiff, hard, etc.) are generally used to describe the soil state.

TEST PRECISION

When tests for maximum and minimum density determinations were first proposed, control samples were sent to a number of government and commercial laboratories for comparison of values. From an analysis of these tests ASTM suggests the following may be obtainable:

Density	Between labs	Same technician
minimum	0.27 to 0.40 kN/m^3	0.07 to 0.14 kN/m^3
maximum	0.38 to 0.71 kN/m^3	0.13 to 0.22 kN/m^3

These are stated as ± values, however, in the author's experience one is most likely to be (+) on minimum densities (i.e., density is too large) and to be low (−) on maximum values (i.e., obtain 16.5 when the value is 17.0 kN/m^3).

The field value is subject to the greatest potential variation. While it is possible to obtain a field density using the methods of Test 10 it is only for that test location. Normal project density variation is likely on the order ± 3 to 5 percent. If one plots the maximum, minimum and the ± density range on the relative density graph provided with this test and on it also plot the field density and ± range the extreme cases makes the relative density of little value since it can vary as much as 50 percent.

Procedure

This is suggested as a group project.

1. Each individual should take a representative sample of the oven-dry cohesionless soil from the stock supply furnished. Be sure to break any remaining lumps prior to use.
2. Obtain a standard compaction mold. Use the same mold for both the maximum- and minimum-density determinations so that you do not have to obtain mold dimensions and compute the volume.

3. Make three trials of maximum density by placing the material in the standard mold in five layers, confining each layer with a round steel block of at least 12 kg, or obtain the assistance of another student to confine the soil using the confining plate shown in Fig. 18-1 while you rap the sides of the mold sharply 15 to 25 times (as specified by instructor) with a rubber mallet. Be sure to record the number of blows and number of layers in your report. After each trial, carefully remix the test soil with that remaining in the container for the next test.

   Alternatively, use the standard ASTM equipment for this test (if allowed and available). If this equipment is used, refer to ASTM D 4253 (or instructor notes) for specific steps.

   Use the average of these densities as the maximum density of the soil. Put this value on the blackboard or where specified by instructor.
4. Next, using the same soil and mold, determine the minimum density. Use Method A, B, C or the author's suggested method depending on the equipment available. In any case do at least three trials (or more if any are more than 10 to 15 g different).

   Average the trials as the minimum density for this soil. Put this value on the blackboard beside your maximum density. Be sure to identify your work so the instructor will know who has not completed the assignment.

Note: There are no data sheets for this project but there is a sheet (see Fig. 18-2) for plotting the maximum, minimum and a field density (or unit weight) so you can ascertain the probable range of D_r for the soil.

Computations

1. Each student is to compute the standard deviation $\overline{\sigma}$ of the maximum and minimum class densities separately[3]. Use $\overline{X}$ = sample mean of the class for the largest and least density values and not the mean of the two populations.

$$\overline{\sigma} = \pm\sqrt{\frac{\sum(\overline{X^2} - X^2)}{N - 1}} \qquad (18\text{-}3)$$

   where N = number of density tests. (If $N > 5$, use for the denominator $N - 1$. If $N \leq 5$, use N for the denominator.)

   X = any density-test values

   Note that the standard deviation is $\pm\sigma$ and this is the likely range of variation in maximum and minimum densities for your laboratory group.
2. Assume a field density that is the average of ρ_{max} and ρ_{min} and has $\pm\overline{\sigma}$ as computed above. Plot these data on one of the relative density graphs from your data set and scale the range in D_r (see Fig. 18-2). Comment on this range.
3. Assuming you are the engineer for a project where this soil is to be used, what would you recommend for quality control to assure that the desired density is obtained? Would you recommend a dry unit weight γ_d (and if so what) or would you recommend a D_r and if so what value?
3. Derive Eq. (18-2) from Eq. (18-1).
4. Comment on ways in which this test (as you did it) might be improved.
5. Comment on use of $N - 1$ instead of N in computing the standard deviation.

[3]The instructor should list all density values and make a copy for each student when all of the laboratory work has been done by the entire class—if the same soil is used for all lab sections.

RELATIVE DENSITY Data Sheet 18

Project: Range of D_r example — Job No. —

Location of Project: — Boring No. — Sample No. —

Description of Soil: Cohesionless — Depth of Sample: —

Tested by: JEB — Date of Testing: ________

To Use: Plot maximum γ or ρ on right and minimum on left; connect with a straight line. Plot field values on both sides and connect. At intersection of two lines drop perpendicular and read D_r.

Example:

$\gamma_{max} = 17.0 \pm 0.4$ kN/m³

$\gamma_{min} = 13.5 \pm 0.4$

$\gamma_{field} = 15.5 \pm 0.3$

Required: Plot data and find possible range of D_r

Solution: From plot $0.43 \leq D_r \leq 0.80$

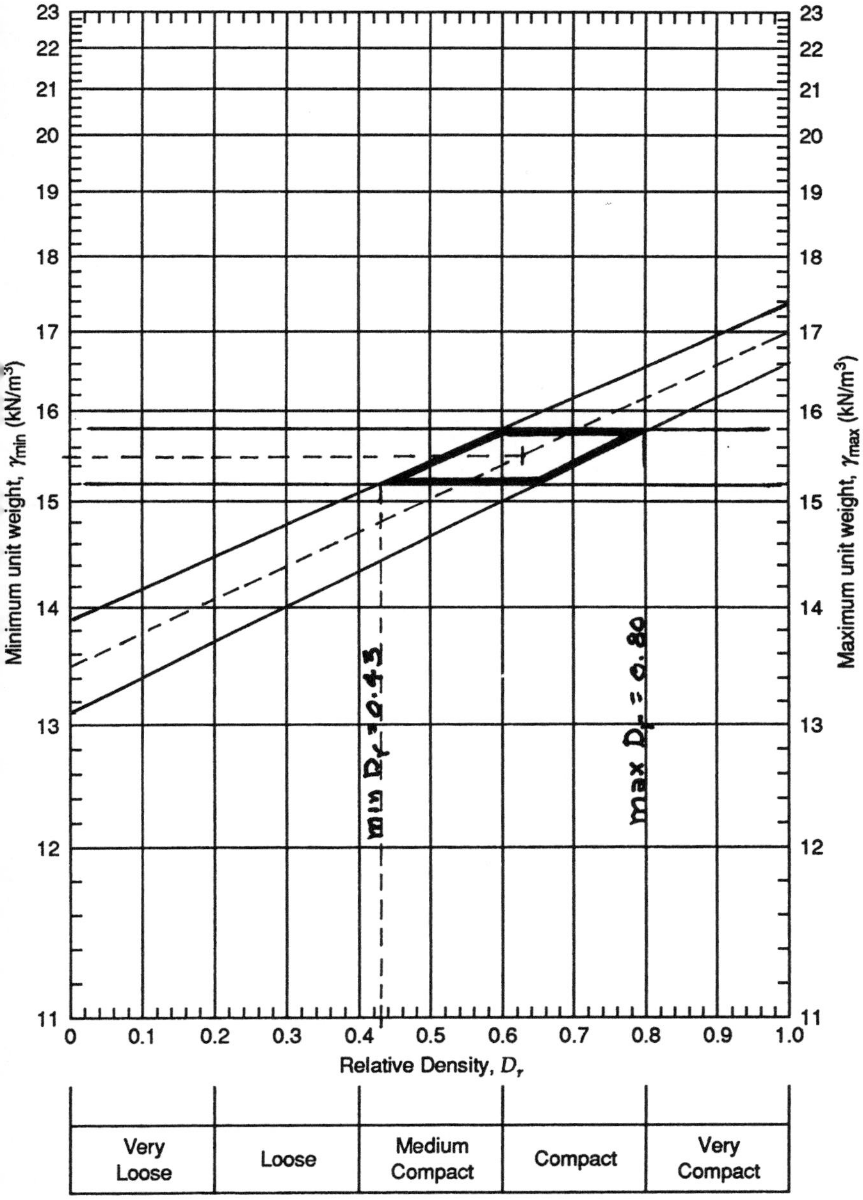

Figure 18-2
Chart for finding relative density range D_r and using sample data given above. Value range plotted solid and given values are dashed. Note that the graph versus computed values may differ by ±0.02, which is negligible.

19: CALIFORNIA BEARING-RATIO (CBR) TEST

References

ASTM D 1883-87
AASHTO T 193-81 (1990)

Objective

To introduce the reader to a method of evaluating the relative quality of subgrade, subbase, and base soils for pavements.

Equipment

CBR equipment (see Fig. 19-1) consisting of 152-mm diam × 178-mm height CBR compaction mold with collar and spacer disk 151-mm diam × 61.4-mm height (or 51-mm height as available)
Compaction rammer (either 24.5 N or 44.5 N as designated by instructor)
Expansion-measuring apparatus with dial gauge reading to 0.01 mm
Surcharge weights as required
Compression machine equipped with CBR penetration piston (49.63-mm diam with cross-sectional area of 19.35 cm^2) and capable of a penetration rate of 1.3 mm/min (0.05 in/min)

General Discussion

The California bearing-ratio test was developed by the California Division of Highways in 1929 as a means of classifying the suitability of a soil for use as a subgrade or base course material in highway construction. During World War II, the U.S. Army Corps of Engineers adopted the test for use in airfield construction.

The CBR test is currently used in pavement design for both roads and airfields. Some Departments of Transportation use the CBR directly. Others convert the CBR value to either the modulus of subgrade reaction k_s or to the Resilient Modulus M_R using empirical relationships. For example, the AASHTO converts CBR to M_R using

$$M_R = 1\,500 \cdot \text{CBR} \quad (\text{lbs/in}^2)$$

$$M_R = 10\,340 \cdot \text{CBR} \quad (\text{kPa})$$

Since the CBR is a percentage and dimensionless, the coefficient includes the units so that $1500 \cdot 6.895 = 10\,340$ kPa and similar.

The laboratory CBR test measures the shearing resistance of a soil under controlled moisture and density conditions. The test yields a bearing-ratio number that is applicable for the state of the soil as tested. The test can also be performed in the field on an in-place soil, but this is beyond the scope of this laboratory manual.

The CBR number (or, simply, the CBR) is obtained as the ratio of the unit stress (in MPa) required to effect a certain depth of penetration of the standard (with an area of 19.35 cm^2) into a compacted specimen of soil at some water content and density to the *standard unit stress* required to obtain the same depth of penetration on a standard sample of crushed stone. In equation form, this is

$$\text{CBR} = \frac{\text{Test unit stress}}{\text{Standard unit stress}} 100 \quad (\%) \qquad (19\text{-}1)$$

Figure 19-1
California Bearing-Ratio (CBR) equipment. Mold, collar, and dial gauge holder for soaking test on left; 24.5-N compaction rammer, base plate, 51-mm-thick spacer disk on base plate. Also shown are the swell plate (dark item in foreground) and surcharge weights to confine sample during soaking.

From this equation it can be seen that the CBR number is a percentage of the standard unit load. In practice, the percentage symbol is dropped and the ratio is simply noted as a number, such as 3, 45, 98. Values of standard unit stress to use in Eq. (19-1) are as follows:

PENETRATION		STANDARD UNIT STRESS	
mm	in	MPa	psi
2.5	0.10	6.9	1 000
5.0	0.20	10.3	1 500
7.5	0.30	13.0	1 900
10.0	0.40	16.0	2 300
12.7	0.50	18.0	2 600

The CBR number is usually based on the load ratio for a penetration of 2.5 mm (0.10 in). If, however, the CBR value at a penetration of 5.0 mm is larger, the test should be redone (ordinarily). If a second test yields also a larger CBR number at 5.0 mm penetration, the CBR for 5.0 mm should be used.

CBR tests are usually made on test specimens at the optimum moisture value for the soil as determined using the standard (or modified) compaction test of Test No. 9. Next, using Method B or D of ASTM D 698 or ASTM D 1557 [for the 152-mm (6-in) diam CBR mold] the specimens are made up using the compaction energy shown[1]:

Method	Blows	Layers	Hammer weight, N
D 698 Method			
B (fine-grained soil)	56	3	24.5
D (coarse soil)	56	3	24.5
D 1557 Method			
B (fine-grained soil)	56	5	44.5
D (coarse soil)	56	5	44.5

[1]For using the standard CBR mold and spacer disk; adjust the number of blows per layer if using the 51-mm thick spacer disk (see data sheet of Fig. 19-6).

Two molds of soil are usually compacted, the first for immediate penetration testing. The second mold is soaked for a period of 96 h with a surcharge approximately equal to the pavement pressure used in the field, *but in no case is the surcharge mass less than 4.54 kg*, and then tested. Swell readings are taken during this period at arbitrary selected times. A penetration test is then made at the end of the soaking period to obtain a CBR value for the soil in a saturated condition.

For the immediate CBR penetration test a surcharge of the same magnitude as for the swell test is placed on the soil sample.

The test on the soaked sample provides:

1. information concerning expected soil expansion beneath the pavement when the soil becomes saturated.
2. an indication of strength loss from field saturation.

Penetration testing is accomplished in a compression machine using a strain rate of 1.3 mm/min. Readings of load vs. penetration are taken at each 0.5 mm of penetration to include the value of 5.0 mm, and then at each 2.5-mm increment thereafter until the total penetration is 12.5 mm (or 0.5 inch).

The CBR number is used to rate the performance of soils primarily for use as bases and subgrades beneath pavements of roads and airfields. The following table gives typical ratings[2].

CBR No.	General rating	Uses	CLASSIFICATION SYSTEM: Unified	CLASSIFICATION SYSTEM: AASHTO
0–3	Very poor	Subgrade	OH, CH, MH, OL	A5, A6, A7
3–7	Poor to fair	Subgrade	OH, CH, MH, OL	A4, A5, A6, A7
7–20	Fair	Subbase	OL, CL, ML, SC, SM, SP	A2, A4, A6, A7
20–50	Good	Base or Subbase	GM, GC, SW, SM, SP, GP	A1*b*, A2-5, A-3, A2-6
> 50	Excellent	Base	GW, GM	A1*a*, A2-4, A3

Many paving-design procedures are published in which one enters a chart with the CBR number and reads directly the thickness of subgrade, base course, and/or flexible pavement thickness based on the expected wheel loads[3]. Sometimes the CBR is converted to a subgrade modulus k or to some other type of strength parameter before entering the paving design charts (one conversion was given earlier).

Procedure

This is suggested as a laboratory group project.

A: LABORATORY WORK

1. Prepare enough soil to compact two CBR molds of soil [approximately 12 kg of fine-grained (−) No. 4 sieve material or 15 kg of 19-mm maximum-size material] at the optimum moisture content of the soil as determined by the appropriate compaction effort (compaction test).[4]

[2]See *The Asphalt Handbook*, The Asphalt Institute, 1970, Chap. 5.

[3]See E. J. Yoder, *Principles of Pavement Design*, 2d. ed., John Wiley & Sons, Inc., New York, 1975, Chaps. 14 and 15. See also, "Development of CBR Flexible Pavement Design Method for Airfields—A Symposium," *Transactions ASCE*, Vol. 115, pp. 453–489 (1950).

[4]For an introductory student laboratory you may use a convenient compacting water content rather than the OMC.

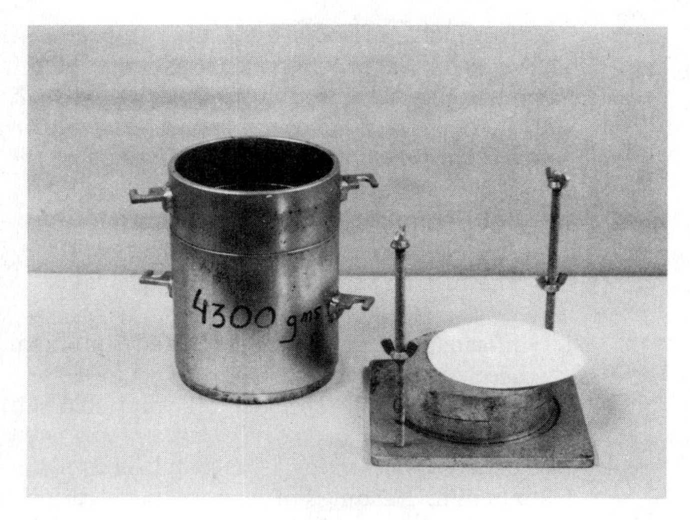

Figure 19-2
Mold, collar, and base plate for preparing a sample. Place spacer disk on perforated base plate and cover with a piece of filter paper.

If it is desired to cure the soil for a more uniform moisture distribution, mix the soil with the necessary amount of water to obtain the requisite percentage of moisture and store in a sealed container for about 12 to 24 h prior to the test.

2. Just before making the compacted molds of soil, take a representative water-content sample (at least 100 g for fine-grained soils). Since compacting the two molds will take some time be sure to keep the soil container covered so the soil does not dry significantly before it is used.
3. Weigh the two molds less base plates and collars.
4. Clamp a mold to the base plate, insert the spacer disk (the 151 × 51 mm solid round disk) in the mold, and cover with a piece of filter paper (refer to Fig. 19-2).
5. Compact the soil according to ASTM D 698 or D 1557 Method B or D for the soil used or *as specified by the instructor.* Strictly, the compaction effort should reproduce the design field density but this is not easy and may require a number of trials.

   Note: If this is done as individual or group projects rather than a lab section project, each group should use a different compaction effort, perhaps as:

   36 blows/3 layers   56 blows/3 layers
   36 blows/4 layers   56 blows/ 4 layers, etc.

6. Remove the collar, and trim the sample smooth and flush with the mold. Patch any holes formed from aggregate displacement when scraping the specimen flush with the mold. Use any soil particles that are smaller than the hole to be filled for patching.
7. Remove the base plate and spacer disk, weigh the mold + compacted soil, and compute the wet mass, density and unit weight and record on your Data Sheet 19*a*.
8. Repeat Steps 4 through 7 for a second mold of soil which will be soaked.
9. Place a piece of filter paper on the base plate, invert the specimen (so the spacer gap is on top), and attach the base plate so the soil is in contact with the filter paper on the base.

For the unsoaked mold of soil, do Steps 10 to 12 next; for soaked samples, proceed to Step 13.

10. Place sufficient slotted weights (but not less than 4.5 kg) on the sample to simulate the required overburden pressure. Be sure this mass is the same as will be the surcharge during soaking of the second sample. Note that a 260-mm concrete pavement slab requires about 11.7 kg of surcharge.

11. Place the specimen in the compression machine and seat the piston using a seating load no greater than 4.5 kg (Fig. 19-3). Set the load dial or load readout to zero and also zero the penetration (or deformation) dial or LVDT readout.
12. Take penetration readings as outlined in the "General Discussion," together with the corresponding load readings and record on Data Sheet 19*b*. Extrude the sample, split and take six moisture samples as follows:

    Two within the top 30 mm of soil
    Two within the bottom 30 mm of soil
    Two at the center of the sample

    Cover until they can be weighed. After weighing remove the lids and put in the oven to determine the compacted sample water content along with the initial water content from Step 2.

For the soaked sample:

13. Place a piece of filter paper on top of the compacted sample and then cover with the perforated plate with adjustable stem. Add sufficient additional slotted weights to obtain the desired surcharge within 2.2 kg but with a total surcharge weight of not less than 4.5 kg. Record the total surcharge mass (be sure to include the perforated plate as part of the surcharge mass).
14. Immerse the mold and weights in a container of water so the water covers the top of the sample to a depth such that any evaporation between readings will not allow the sample to dehydrate.

Figure 19-3
Setup for CBR penetration test. The enlargement on the penetration piston produces the standard 19.35-cm^2 area required. There are slots or holes in the surcharge weights (left on during penetration) so that piston can make contact with the soil.

Next attach the dial gauge (reading to 0.01 mm) in its holding bracket (see Figs. 19-4 and 19-5). Mark the mold where the bracket is placed so that it can be re-placed properly between readings—it may be used for more than one mold. The stem of the perforated plate is adjustable so the dial gauge can be initially set to zero for all of the tests and readings will be valid as long as the bracket is put in the same place.

15. Set the swell gauge to zero by adjusting the stem on the perforated plate and record the time of the start of the test. Take readings at 0, 1, 2, 4, 8, 12, 24, 36, 48, 72, and 96 h of elapsed time. The swell test may be terminated after 48 h if the swell-dial reading has been constant for the last 24 h.

Figure 19-4
Setup for the soaking test. Water in the laboratory sink just covers the surcharge plates resting on the swell plate. The swell plate extension rod makes contact with the dial gauge extension to measure swell. A standard dial gauge frame is used here. Its position on the CBR mold should be marked so that it can be removed (for use in measuring swell of other tests in progress) and replaced to the same position.

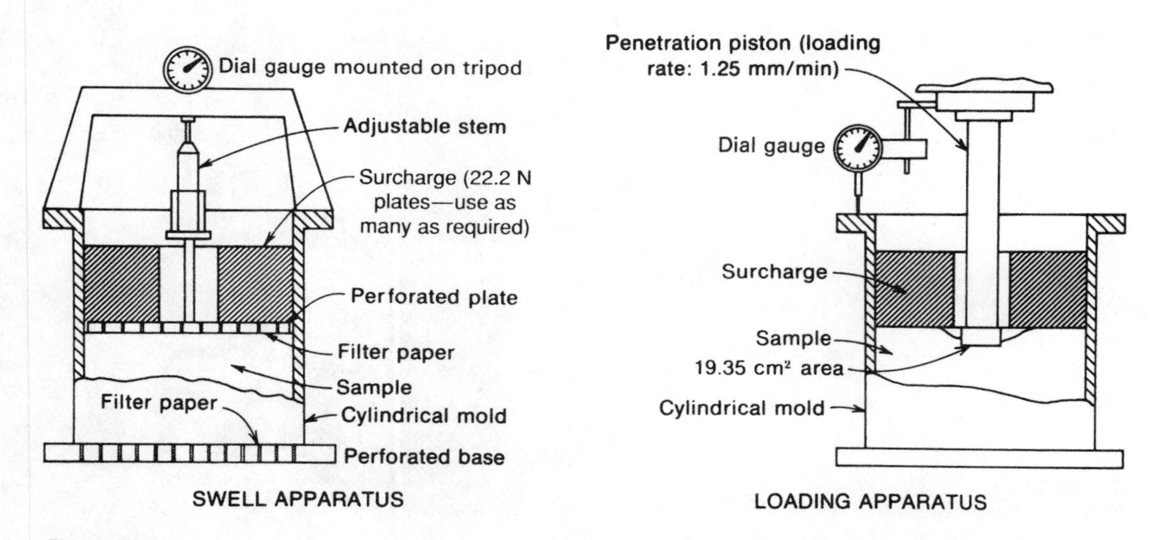

Figure 19-5
Line details of the CBR test. Use 2.3 kg (22.2-N) surcharge plate weights so that they can be easily weighed on the 2600 g laboratory balances.

16. At the end of the soaking period (96 h or other), remove the sample and let it drain for 15 min. Blot the free surface water from the top surface of the sample with paper towels.
17. Weigh the soaked sample including the mold and record on your Data Sheet 19*a*.
18. Do Steps 10 to 12 (i.e., take penetration and load readings and water content samples). Record these data on Data Sheet 19*b* similarly to Fig. 19-7.

B. CALCULATIONS AND DATA PRESENTATION

1. Plot a curve of penetration resistance (the ordinate) in kilopascals (kPa) vs. penetration in millimeters for both the freshly compacted and the soaked samples. If the curve is not essentially linear through the origin, extend a line from the straight-line portion to intersect the abscissa. The difference between this value and zero penetration is a correction to apply to compute the CBR value.

   Curves for both samples (soaked and unsoaked) should be plotted on the same graph and clearly identified, together with the curve correction values, so that one may readily observe the effect of soaking the sample. Refer to Figs. 19-6 to 19-8 for computations and presentation of data. Use a sheet of graph paper from the data sheet section.
2. Obtain the penetration resistance for 2.5 and 5.0 mm from the curve (using corrections from Step 1 above, if necessary) and compute the two CBR numbers using Eq. (19-1).

   Note: For student laboratory work, if the CBR at 5.0 mm is larger than at 2.5 mm penetration, the test need not be redone but *both CBR values should be reported.*
3. Compute the average water contents and dry densities and unit weights of both samples before soaking and of the soaked sample after soaking.
4. Compute the percentage of swell based on the nominal initial height of the sample. Plot a curve of percent swell (ordinate) vs. elapsed time on a separate sheet of graph paper.
5. The report should compare the two CBR values, present a summary of the water contents properly identified, and include the required curves. Discuss the significance of a large or small CBR value, any changes in CBR with soaking, and any swell that occurred.

   How might the swell be reduced (if there is a significant amount) or eliminated for this soil?

   Be sure your report includes classification data and the AASHTO classification of this soil.

BEARING RATIO TEST Data Sheet 19a

Project: CBR Test — Job No.: —

Location of Project: Soil Laboratory

Description of Soil: Silty blue Clay A-4(8) $w_L = 28.4\%$ $w_P = 19.8\%$

Tested By: JEB & RGL — Date of Testing: 6/16/19--

Compaction Energy: Rammer 24.5 N — No. of layers 5 — Blows/layer 40

w at compaction 18.7 % — Mold diam. 15.2 cm — Ht. of soil 12.7 cm

Vol. 2305 cm³ — used 5 cm spacer disk

Swell Data — Note: both molds same size

Starting time and date	Elapsed time	Mold no. 1, Surcharge 44.5 N: Dial reading[a] (×0.01 mm)	$\% = \frac{S}{H}(100)$	Mold no. 3, Surcharge ___ N: Dial reading (×___)	$\% = \frac{S}{H}(100)$	Mold no. ___, Surcharge ___ N: Dial reading (×___)	$\% = \frac{S}{H}(100)$
6/16 10:30 a.m.	0h	0	0	Not Soaked			
	1h	0.047	0.037%				
	2h	0.071	0.056				
	4h	0.108	0.085				
	14	0.140	0.110				
6/17 10:30 a.m.	24	0.190	0.150				
12 midn.	37.5	0.272	0.214				
6-18 12 noon	49.5	0.300	0.236				
12 mid	61.5	0.325	0.256				
6-19 12 noon	73.5	0.335	0.264				
6-20 1:00 P.m.	98.5	0.355	0.280				

[a] load dial units or direct load cell readings.

After Soaking

Mold No.	1	3	
Surcharge, N	44.5		
Initial mass of wet soil + mold + base plate	12056 g.	12095 g	
Final mass of wet soil + mold + base plate	12130	-	
Mass of mold + base plate	7352	7374	
Initial mass of wet soil, M_i	4704 g	4721 g	
Initial wet density, ρ_i	2.041 g/cm³	2.048 g/cm³	
Mass of water absorbed, M_w	74 g		
% water absorbed = $M_w/M_s \times 100$	1.87 % →	$\frac{74 \times 100}{4704/1.187}$ (1 + w)	

Note: Insert units in column headings as necessary.

Figure 19-6
Data from a CBR swell test.

BEARING RATIO TEST Data Sheet 19*b*

Project CBR Test Job No. —

Test Information A-4(8) $w_L = 28.4$ $w_P = 19.8$ CBR Piston $A_p = 19.35\ cm^2$

Tested By JEB & RGL Date of Testing 6/16/19 -- unsoaked 6/20/19 -- soaked

CBR Test Load Data

Penetration, mm	(~~soaked~~, unsoaked) Mold no. 3 Surcharge 44.5 N		(soaked, ~~unsoaked~~) Mold no. 1 Surcharge 44.5 N		(soaked, unsoaked) Mold no. ____ Surcharge ____ N	
	Piston load dial reading (5.678 N/d)	Stress, kPa	Piston load dial reading (5.678)	Stress, kPa	Piston load dial reading (____)	Stress, kPa
0.000	0	0	0	0		
0.5	100	293	33	97	$\sigma = \frac{DR(5.678)(10)}{19.35}$	
1.0	154	452	57	167		
1.5	188	552✓	78	229	$\sigma = \frac{188(5.678)(10)}{19.35}$	
2.0	222	651	96	282	= 552✓ kPa	
2.5	251	737	113	332		
3.0	277	813	131	384		
4.0	301	883	149	437		
5.0	347	1018	181	531		
6.0	388	1138	204	599		
7.5	429	1260	249	731		
9.0	494	1450	274	804		
10.0	525	1540	290	851		
12.5	635	1863	332	974		
Actual water content, *w*% (*use water content sheets*)	Top $\frac{1}{3}$	18.1%	Full sample oven-dried			
	Middle $\frac{1}{3}$	18.5				
	Bottom $\frac{1}{3}$	19.1				
	Average	18.6% (18.7)		21.5%		

(Annotation: "Dial readings" with arrows pointing to the piston load dial reading columns.)

Wet density, ρ_{wet} = 2.048 ← P 193 → 2.041 ____

Dry density, ρ_d = 1.727 g/cm³ 1.680 = $\frac{2.041}{1+0.215}$ ____

Note: Insert units in column headings as necessary.

Figure 19-7
Load-penetration data for both an unsoaked and a soaked CBR test as normally performed.

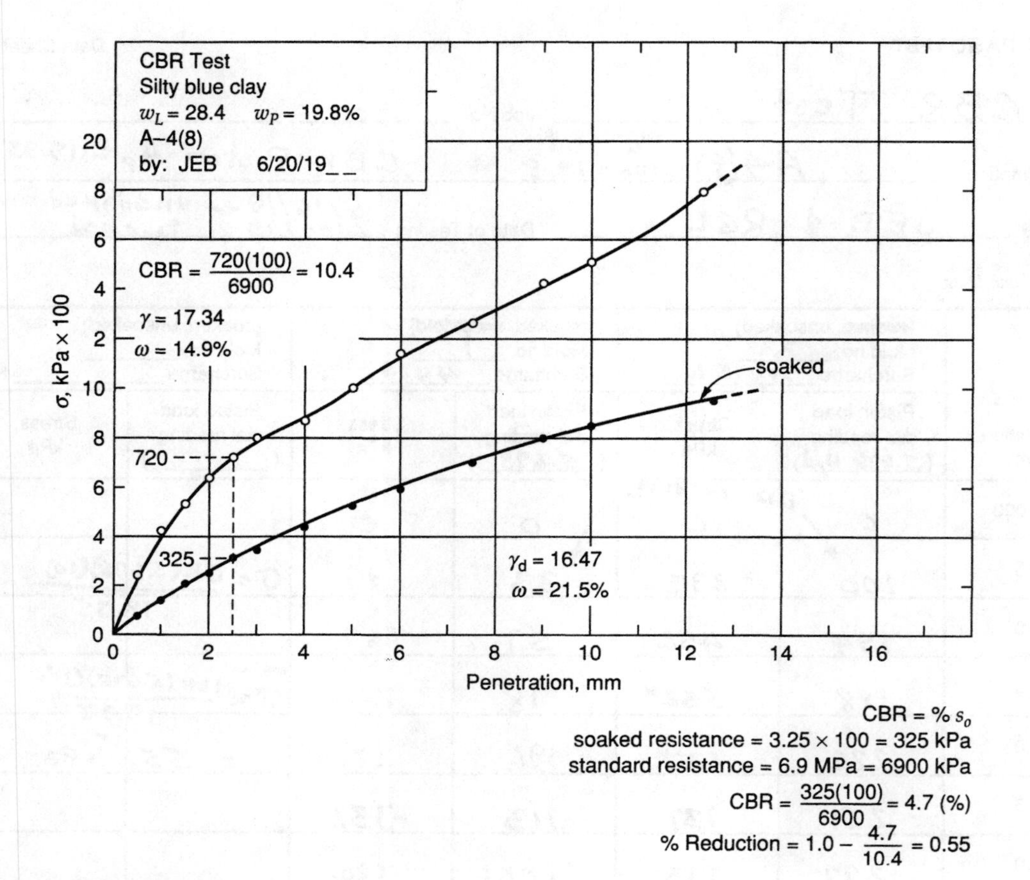

CBR = % s_o

soaked resistance = 3.25 × 100 = 325 kPa

standard resistance = 6.9 MPa = 6900 kPa

$$\text{CBR} = \frac{325(100)}{6900} = 4.7\ (\%)$$

$$\%\ \text{Reduction} = 1.0 - \frac{4.7}{10.4} = 0.55$$

Figure 19-8
Plot of load vs. penetration to compute the CBR numbers for this soil. Soaking the soil has reduced the CBR over 50 percent.

20: FLOW-NET CONSTRUCTION USING AN ELECTRICAL ANALOGY

References

Herbert, R., and K. R. Rushton (1966), "Ground-Water Flow Studies by Resistance Networks," *Geotechnique*, London, Vol. 16, No. 1, March, pp. 53–75.
Lane, E. W., F. B. Campbell, and W. H. Price (1934), "The Flow Net and the Electric Analogy," *Civil Engineering, ASCE*, October, pp. 510–514.
Selim, M. A. (1947), "Dams on Porous Media," *Transactions, ASCE*, Vol. 112, pp. 488–505.
Zanger, C. N. (1953), *Theory and Problems of Water Percolation*, U. S. Bureau of Reclamation, Engineering Monograph No. 8, April, 76 pages.

Objective

To introduce the reader to the concept that the flow of a fluid through a porous medium is similar (or analogous) to the flow of an electric current through a current-conducting medium.

Equipment

Voltmeter (VM) or Wheatstone bridge (Fig. 20-1)
DC voltage supply (with current output in the mA range)
Teledeltos paper[1]
Scissors
Silver paint to make electrodes[2]

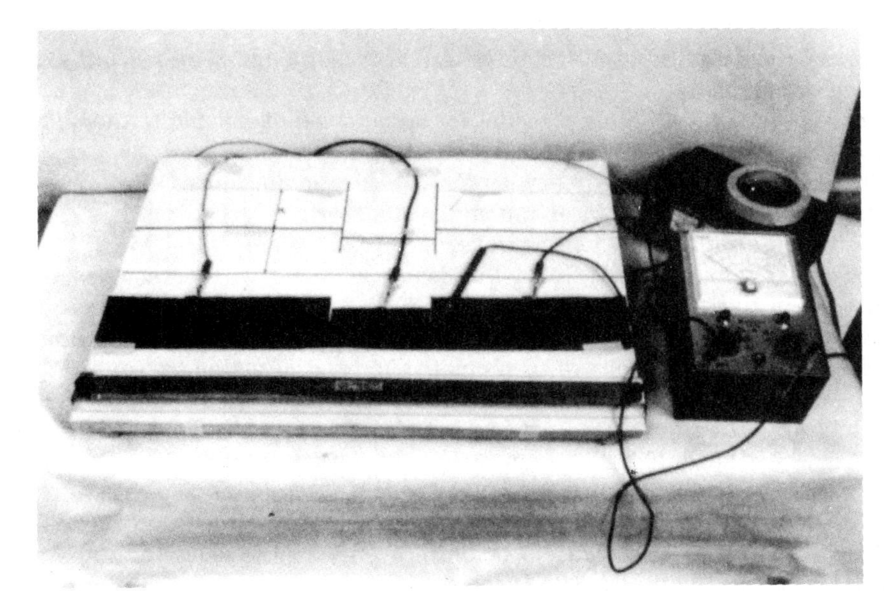

Figure 20-1
Electrical analogy setup. Cut-out in Teledeltos paper represents the conditions shown in the drawing mounted above the cutout. Notches cut in paper represent sheet-pile penetration. Voltage is impressed from both sides to the central portion to simulate water around a sheet-pile cofferdam.

[1]Teledeltos paper may be obtained from Western Union Telegraph Co., 1 Lake Street, Upper Saddle River, NJ 07458. You may be able to obtain some from your electrical engineering department.

[2]May be obtained from Newark Electronics Corporation—a nationwide electronics distributor.

General Discussion

A flow net is a graphical solution of the Laplace equation

$$k_x \frac{\partial^2 h}{\partial x^2} + k_y \frac{\partial^2 h}{\partial y^2} = 0 \tag{20-1}$$

and the construction of the net is a very time-consuming task if done accurately. The solution becomes even more difficult if the coefficients of permeability k_x and k_y in Eq. (20-1) are different in the x and y directions. Construction difficulties further increase if the soil is stratified or boundary conditions are irregular.

It has been found that the flow of an electric current from a high to a low potential is analogous to the flow of water caused by a differential head. From this similarity it follows that one can construct the shape of the porous soil mass out of an electrical conducting material and impress a differential voltage at locations to simulate the differential head of water between the two (or more) points. The voltages, as obtained with an electric probe at various locations within the model, will be in some ratio to the water head (or pressure potential) at that point. A locus of points of equal voltage produces a line of constant head or an equipotential line. If lines of equipotential are found, finishing the flow net by sketching in flow lines that intersect at right angles and form approximate squares is made somewhat easier.

Electrical conducting models can be constructed of:

1. Wire mesh (as used for the No. 200 sieve or other fine mesh).
2. Metal plates or metal foil.
3. Fine sand that has been treated to become electrically nonconductive, built into a model, and then saturated with a water solution containing an electrolyte. This may require using ac voltage to avoid polarizing parts of the model when the probe is used.
4. Teledeltos paper (or any other electrically conducting paper of commercial or laboratory manufacture).

Other model materials may be used for an electrical analogy, but Items 1 and 4 above are probably the most common.

Note that it is possible to build a model for both two-dimensional (either plan or elevation) flow and three-dimensional flow.

Different coefficients of permeability, as for stratification or lenses of different materials, can be simulated in a model by inserting pieces of material of greater or less conductivity depending on the user's ingenuity and perseverance.

Several simple examples using Teledeltos (same as using wire screen) paper cutouts will be illustrated in the scope of this text. In Fig. 20-2*a*, the existing soil mass is shown on an impervious base (a boundary condition). Figure 20-2*b* gives the corresponding cutout and locations of the impressed voltage to simulate the head of water shown in Fig. 20-2*a*.

The VM is set to read an impressed voltage between electrodes A and B of the cutout of, say, 10 volts (V) by an appropriate adjustment of the dc power supply. This is analogous to the potential water head of 13.5 shown in Fig. 20-2*a*; thus each 1.35 m of water head corresponds to a differential voltage of 1 V on the VM scale. It is possible to use a one-to-one voltage-to-water head analogy, but this may require voltages high enough to be uncomfortable to work with or to cause sparking—even with the very low amperages that the dc power supply should output. The current supplied by the power source should be in the milliampere range.

Now, to find an equipotential line (line of equal total head), it is necessary to take a probe from the "low" side of the VM and find a location, such as C in the cutout of Fig. 20-2*b*, which corresponds to, say, 9 V (we started with 10 V in this example to represent 13.5 m of pressure head). The remaining head potential in the soil mass represented by

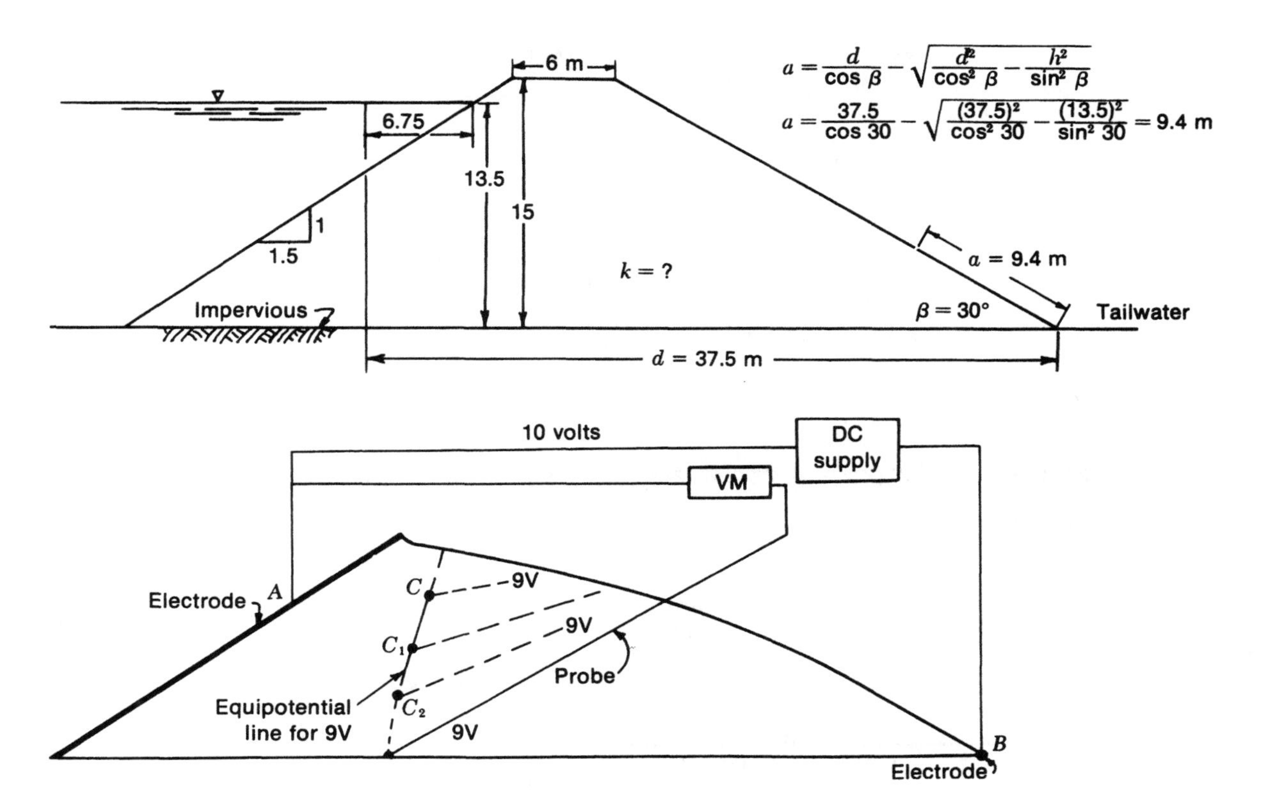

Figure 20-2
Pervious earth dam on an impervious base.

9 volts is calculated in this way:

If 10 v = 13.5 m pressure head

1 v = 1.35 m

At 9v the remaining head is:

$$h \text{ remaining} = (10 - 1) \times 1.35 \text{ m}$$
$$= 9 \times 1.35 \text{ m} = 12.15 \text{ m}$$

If one finds a series of points such as C for which the voltmeter reads 9 V at each location, the locus of points so established is an equipotential (or constant head) line for a remaining water head of 12.15 m of water. A similar analysis for 8 V, 7 V, 6 V, and so on can be made, yielding equipotential lines for 10.80 m, 9.45 m, 8.10 m, and so forth remaining water head.

It is difficult to obtain the flow-path boundaries to complete the flow net using this technique. However with the equipotential line fairly accurately drawn, the sketching in of flow paths to make squares to satisfy Eq. (20-1) is not as difficult as a strict "trial and correction" process.

Figure 20-3*a* and *b* represents an existing sheet-pile wall system and the corresponding cutout for use in the electrical analogy solution to establish equipotential lines.

In Fig. 20-3*b* the thin slot is an electric barrier simulating the impervious sheet-pile wall. As before, the impressed voltage from A to B simulates the differential head of water (of 6 m in this case). Notice that the effect of tailwater is to establish the differential head across the seepage path of the system. Again, by probing at a particular voltage, one

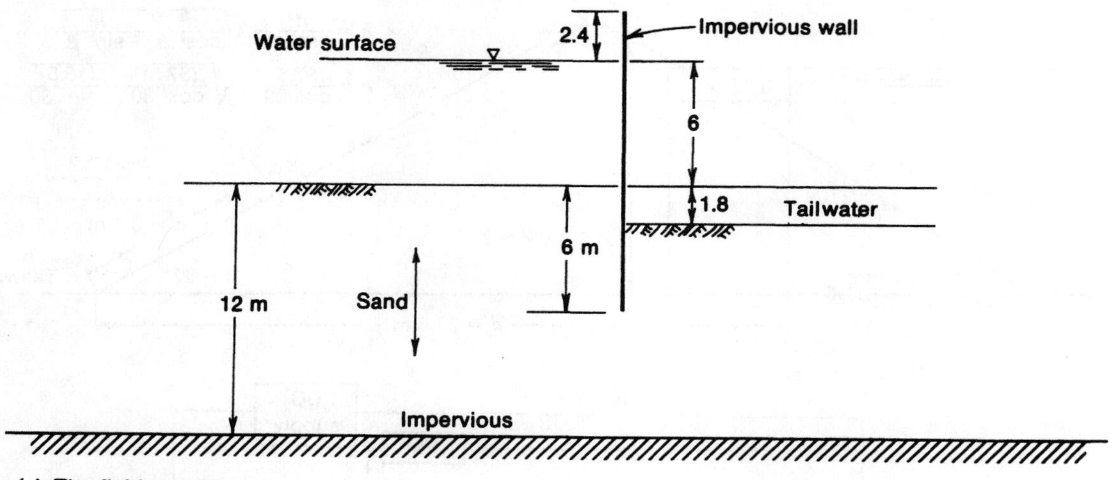

(*a*) The field conditions.

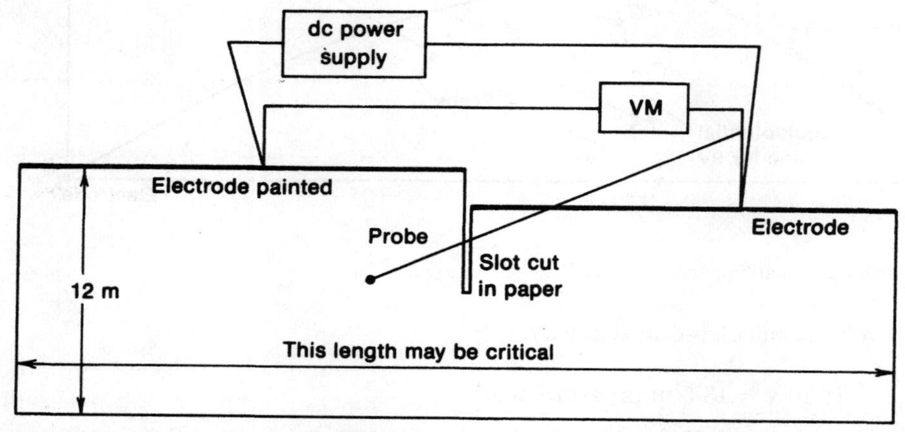

(*b*) The cut out made of Teledeltos paper with electrodes painted in proper location together with impressed voltage and VTVM.

Figure 20-3
Single sheep-pile wall for flow new analysis.

establishes a locus of points for an equipotential line. Flow paths should be sketched in by eye using the concept of establishing squares to complete the flow net for the system.

For problems of the type shown in Figs. 20-3 and 20-4, the upstream and downstream boundary conditions (large distance of $L \to \infty$) can be approximated by using a model length three to six times the height H or stratum thickness. The proper length can be checked by observing how the equipotential lines intersect into the bottom boundary. That is, if the equipotential lines do not intersect off a smooth curve and at nearly right angles, the model length is not adequate for the thickness of the pervious soil stratum.

A pervious sheet-pile wall (and most are) may be simulated by using a slotted rather than a full-slot wall cutout. If the soil is stratified, or its coefficient of permeability k changes, this may be simulated by increasing or decreasing the conductance of the paper by adding (add a light coat of paint) or removing (punch holes) electrolyte from the cutout in the appropriate zone(s). These techniques are "trial and adjustment" but may be used to indicate possible ranges of expected water flow—especially in view of the fact that considerable uncertainty exists in determining the coefficient of permeability of the soil.

Teledeltos paper is an excellent material for use as an electrical-analogy cutout. It is a paper with a graphite coating, and it has been found that the conductivity may vary slightly

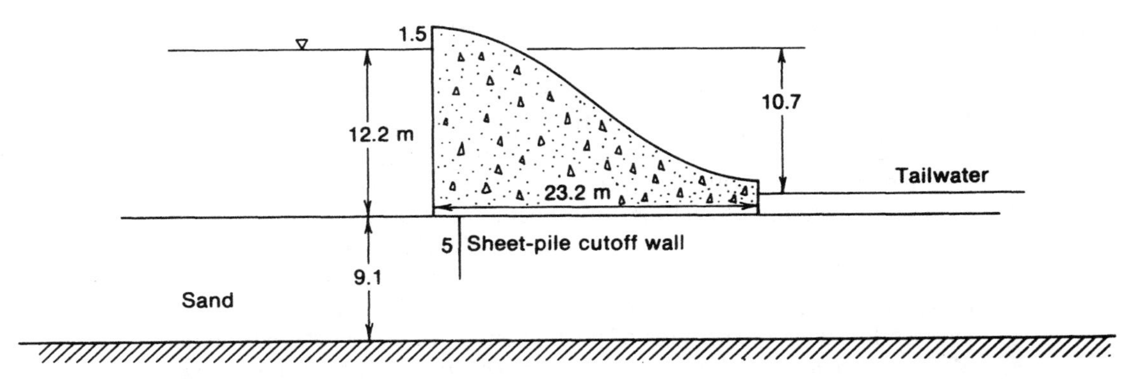

(a) Field conditions.

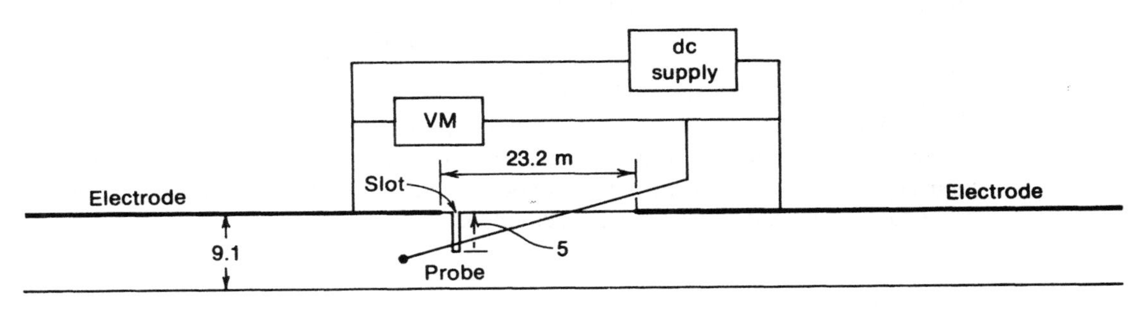

(b) The cut out made from Teledeltos paper with electrodes painted in proper location together with impressed voltage and VTVM.

Figure 20-4
Concrete dam on pervious foundation with an upstream sheet-pile cut-off wall.

between x and y directions. However, for the solution of most soil hydraulic problems, any differences in conductivity are of academic importance. This is especially so when one considers the degree of precision associated with the determination of the coefficient of permeability of the soil.

In order to simulate water in contact at more than a point at any boundary, it is necessary to paint an electrode of appropriate length on the paper cutout so that the electric current simulates the water on entering and exiting the soil mass. The electrodes should be highly conductive with respect to the paper, and silver paint (silver base) is recommended for this purpose. One might clamp metal strips to the paper cutout, but this procedure usually results in an awkward working model, and the silver paint is preferred. The electrodes should be painted as straight as possible and should be formed as a relatively narrow strip because of the reduction factor already involved in the small size of the model with respect to the field simulation.

The Teledeltos paper will possess erratic conducting qualities if it becomes perforated, creased, or crumpled. Therefore, it should be carefully handled and stored.

If $k_x \neq k_y$ recall from flow-net theory that the model must be modified (or one of the dimensions transformed). The transformation for either x or y (but not both) is done using the following relationships:

$$x' = x\sqrt{\frac{k_x}{k_y}} \qquad \text{or} \qquad y' = y\sqrt{\frac{k_y}{k_x}}$$

This computation is made so that when one of the dimensions is transformed, squares can be sketched on the model when constructing the flow net. For the electrical-analogy scale cutout, the dimensions should also be adjusted, using the appropriate transformation equation above prior to cutting the model.

Procedure

1. Each student will make a scale cutout using Fig. 20-1*p*, 20-2*p*, or 20-3*p*, as designated by the instructor, with the Teledeltos paper furnished.
2. Carefully paint the electrodes on the cutout using the silver electrode paint furnished. Use electrode locations that are appropriate to determine equipotential lines.
3. Impress a convenient voltage (a multiple of 10 is recommended) across the electrodes of the cutout. Use the probe from the VM to locate at least 10 equipotential lines on the model. Lightly mark the locus of points by circling them with a pencil, *being extremely careful not to perforate the paper cutout.* Remember that holes will change the conductivity (coefficient of permeability) of the electrical model.
4. Make a tracing of the cutout, ink in the equipotential line locations, and complete the flow-net construction by sketching the flow paths freehand. Compute the seepage quantity per meter of structure per day.
5. Redo the assigned problem of Step 1 above using free-hand work for both equipotential and flow-path lines. Do not be highly concerned with making an accurate drawing; merely make a sketch (to scale, of course) that does not "look too bad" and compute the seepage per meter of structure per day as in Step 4.
6. Compare the results of Steps 4 and 5 and comment on any significant differences. In your report be sure to include your paper cutout mounted on a sheet of heavy paper or cardboard so that it does not get torn or lost.

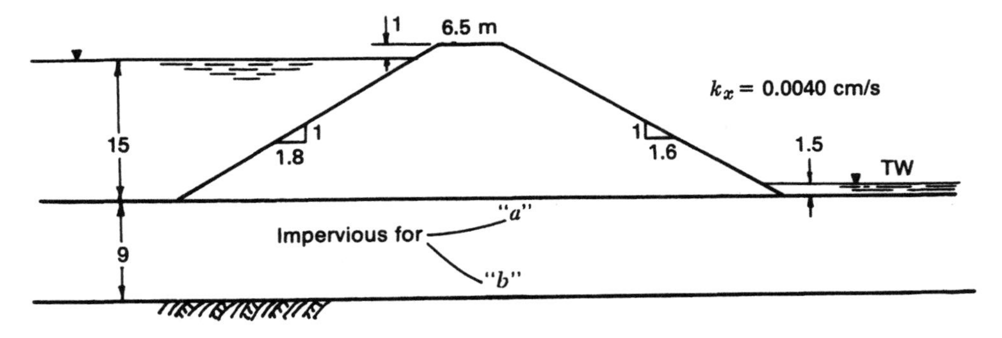

Figure 20-1p (not to scale)
(*a*) Find the seepage quantity in m³/m/day for the structure shown if it is on an *impervious base*. (*b*) Find the seepage quantity in m³/m/day if the base is *pervious* and both dam and base have the same k. (*c*) Redo (a) if $k_y = 0.25k_x$, where k_x = value shown on Fig. 20-1p.

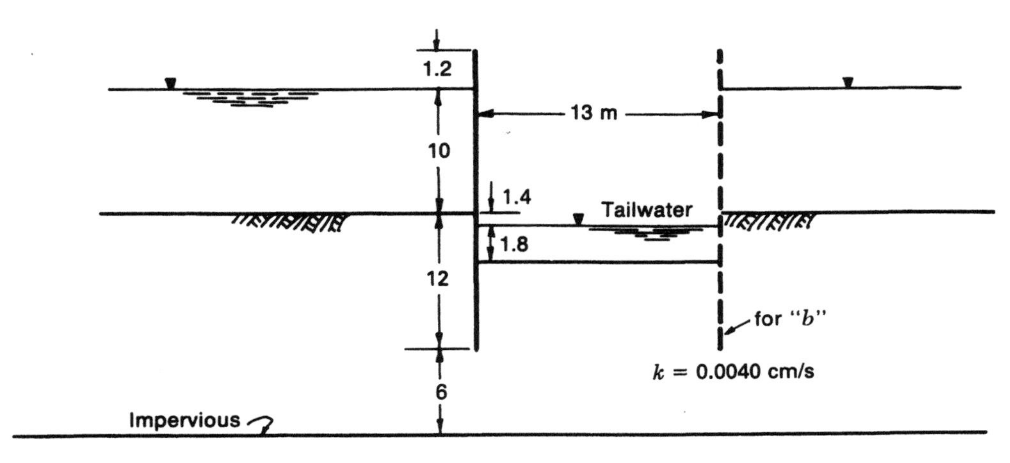

Figure 20-2p (not to scale)
(*a*) Find the seepage quantity in m³/m/day for the sheet-pile wall on the left side only. (*b*) Find the seepage quantity per meter of cofferdam perimeter per day for the above if both sheet-pile walls exist.

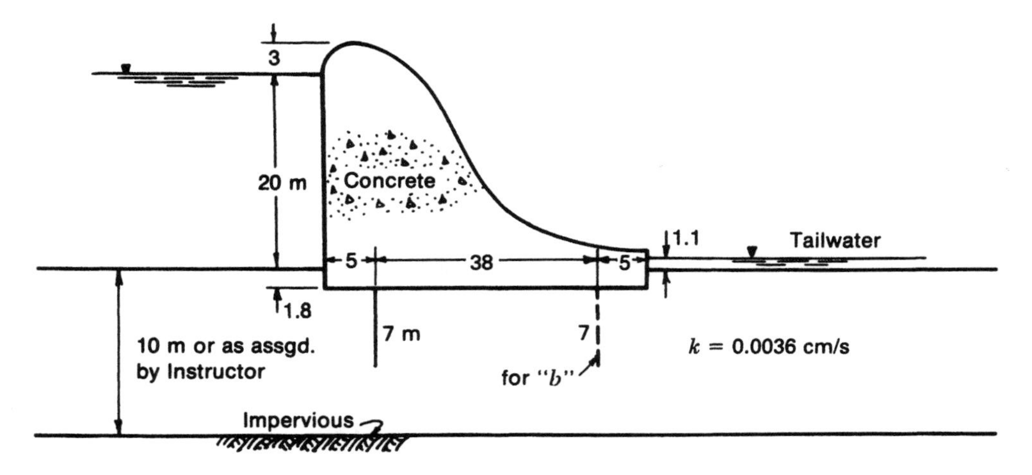

Figure 20-3p (not to scale)
(*a*) Find the seepage in m³/m/day for the dam shown with the sheet-pile cutoff wall at the *upstream* location (the heavy lined wall). (*b*) Find the seepage per day if the sheet-pile cutoff wall is at the *downstream* location (dashed line). (*c*) Find the seepage per day if the sheet-pile cutoff walls are at *both* locations.

21: VOLUMETRIC-GRAVIMETRIC RELATIONSHIPS

References

Archimedes' principle that a body submerged in a liquid will displace its own volume, as found in any introductory physics textbook.

Objective

To introduce the student to the concept of dry and saturated density, unit weights, void ratio e, soil structure (state), and a method of computing the specific gravity of a coarse-grained material.

Equipment

Large straight-sided container[1] of known volume
Steel straightedge
Quantity of oven-dry sandy gravel (say, 2 to 20 mm diameters)
Balance with about 20 kg capacity
Graduated cylinder (500 to 1000 mL)

Figure 21-1
Equipment for volumetric-gravimetric relationships. Filled battery jar on 20-kg balance, and 1000-mL graduated cylinder with water to pour into battery jar of gravel after initially weighing for dry density. Volume and mass of the battery jar have already been determined to save time.

Procedure

This is suggested as a class demonstration.

1. Each group carefully weigh the empty volume container and fill with soil, carefully strike the surface flush, and weigh. If the volume of soil changes due to weighing vibrations,

[1]The author suggests using a battery jar of dimensions 10 cm diam × 15 cm height (vol approximately 1200 mL) or, preferably, 15 cm diam × 15 cm height (vol. approximately 2400 mL) and readily available from laboratory supply companies. One may use round kitchen coffee or sugar storage containers if they are transparent so that observation can be made of soil state and entrapped air bubbles.

you have introduced a state (soil structure) change. Add additional soil, reweigh, and/or repeat as necessary until weighing vibrations do not cause further state change. Record the final mass of soil + container on your Data Sheet 21 in the space provided.

Record the volume of the container as V_c (compute if necessary). Compute the mass of dry soil used to fill the container and record as M_{ds}.

2. Fill the graduated cylinder with temperature-stabilized (approximately 20°C) tap water (let tap run sufficiently that cold water is provided). Carefully pour the water into the container of soil, by pouring around the edges and slowly, so as to avoid as much as possible trapped air bubbles in the soil voids.

   Fill the volume container level full with water but do not overfill so that no water is lost. Also be careful to fill level full and not "rounded" full from the meniscus effect. Observe the sides of the container as you fill it and if air bubbles appear to be trapped, gently rock the container or lightly tap the container at the bubble to dislodge it. Be very careful not to introduce a soil-state change at this point.

   Record the volume of water used to fill the soil container as V_w.

3. Carefully place the filled container on the balance (you may wish to place the container partially filled with water on the balance, then finish filling, being careful to keep the container level so that it is completely filled with water at the end of the operation).

   Record the mass of soil + water + container. Compare the difference in mass with the milliliters of water added. The difference should only be 2 to 5 grams; if more than this, recheck the graduated cylinder and/or weighing. Now compute the volume of water in the soil voids as V_w and record on your data sheet. This will also be the *approximate* volume of soil voids V_v.

Computations

1. Compute the dry density ρ_d using Eq. (I-6) as

$$\rho_d = \frac{M_{ds}}{V_c} \text{ g/cm}^3$$

The dry unit weight is computed as

$$\gamma_d = \rho_d \times 9.807 \text{ kN/m}^3$$

2. Compute the saturated density ρ_{sat} using Eq. (I-5) as

$$\rho_{sat} = \frac{M_{ds} + M_w}{V_c} \text{ g/cm}^3$$

Similarly the saturated unit weight is computed as

$$\gamma_{sat} = \rho_{sat} \times 9.807 \text{ kN/m}^3$$

3. Compute the void ratio e using Eq. (I-1) as follows:
   a. First compute the volume of soil solids V_s by direct reasoning:

$$V_s = V_c - V_w \quad \text{cm}^3$$

   b. Next compute the volume of voids (also by direct reasoning) since the voids volume must be the volume of water poured into the container or

$$V_v = V_w \quad \text{cm}^3$$

c. Now compute the void ratio as

$$e = \frac{V_v}{V_s}$$

4. Compute the approximate specific gravity of the soil solids G_s using Eq. (I-8) to give

$$G_s = \frac{M_{\text{ds}}}{V_s \rho_w}$$

Refer to Test 7, which gives a short table for the normal range in G_s for selected soils, and compare the tabular value with the value you have just computed.

The Report

1. Complete Data Sheet No. 21 and include a neat block diagram similar to Fig. I-1 showing all volumes and masses.
2. Comment on possible errors in computing G_s in this manner. Why has the soil size been limited?
3. Comment on whether any test improvements might be made, such as:

   Would it be better to use a few drops of wetting agent in the water to reduce the formation of air bubbles on soil or container sides?
   How does the size of the volume container affect results?
   What is the error introduced by not using distilled and/or deaerated water at 4°C instead of tap water at about 20°C?

4. Derive the constant 9.807 used to convert g/cm^3 to kN/m^3.
5. Put the above information in a folder with a cover page, and submit as assigned.

22: UNIT WEIGHT OF COHESIVE SOILS

References

See Experiment No. 21.
ASTM D-2937 [for information only as this test method is not yet (see footnote 2, page 241) a standardized procedure].

Objective

To introduce the reader to a rapid method of determining the unit density (or unit weight) of a *wet* cohesive soil.

Equipment

Container of known volume [such as a 1-quart (or 1-liter) canning jar or a discarded mayonnaise jar]
Balance weighing to 0.1 g
Graduated cylinder (500 mL or larger)

Figure 22-1
Equipment for a simple and rapid determination o. unit density of a cohesive soil. Volume of glass jar has been determined (969 cm^3) to save lab time. The 2.6-kg balances shown are usually adequate for weighing. Use a graduated cylinder of no less than 500 mL for this size volumetric jar. Use a soil sample large enough that the smallest division of the graduated cylinder does not materially affect results (or use differences in mass of the graduated cylinder before and after filling the volumetric jar for volume of water added to jar).

General

This procedure can be used to obtain the displaced volume of any cohesive soil that will not slake apart in water for the 30 s to 1 min required to perform the test. This procedure is not applicable for porous materials or very dry soils unless they are very dense, as any appreciable water adsorption via capillary action or otherwise will affect the results. For a very large number of soils this procedure will be quite adequate, rapid, and economical. It is applicable for soils located at a depth in the ground; for shallow soils the methods of Test 10 may be used.

Procedure

This can be an individual or group project or class demonstration.

1. Determine the volume of the volumetric container as V_c.
2. Obtain a tube sample (if available) of cohesive soil[1] and from it cut one or more test samples that do not contain any obvious void spaces.
3. Fill the graduated cylinder with temperature-stabilized tap water (approx. 20°C by letting tap run for a period of time or previously collecting the water and letting it stabilize to room temperature).
4. Carefully weigh the sample to obtain its wet mass M_{ws}. For best results the wet test samples should have a mass between 400 and 700 g.

   Now carefully place the sample in the volumetric container without fracturing it.

   Next *quickly* fill the volumetric container containing the soil sample with water from the graduated cylinder. Now quickly empty the water from the volumetric container, remove the soil sample, blot its surface dry with paper towels, and reweigh. If the initial and final sample weights are within 1 or 2 g, the test is quite satisfactory—more than this will require judgment of test acceptance. If the test is satisfactory, record the volume of water used from the graduated cylinder as V_w.

   Compute the volume of the soil sample as

$$V_{ws} = V_c - V_w$$

5. Place the complete wet soil sample in a dish and oven-dry overnight to obtain the dry soil mass M_{ds} (unless the water content is already known).
6. Compute the wet density of the soil as follows:

$$\rho_{wet} = \frac{M_{ws}}{V_{ws}}$$

   The dry density after oven-drying (with no allowance for any volume reduction from drying) to obtain M_{ds} is

$$\rho_d = \frac{M_{ds}}{V_{ws}}$$

   or if the water content w is known use the wet density to compute

$$\rho_d = \frac{\rho_{wet}}{1 + w}$$

   Next compute the wet and dry *unit weights* as

$$\gamma_i = \rho_i \times 9.807 \text{ kN/m}^3$$

[1]Where tube samples are not readily available in student soil laboratories, a set of cohesive samples may be made up using the Harvard miniature compaction equipment. This produces samples of known volume and density to establish validity of method.

The Report

1. Complete Data Sheet No. 22, and show wet and dry unit weights.
2. Comment on test limitations such as
   a. When test is not applicable (such as dry or wet soil state)
   b. Whether some other liquid would be better to use than water
   c. Size of sample and/or alternative sample preparation (grease coating, some other coating[2], etc.)
3. Put the project data in a folder with a cover sheet and submit as required.

[2]ASTM is currently working on a "standard" for this procedure, the essential difference being that the soil sample is to be coated with wax. This certainly introduces considerable additional work into the test for a questionable increase in precision. For tube samples all one obtains is the density for that sample recovered from that location.

WATER CONTENT DETERMINATION

Data Sheet 1

Project ______________________ Job No. ______________________

Location of Project ______________________

Description of Soil ______________________

Tested By ______________________ Date of Testing ______________________

Date of Weighing ______________________

Boring no.					
Container no. (cup)					
Mass of cup + wet soil					
Mass of cup + dry soil					
Mass of cup					
Mass of dry soil, M_s					
Mass of water, M_w					
Water content, w%					

Boring no.					
Container no. (cup)					
Mass of cup + wet soil					
Mass of cup + dry soil					
Mass of cup					
Mass of dry soil, M_s					
Mass of water, M_w					
Water content, w%					

$$w = \frac{M_w}{M_s}(100)\ \%$$

WATER CONTENT DETERMINATION Data Sheet 1

Project ________________ Job No. ________________

Location of Project ________________

Description of Soil ________________

Tested By ________________ Date of Testing ________________

Date of Weighing ________________

Boring no.					
Container no. (cup)					
Mass of cup + wet soil					
Mass of cup + dry soil					
Mass of cup					
Mass of dry soil, M_s					
Mass of water, M_w					
Water content, w%					

Boring no.					
Container no. (cup)					
Mass of cup + wet soil					
Mass of cup + dry soil					
Mass of cup					
Mass of dry soil, M_s					
Mass of water, M_w					
Water content, w%					

$w = \frac{M_w}{M_s}(100)\ \%$

WATER CONTENT DETERMINATION Data Sheet 1

Project ______________________ Job No. ______________

Location of Project ______________________

Description of Soil ______________________

Tested By ______________________ Date of Testing ______________

Date of Weighing ______________

Boring no.					
Container no. (cup)					
Mass of cup + wet soil					
Mass of cup + dry soil					
Mass of cup					
Mass of dry soil, M_s					
Mass of water, M_w					
Water content, w%					

Boring no.					
Container no. (cup)					
Mass of cup + wet soil					
Mass of cup + dry soil					
Mass of cup					
Mass of dry soil, M_s					
Mass of water, M_w					
Water content, w%					

$w = \frac{M_w}{M_s}(100)\ \%$

WATER CONTENT DETERMINATION Data Sheet 1

Project ______________________ Job No. ______________

Location of Project ______________________

Description of Soil ______________________

Tested By ______________________ Date of Testing ______________

Date of Weighing ______________

Boring no.					
Container no. (cup)					
Mass of cup + wet soil					
Mass of cup + dry soil					
Mass of cup					
Mass of dry soil, M_s					
Mass of water, M_w					
Water content, w%					

Boring no.					
Container no. (cup)					
Mass of cup + wet soil					
Mass of cup + dry soil					
Mass of cup					
Mass of dry soil, M_s					
Mass of water, M_w					
Water content, w%					

$w = \frac{M_w}{M_s}(100)\ \%$

WATER CONTENT DETERMINATION Data Sheet 1

Project ______________________ Job No. ______________________

Location of Project ______________________

Description of Soil ______________________

Tested By ______________________ Date of Testing ______________________

Date of Weighing ______________________

Boring no.					
Container no. (cup)					
Mass of cup + wet soil					
Mass of cup + dry soil					
Mass of cup					
Mass of dry soil, M_s					
Mass of water, M_w					
Water content, w%					

Boring no.					
Container no. (cup)					
Mass of cup + wet soil					
Mass of cup + dry soil					
Mass of cup					
Mass of dry soil, M_s					
Mass of water, M_w					
Water content, w%					

$$w = \frac{M_w}{M_s} (100) \%$$

WATER CONTENT DETERMINATION Data Sheet 1

Project ______________________ Job No. ______________

Location of Project ______________________

Description of Soil ______________________

Tested By ______________________ Date of Testing ______________

Date of Weighing ______________

Boring no.					
Container no. (cup)					
Mass of cup + wet soil					
Mass of cup + dry soil					
Mass of cup					
Mass of dry soil, M_s					
Mass of water, M_w					
Water content, w%					

Boring no.					
Container no. (cup)					
Mass of cup + wet soil					
Mass of cup + dry soil					
Mass of cup					
Mass of dry soil, M_s					
Mass of water, M_w					
Water content, w%					

$w = \frac{M_w}{M_s}(100)\ \%$

Project and Location

Foreman ____________ Helper ____________ Water Level ________ at ________ hrs. ________ Job No. ________

Boring Unit ____________ Casing Used ____________ Date ________

Surface Elevation ____________ Split Spoon Size ________ Wt. of Hammer ________ Sht ______ of ______

Boring No. ____________ Weather ____________ Temp. ________

Sample no.	Elevation		Penetration test				Length recovered m, ft	Field strength q_u kPa, Tsf	Sample method (SS, ST, Aug, etc)		Remarks
	From	To	N for: 150 mm 6 in			Penetration N					Sample description
			1st	2nd	3rd						

Abbreviations

F.T. — Fish-tail
W.O. — Wash out
S.T. — Shelby tube
S.S. — Split spoon
D.B. — Diamond bit
C. — Core
R.B. — Rock bit

Project and Location

Foreman ______________ Helper ______________ Water Level ________ at ________ hrs. ________ Job No. ________

Boring Unit ______________ Casing Used ______________ Date ________

Surface Elevation ______________ Split Spoon Size ________ Wt. of Hammer ________ Sht ____ of ____

Boring No. ______________ Weather ______________ Temp. ________

Sample no.	Elevation		Penetration test				Length recovered m, ft	Field strength q_u kPa, Tsf	Sample method (SS, ST, Aug, etc)		Remarks
	From	To	N for: 150 mm 6 in — 1st	2nd	3rd	Penetration N					Sample description

Abbreviations

- F.T. — Fish-tail
- W.O. — Wash out
- S.T. — Shelby tube
- S.S. — Split spoon
- D.B. — Diamond bit
- C. — Core
- R.B. — Rock bit

ATTERBERG LIMITS DETERMINATION Data Sheet 3

Project __________ Job No. __________

Location of Project __________ Boring No. __________ Sample No. __________

Description of Soil __________

Depth of Sample __________ Tested By __________ Date __________

Liquid Limit Determination

Can no.					
Mass of wet soil + can					
Mass of dry soil + can					
Mass of can					
Mass of dry soil					
Mass of moisture					
Water content, w%					
No. of blows N					
Penetration D, mm					

Water content, w%

10 15 20 25 30 40 50 60 80 100

No. of blows, N

Flow index F_i = __________

Liquid limit = __________

Plastic limit = __________

Plasticity index I_P = __________

Plastic Limit Determination

Can no.					
Mass of wet soil + can					
Mass of dry soil + can					
Mass of can					
Mass of dry soil					
Mass of moisture					
Water content, w% = w_P					

ATTERBERG LIMITS DETERMINATION

Data Sheet 3

Project ______________________ Job No. ______________________

Location of Project ______________________ Boring No. ____________ Sample No. ____________

Description of Soil ______________________

Depth of Sample ______________ Tested By ______________ Date ______________

Liquid Limit Determination

Can no.					
Mass of wet soil + can					
Mass of dry soil + can					
Mass of can					
Mass of dry soil					
Mass of moisture					
Water content, w%					
No. of blows N					
Penetration D, mm					

Water content, w%

10 15 20 25 30 40 50 60 80 100

No. of blows, N

Flow index F_i = ____________

Liquid limit = ____________

Plastic limit = ____________

Plasticity index I_P = ____________

Plastic Limit Determination

Can no.					
Mass of wet soil + can					
Mass of dry soil + can					
Mass of can					
Mass of dry soil					
Mass of moisture					
Water content, $w\% = w_P$					

ATTERBERG LIMITS DETERMINATION Data Sheet 3

Project ______________________ Job No. ______________________

Location of Project ______________________ Boring No. __________ Sample No. __________

Description of Soil ____

Depth of Sample ______________ Tested By ______________ Date ______________

Liquid Limit Determination

Can no.					
Mass of wet soil + can					
Mass of dry soil + can					
Mass of can					
Mass of dry soil					
Mass of moisture					
Water content, w%					
No. of blows N					
Penetration D, mm					

Water content, w%

10 15 20 25 30 40 50 60 80 100

No. of blows, N

Flow index F_i = ______

Liquid limit = ______

Plastic limit = ______

Plasticity index I_P = ______

Plastic Limit Determination

Can no.					
Mass of wet soil + can					
Mass of dry soil + can					
Mass of can					
Mass of dry soil					
Mass of moisture					
Water content, $w\% = w_P$					

ATTERBERG LIMITS DETERMINATION Data Sheet 3

Project ______________________ Job No. ______________________

Location of Project ______________________ Boring No. __________ Sample No. __________

Description of Soil ____

Depth of Sample ______________ Tested By ______________ Date ______________

Liquid Limit Determination

Can no.					
Mass of wet soil + can					
Mass of dry soil + can					
Mass of can					
Mass of dry soil					
Mass of moisture					
Water content, w%					
No. of blows N					
Penetration D, mm					

Water content, w%

10 15 20 25 30 40 50 60 80 100

No. of blows, N

Flow index F_i = __________

Liquid limit = __________

Plastic limit = __________

Plasticity index I_P = __________

Plastic Limit Determination

Can no.					
Mass of wet soil + can					
Mass of dry soil + can					
Mass of can					
Mass of dry soil					
Mass of moisture					
Water content, $w\% = w_P$					

RINKAGE LIMIT TEST

Data Sheet 4

ject ______________________ Job No. ______________________

ation of Project ______________________ Boring No. ____________ Sample No. ____________

scription of Soil ______________________ Depth of Sample ____________

sted By ______________________ Date of Testing ______________________

Mass of coated dish + wet soil = ____________ g (a)

Mass of coated dish + dry soil = ____________ g (b)

Mass of coated dish, M_d = ____________ g (c) dish # = ______

Mass of dry soil, M_s = ____________ g (d) (b – c)

Total mass of water, M_w = ____________ g (e) (a – b)

nitial water content, w_o = ____________ % (f) (e/d·100)

a for volume of soil cake

ial cake volume (wet) *Wax Method*

Mass coat dish + Hg, M_{dm} = ____________ g (g) M_{dw}

Mass of coat dish, M_d = ____________ g (h) M_d

Vol. of shrink dish V_o = (g – h)/ρ = ______________________ cm³

cake volume (final)

Mass of vol. cup + Hg, M_{vcm} = ____________ g (i) M_{bw}

Mass of vol. cup + Hg – soil, M_{vcms} = ____________ g (j) M_{bws}

Mass of displaced Hg (i – j) = ____________ g (k) M_s

Vol. of soil cake V_f (with Hg = k/ρ) = ______________________ cm³

Vol. of soil cake V_f (w/wax = i + k – j) = ______________________ cm³

$w_s = w_o - \frac{(V_o - V_f)\rho_w}{M_s} \times 100$ = ______________________

$SR = \frac{M_s}{\rho_w V_f}$ = ______________________

RINKAGE LIMIT TEST

Data Sheet 4

ject ______________________ Job No. ______________________

cation of Project ______________________ Boring No. ____________ Sample No. ____________

scription of Soil ______________________ Depth of Sample ____________

sted By ______________________ Date of Testing ______________________

Mass of coated dish + wet soil = ____________ g (a)

Mass of coated dish + dry soil = ____________ g (b)

Mass of coated dish, M_d = ____________ g (c) dish # = ______

Mass of dry soil, M_s = ____________ g (d) (b – c)

Total mass of water, M_w = ____________ g (e) (a – b)

Initial water content, w_o = ____________ % (f) (e/d·100)

ta for volume of soil cake

tial cake volume (wet) *Wax Method*

Mass coat dish + Hg, M_{dm} = ____________ g (g) M_{dw}

Mass of coat dish, M_d = ____________ g (h) M_d

Vol. of shrink dish $V_o = (g - h)/\rho$ = ______________________ cm^3

y cake volume (final)

Mass of vol. cup + Hg, M_{vcm} = ____________ g (i) M_{bw}

Mass of vol. cup + Hg – soil, M_{vcms} = ____________ g (j) M_{bws}

Mass of displaced Hg (i – j) = ____________ g (k) M_s

Vol. of soil cake V_f (with Hg = k/ρ) = ______________________ cm^3

Vol. of soil cake V_f (w/wax = i + k – j) = ______________________ cm^3

$w_s = w_o - \frac{(V_o - V_f)\rho_w}{M_s} \times 100$ = ______________________

$SR = \frac{M_s}{\rho_w V_f}$ = ______________________

GRAIN SIZE ANALYSIS–MECHANICAL Data Sheet 5*a*

Project ______________________________ Job No. __________

Location of Project ______________ Boring No. __________ Sample No. __________

Description of Soil ______________________________ Depth of Sample __________

Tested By ______________________ Date of Testing ______________________

Soil Sample Size (ASTM D1140-54)

Nominal diameter of largest particle	Approximate minimum mass of sample, g
No. 10 sieve	200
No. 4 sieve	500
3/4 in.	1500

Mass of dry sample + dish	
Mass of dish	
Mass of dry sample, M_s	

Sieve analysis and grain shape

Sieve no.	Diam. (mm)	Mass retained	% retained	% passing

% passing = 100 − Σ % retained.

GRAIN SIZE ANALYSIS–MECHANICAL Data Sheet 5*a*

Project ______________________________ Job No. __________

Location of Project ______________ Boring No. ________ Sample No. ________

Description of Soil ______________________ Depth of Sample ________

Tested By ______________________ Date of Testing ______________________

Soil Sample Size (ASTM D1140-54)

Nominal diameter of largest particle	Approximate minimum mass of sample, g
No. 10 sieve	200
No. 4 sieve	500
3/4 in.	1500

Mass of dry sample + dish	
Mass of dish	
Mass of dry sample, M_s	

Sieve analysis and grain shape

Sieve no.	Diam. (mm)	Mass retained	% retained	% passing

% passing = 100 − Σ % retained.

GRAIN SIZE ANALYSIS–MECHANICAL

Data Sheet 5*a*

Project ______________________ Job No. __________

Location of Project ______________ Boring No. __________ Sample No. __________

Description of Soil ______________________ Depth of Sample __________

Tested By ______________ Date of Testing ______________

Soil Sample Size (ASTM D1140-54)

Nominal diameter of largest particle	Approximate minimum mass of sample, g
No. 10 sieve	200
No. 4 sieve	500
3/4 in.	1500

Mass of dry sample + dish	
Mass of dish	
Mass of dry sample, M_s	

Sieve analysis and grain shape

Sieve no.	Diam. (mm)	Mass retained	% retained	% passing

% passing = 100 − Σ % retained.

GRAIN SIZE DISTRIBUTION

Data Sheet 5*b*

Project ________________ Job No. ________

Location of Project ________ Boring No. ________ Sample No. ________

Tested By ________ Date of Testing ________

Gravel | Sand (Course to medium; Fine) | Silt | Clay

U.S. standard sieve sizes

No. 4 — No. 10 — No. 20 — No. 40 — No. 100 — No. 200

Percent finer: 100, 80, 60, 40, 20, 0

Particle diameter, mm: 19, 4.75, 1, 0.840, 0.420, 0.150, 0.1, 0.075, 0.01, 0.002, 0.001

Visual soil description ________________

____

Soil classification ________________

System ________________

GRAIN SIZE DISTRIBUTION

Data Sheet 5*b*

Project ______________________ Job No. __________

Location of Project ______________ Boring No. ______________ Sample No. __________

Tested By ______________ Date of Testing ______________

Gravel	Sand		Silt	Clay
	Course to medium	Fine		

U.S. standard sieve sizes

No. 4, No. 10, No. 20, No. 40, No. 100, No. 200

Percent finer: 0, 20, 40, 60, 80, 100

Particle diameter, mm: 19, 4.75, 1, 0.840, 0.420, 0.150, 0.1, 0.075, 0.01, 0.002, 0.001

Visual soil description ______________________

____

Soil classification ______________________

System ______________________

GRAIN SIZE DISTRIBUTION

Data Sheet 5*b*

Project ______________________ Job No. __________

Location of Project ______________ Boring No. ______________ Sample No. __________

Tested By ______________ Date of Testing ______________

Gravel	Sand		Silt	Clay
	Course to medium	Fine		

U.S. standard sieve sizes

No. 4, No. 10, No. 20, No. 40, No. 100, No. 200

Percent finer: 0, 20, 40, 60, 80, 100

Particle diameter, mm: 19, 4.75, 1, 0.840, 0.420, 0.150, 0.1, 0.075, 0.01, 0.002, 0.001

Visual soil description ______________________

____

Soil classification ______________________

System ______________________

GRAIN SIZE DISTRIBUTION

Data Sheet 5*b*

Project ______________________ Job No. ________

Location of Project ________ Boring No. ________ Sample No. ________

Tested By ________ Date of Testing ________

Visual soil description ________________________

____

Soil classification ________________________

System ________________________

GRAIN SIZE DISTRIBUTION Data Sheet 5*b*

Project ______________________ Job No. __________

Location of Project ______________ Boring No. ______________ Sample No. __________

Tested By ______________ Date of Testing ______________

Gravel	Sand		Silt	Clay
	Course to medium	Fine		

U.S. standard sieve sizes: No. 4, No. 10, No. 20, No. 40, No. 100, No. 200

Percent finer: 0, 20, 40, 60, 80, 100

Particle diameter, mm: 19, 4.75, 1, 0.840, 0.420, 0.150, 0.1, 0.075, 0.01, 0.002, 0.001

Visual soil description ______________________

____

Soil classification ______________________

System ______________________

GRAIN SIZE ANALYSIS–HYDROMETER METHOD Data Sheet 6

Project ______________________ Job No. ______________________

Location of Project ______________________ Boring No. __________ Sample No. __________

Description of Soil ______________________ Depth of Sample __________

Tested By ______________________ Date of Testing ______________________

Hydrometer analysis

General Data: Hydrometer type __________ Zero correction __________ Meniscus __________

Dispersing agent ______________________ Amount used ______________________

G_s of solids = ______________ CF a = ______________ w (if air-dry) = __________%

Mass soil (wet, dry) = __________ g % Finer[1] = __________ % Control sieve no. __________

Date	Time of reading	Elapsed time, min	Temp., °C	Actual Hyd. reading, R_a	Corr. Hyd. reading, R_c	Act/Adj % Finer	Hyd. Corr. only for meniscus, R	L from Table 6-5	$\frac{L}{t}$	K from Table 6-4	D, mm

[1]mark out wet or dry; % Finer if applicable; w for air-dry soil only.

$R_c = R_{\text{actual}} - \text{zero correction} + C_T$ Act % finer $= R_c(a)/M_s$ $D = K\sqrt{L/t}$

GRAIN SIZE ANALYSIS–HYDROMETER METHOD Data Sheet 6

Project ________________________ Job No. ________________________

Location of Project ________________________ Boring No. ____________ Sample No. ____________

Description of Soil ________________________ Depth of Sample ____________

Tested By ________________________ Date of Testing ________________________

Hydrometer analysis

General Data: Hydrometer type ____________ Zero correction ____________ Meniscus ____________

Dispersing agent ________________________ Amount used ________________________

G_s of solids = ________________ CF a = ____________ w (if air-dry) = ____________%

Mass soil (wet, dry) = ____________ g % Finer[1] = ____________ % Control sieve no. ____________

Date	Time of reading	Elapsed time, min	Temp., °C	Actual Hyd. reading, R_a	Corr. Hyd. reading, R_c	Act/Adj % Finer	Hyd. Corr. only for meniscus, R	L from Table 6-5	$\frac{L}{t}$	K from Table 6-4	D, mm

[1]mark out wet or dry; % Finer if applicable; w for air-dry soil only.

$R_c = R_{\text{actual}} - \text{zero correction} + C_T$ Act % finer $= R_c(a)/M_s$ $D = K\sqrt{L/t}$

SPECIFIC GRAVITY OF SOIL SOLIDS (G_s)

Data Sheet 7

Project ______ Job No. ______

Location of Project ______ Boring No. ______ Sample No. ______

Description of Soil ______ Depth of Sample ______

Tested By ______ Date of Testing ______

fl. = volumetric flask

Test no.				
Vol. of flask at 20°C				
Method of air removal[a]				
Mass fl. + water + soil = M_{bws}				
Temperature, °C				
Mass fl. + water[b] = M_{bw}				
Dish no.				
Mass dish + dry soil				
Mass of dish				
Mass of dry soil = M_s				
$M_w = M_s + M_{bw} - M_{bws}$				
$\alpha = \rho_T/\rho_{20°C}$				
$G_s = \alpha\ M_s/M_w$				

[a]Indicate vacuum or aspirator for air removal.

[b]M_{bw} is the mass of the flask filled with water at same temp. ± 1°C as for M_{bws} or value from calibration curve at T of M_{bws}.

Remarks ______

______

Average specific gravity of soil solids (G_s) = ______

SPECIFIC GRAVITY OF SOIL SOLIDS (G_s) Data Sheet 7

Project ______________________ Job No. ______________________

Location of Project ______________________ Boring No. __________ Sample No. __________

Description of Soil ______________________ Depth of Sample ______________________

Tested By ______________________ Date of Testing ______________________

fl. = volumetric flask

Test no.				
Vol. of flask at 20°C				
Method of air removal[a]				
Mass fl. + water + soil = M_{bws}				
Temperature, °C				
Mass fl. + water[b] = M_{bw}				
Dish no.				
Mass dish + dry soil				
Mass of dish				
Mass of dry soil = M_s				
$M_w = M_s + M_{bw} - M_{bws}$				
$\alpha = \rho_T/\rho_{20°C}$				
$G_s = \alpha\ M_s/M_w$				

[a]Indicate vacuum or aspirator for air removal.
[b]M_{bw} is the mass of the flask filled with water at same temp. ± 1°C as for M_{bws} or value from calibration curve at T of M_{bws}.

Remarks ____

____

Average specific gravity of soil solids (G_s) = ______________________

SOIL CLASSIFICATION Data Sheet 8

Project ______________________________

Location of Project ______________________________

Tested By ______________________ Date ______________

Soil Identification (Container No., etc.) ______________________

Sieve Analysis (washed, dry)

Sieve	% Passing	Index Properties
No. 4	________	w_L = ________
No. 10	________	w_P = ________
	________	I_P = ________
No. 40	________	________
	________	C_U = ________ } if (–) #200 ≤ 12
No. 100	________	C_C = ________ } Plot curve
No. 200	________	

Classification (write descript.) ______________________

Symbols: AASHTO: ______________ Unified: ______________

Soil Identification (Container No., etc.)

Sieve Analysis (washed, dry)

Sieve	% Passing	Index Properties
No. 4	________	w_L = ________
No. 10	________	w_P = ________
	________	I_P = ________
No. 40	________	________
	________	C_U = ________ } if (–) #200 ≤ 12
No. 100	________	C_C = ________ } Plot curve
No. 200	________	

Classification (write descript.) ______________________

Symbols: AASHTO: ______________ Unified: ______________

SOIL CLASSIFICATION Data Sheet 8

Project ______________________________

Location of Project ______________________________

Tested By ______________________ Date ______________

Soil Identification (Container No., etc.) ______________________________

Sieve Analysis (washed, dry)

Sieve	% Passing	Index Properties
No. 4	________	w_L = ________
No. 10	________	w_P = ________
	________	I_P = ________
No. 40	________	________
	________	C_U = ________ } if (–) #200 ≤ 12
No. 100	________	C_C = ________ } Plot curve
No. 200	________	

Classification (write descript.) ______________________________

Symbols: AASHTO: ______________ Unified: ______________

Soil Identification (Container No., etc.)

Sieve Analysis (washed, dry)

Sieve	% Passing	Index Properties
No. 4	________	w_L = ________
No. 10	________	w_P = ________
	________	I_P = ________
No. 40	________	________
	________	C_U = ________ } if (–) #200 ≤ 12
No. 100	________	C_C = ________ } Plot curve
No. 200	________	

Classification (write descript.) ______________________________

Symbols: AASHTO: ______________ Unified: ______________

SOIL CLASSIFICATION Data Sheet 8

Project ______________________________

Location of Project ______________________________

Tested By ______________________ Date ______________

Soil Identification (Container No., etc.) ______________________

Sieve Analysis (washed, dry)

Sieve	% Passing	Index Properties
No. 4	______	w_L = ______
No. 10	______	w_P = ______
	______	I_P = ______
No. 40	______	______
	______	C_U = ______ } if (–) #200 ≤ 12 Plot curve
No. 100	______	C_C = ______ }
No. 200	______	

Classification (write descript.) ______________________

Symbols: AASHTO: ______________ Unified: ______________

Soil Identification (Container No., etc.)

Sieve Analysis (washed, dry)

Sieve	% Passing	Index Properties
No. 4	______	w_L = ______
No. 10	______	w_P = ______
	______	I_P = ______
No. 40	______	______
	______	C_U = ______ } if (–) #200 ≤ 12 Plot curve
No. 100	______	C_C = ______ }
No. 200	______	

Classification (write descript.) ______________________

Symbols: AASHTO: ______________ Unified: ______________

SOIL CLASSIFICATION Data Sheet 8

Project ______________________________

Location of Project ______________________________

Tested By ______________________ Date ______________

Soil Identification (Container No., etc.) ______________________________

Sieve Analysis (washed, dry)

Sieve	% Passing
No. 4	______
No. 10	______
	______
No. 40	______
	______
No. 100	______
No. 200	______

Index Properties

w_L = ______

w_P = ______

I_P = ______

______

C_U = ______

C_C = ______

} if (–) #200 ≤ 12 Plot curve

Classification (write descript.) ______________________________

Symbols: AASHTO: ______________ Unified: ______________

Soil Identification (Container No., etc.)

Sieve Analysis (washed, dry)

Sieve	% Passing
No. 4	______
No. 10	______
	______
No. 40	______
	______
No. 100	______
No. 200	______

Index Properties

w_L = ______

w_P = ______

I_P = ______

______

C_U = ______

C_C = ______

} if (–) #200 ≤ 12 Plot curve

Classification (write descript.) ______________________________

Symbols: AASHTO: ______________ Unified: ______________

COMPACTION TEST

Data Sheet 9

Project ______________________ Job No. ______________________

Location of Project ______________________ Boring No. ____________ Sample No. ____________

Description of Soil ______________________

Test Performed By ______________________ Date of Test ______________________

Blows/Layer ____________ No. of Layers ____________ Mass of Rammer ______ N

Mold dimensions: Diam. ____________ cm Ht. ____________ cm Vol. ____________ cm^3

Water Content Determination

Sample no.	1		2		3		4		5		6	
Moisture can no.												
Mass of cup + wet soil												
Mass of cup + dry soil												
Mass of water												
Mass of cup, g												
Mass of dry soil, g												
Water content, w%												

Density and Unit Weight: $\rho = M_{ws}/\text{vol}$ g/cm^3; $\gamma_{wet} = \rho \times 9.807$ kN/m^3

Assumed water content						
Water content, w%						
Mass of soil + mold, g						
Mass of mold, g						
Mass of soil in mold, M_{ws}						
Wet unit weight, kN/m^3						
Dry unit weight γ, kN/m^3						

Dry unit weight, kN/m^3

Water content, w%

Optimum moisture = ____________ % Maximum dry unit weight, γ_d = ____________ kN/m^3

COMPACTION TEST Data Sheet 9

Project ______________________ Job No. ______________________

Location of Project ______________________ Boring No. ____________ Sample No. ____________

Description of Soil ____

Test Performed By ______________________ Date of Test ______________________

Blows/Layer ______________ No. of Layers ______________ Mass of Rammer ______ N

Mold dimensions: Diam. ____________ cm Ht. ____________ cm Vol. ____________ cm^3

Water Content Determination

Sample no.	1		2		3		4		5		6	
Moisture can no.												
Mass of cup + wet soil												
Mass of cup + dry soil												
Mass of water												
Mass of cup, g												
Mass of dry soil, g												
Water content, w%												

Density and Unit Weight: $\rho = M_{ws}/\text{vol}$ g/cm^3; $\gamma_{wet} = \rho \times 9.807$ kN/m^3

Assumed water content						
Water content, w%						
Mass of soil + mold, g						
Mass of mold, g						
Mass of soil in mold, M_{ws}						
Wet unit weight, kN/m^3						
Dry unit weight γ, kN/m^3						

Dry unit weight, kN/m^3

Water content, w%

Optimum moisture = ____________ % Maximum dry unit weight, γ_d = ____________ kN/m^3

COMPACTION TEST Data Sheet 9

Project ______________________ Job No. ______________________

Location of Project ______________________ Boring No. ____________ Sample No. ____________

Description of Soil ______________________

Test Performed By ______________________ Date of Test ______________________

Blows/Layer ____________ No. of Layers ____________ Mass of Rammer ______ N

Mold dimensions: Diam. ____________ cm Ht. ____________ cm Vol. ____________ cm^3

Water Content Determination

Sample no.	1		2		3		4		5		6	
Moisture can no.												
Mass of cup + wet soil												
Mass of cup + dry soil												
Mass of water												
Mass of cup, g												
Mass of dry soil, g												
Water content, w%												

Density and Unit Weight: $\rho = M_{ws}/\text{vol}$ g/cm^3; $\gamma_{wet} = \rho \times 9.807$ kN/m^3

Assumed water content						
Water content, w%						
Mass of soil + mold, g						
Mass of mold, g						
Mass of soil in mold, M_{ws}						
Wet unit weight, kN/m^3						
Dry unit weight γ, kN/m^3						

Dry unit weight, kN/m^3

Water content, w%

Optimum moisture = ____________ % Maximum dry unit weight, γ_d = ____________ kN/m^3

FIELD DENSITY TEST (Sand-cone, Balloon) Data Sheet 10*a*

Project ______________________ Job No. ______________

Location of Project ______________________

Description of Soil ______________________

Test Performed By ______________________ Date of Test ______________

Laboratory Data from Field Test

Sand-cone method

Mass of wet soil + can ______________

Mass of can ______________

Mass of wet soil, M' ______________

Mass of wet soil + pan ______________

Mass of dry soil + pan ______________

Mass of pan ______________

Mass of dry soil ______________

Water content, w% ______________

Balloon method

Mass of wet soil + can ______________

Mass of can ______________

Mass of wet soil, M' ______________

Mass of wet soil + pan ______________

Mass of dry soil + pan ______________

Mass of pan ______________

Mass of dry soil ______________

Water content, w% ______________

Field Data

Sand-cone method

Type of sand used ______________

Density of sand, ρ_{sand} = ______________ g/cm^3

Mass of jug + cone before use ______________ g

Mass of jug + cone after use ______________ g

Mass of sand used (hole + cone) ______________ g

Mass of sand in cone (from calib.) ______________ g

Mass of sand in hole, M ______________ g

Vol. of hole, $V_h = M/\rho_{sand}$ = ______________ cm^3

Balloon method

Correction factor CF = ______________

Final scale reading ______________ cm^3

Initial scale reading ______________ cm^3

Vol. of hole, V'_h ______________ cm^3

Vol. of hole = V'_h (CF) ______________ cm^3

$\rho_{wet} = M'/V_h$ = ______________ g/cm^3

Unit Weight of Soil: Wet $\gamma_{wet} = \rho_{wet} \times 9.807$ = ______________ kN/m^3

Dry $\gamma_{dry} = \gamma_{wet}/(1 + w)$ = ______________ kN/m^3

FIELD DENSITY TEST (Sand-cone, Balloon)

Data Sheet 10*a*

Project ______________________ Job No. ______________

Location of Project ______________________

Description of Soil ______________________

Test Performed By ______________________ Date of Test ______________

Laboratory Data from Field Test

Sand-cone method

Mass of wet soil + can ______________

Mass of can ______________

Mass of wet soil, M' ______________

Mass of wet soil + pan ______________

Mass of dry soil + pan ______________

Mass of pan ______________

Mass of dry soil ______________

Water content, w% ______________

Balloon method

Mass of wet soil + can ______________

Mass of can ______________

Mass of wet soil, M' ______________

Mass of wet soil + pan ______________

Mass of dry soil + pan ______________

Mass of pan ______________

Mass of dry soil ______________

Water content, w% ______________

Field Data

Sand-cone method

Type of sand used ______________

Density of sand, ρ_{sand} = ______________ g/cm^3

Mass of jug + cone before use ______________ g

Mass of jug + cone after use ______________ g

Mass of sand used (hole + cone) ______________ g

Mass of sand in cone (from calib.) ______________ g

Mass of sand in hole, M ______________ g

Vol. of hole, $V_h = M/\rho_{sand}$ = ______________ cm^3

Balloon method

Correction factor CF = ______________

Final scale reading ______________ cm^3

Initial scale reading ______________ cm^3

Vol. of hole, V'_h ______________ cm^3

Vol. of hole = V'_h (CF) ______________ cm^3

$\rho_{wet} = M'/V_h$ = ______________ g/cm^3

Unit Weight of Soil: Wet $\gamma_{wet} = \rho_{wet} \times 9.807$ = ______________ kN/m^3

Dry $\gamma_{dry} = \gamma_{wet}/(1 + w)$ = ______________ kN/m^3

FIELD DENSITY TEST (cont'd) Data Sheet 10*b*

Name ______________________________ Date of Testing ______________________________

Calibration Data

I. *Sand-cone method*

A. Sand density determination

Sand used ______________________________

Type of vol. measure ______________________________ Vol., V_m ______________ cm^3

Mass of sand to fill vol. measure: Trial no. 1 ______________________

Trial no. 2 ______________________

Trial no. 3 ______________________

Average mass M_a = ______________________

Density of sand, $\rho_{sand} = M_a/V_m$ = ______________________________ g/cm^3

B. Mass of sand to fill cone

Mass of filled jug + cone = ______________

Mass after trial No. 1 = ______________ Mass used = ______________

Mass after trial No. 2 = ______________ Mass used = ______________

Mass after trial No. 3 = ______________ Mass used = ______________

Average mass to fill cone = ______________ g

II. *Volumeasure (balloon apparatus) calibration*

Type of container used ______________________________

Vol. of container, V_c = ______________ cm^3

Initial reading ______________

Reading after trial No. 1 ______________ ; Change in vol. ______________ cm^3

Reading after trial No. 2 ______________ ; Change in vol. ______________ cm^3

Reading after trial No. 3 ______________ ; Change in vol. ______________ cm^3

Average ΔV ______________ cm^3

Correction factor CF = $V_c/\Delta V$ = ______________

[Note, if correction factor is less than ± 0.002, neglect it.]

FIELD DENSITY TEST (cont'd) Data Sheet 10*b*

Name ______________________ Date of Testing ______________________

Calibration Data

I. *Sand-cone method*

A. Sand density determination

Sand used ______________________

Type of vol. measure ______________________ Vol., V_m __________ cm^3

Mass of sand to fill vol. measure: Trial no. 1 ______________

Trial no. 2 ______________

Trial no. 3 ______________

Average mass M_a = ______________

Density of sand, $\rho_{sand} = M_a/V_m$ = ______________ g/cm^3

B. Mass of sand to fill cone

Mass of filled jug + cone = ______________

Mass after trial No. 1 = ______________ Mass used = ______________

Mass after trial No. 2 = ______________ Mass used = ______________

Mass after trial No. 3 = ______________ Mass used = ______________

Average mass to fill cone = ______________ g

II. *Volumeasure (balloon apparatus) calibration*

Type of container used ______________________

Vol. of container, V_c = ______________ cm^3

Initial reading ______________

Reading after trial No. 1 ______________ ; Change in vol. ______________ cm^3

Reading after trial No. 2 ______________ ; Change in vol. ______________ cm^3

Reading after trial No. 3 ______________ ; Change in vol. ______________ cm^3

Average ΔV ______________ cm^3

Correction factor CF = $V_c/\Delta V$ = ______________

[Note, if correction factor is less than ± 0.002, neglect it.]

COEFFICIENT OF PERMEABILITY (Constant Head, Falling Head) Data Sheet 11, 12

Project ______________________ Job No. ______________________

Location of Project ______________________

Description of Soil ______________________

Tested by ______________________ Date of Testing ______________________

Sample Dimensions: Diam. ________ cm; Area ________ cm^2

Mass soil + pan Init. ________ g Ht. ________ cm

Mass soil + pan Final ________ g Vol. ________ cm^3

Mass of Sample ________ g Density, ρ ________ g/cm^3

Constant Head

$h =$ ________ cm

Test data / *Test data used*

Test No.	t, s	Q, cm^3	T, °C	Test No.	t, s	Q, cm^3	T, °C
1							
2							
3							
4							
				Average[a]			

$k_T = \frac{QL}{Aht} =$ ________ $\alpha = \eta_t/\eta_{20} =$ ________

$=$ ________ cm/s $k_{20} = \alpha k_T =$ ________ cm/s

Falling Head

Standpipe = [burette, other (specify)] ________ Area standpipe, $a =$ ________ cm^2

Test data[b] / *Test data used*

Test no.	h_1, cm	h_2, cm	t, s	Q_{in}, cm^3	Q_{out}, cm^3	T, °C	Test no.	h_1, cm	h_2, cm	t, s	T, °C
1											
2											
3											
4											
							Average				

$\alpha = \eta_T/\eta_{20} =$ ________

$k_T = \frac{aL}{At} \ln \frac{h_1}{h_2} =$ ________ $=$ ________ cm/s

$k_{20} = \alpha k_T =$ ________ $=$ ________ cm/s

[a]Use averaged values only if there is a small difference in test temperature, say, 1–2°C.
[b]Simplify by using the same h_1 and h_2 each time, so you can average k.

COEFFICIENT OF PERMEABILITY (Constant Head, Falling Head) Data Sheet 11, 12

Project ______________________ Job No. ______________________

Location of Project ______________________

Description of Soil ______________________

Tested by ______________________ Date of Testing ______________________

Sample Dimensions: Diam. __________ cm; Area __________ cm^2

Mass soil + pan Init. __________ g Ht. __________ cm

Mass soil + pan Final __________ g Vol. __________ cm^3

Mass of Sample __________ g Density, ρ __________ g/cm^3

Constant Head

$h =$ __________ cm

Test data / *Test data used*

Test No.	t, s	Q, cm^3	T, °C	Test No.	t, s	Q, cm^3	T, °C
1							
2							
3							
4							
				Average[a]			

$k_T = \frac{QL}{Aht} =$ __________ $\alpha = \eta_t/\eta_{20} =$ __________

$=$ __________ cm/s $k_{20} = \alpha k_T =$ __________ cm/s

Falling Head

Standpipe = [burette, other (specify)] __________ Area standpipe, $a =$ __________ cm^2

Test data[b] / *Test data used*

Test no.	h_1, cm	h_2, cm	t, s	Q_{in}, cm^3	Q_{out}, cm^3	T, °C	Test no.	h_1, cm	h_2, cm	t, s	T, °C
1											
2											
3											
4											
							Average				

$\alpha = \eta_T/\eta_{20} =$ __________

$k_T = \frac{aL}{At} \ln \frac{h_1}{h_2} =$ __________ $=$ __________ cm/s

$k_{20} = \alpha k_T =$ __________ $=$ __________ cm/s

[a]Use averaged values only if there is a small difference in test temperature, say, 1–2°C.
[b]Simplify by using the same h_1 and h_2 each time, so you can average k.

COEFFICIENT OF PERMEABILITY (Constant Head, Falling Head) Data Sheet 11, 12

Project ______________________ Job No. ______________________

Location of Project ______________________

Description of Soil ______________________

Tested by ______________________ Date of Testing ______________________

Sample Dimensions: Diam. __________ cm; Area __________ cm^2

Mass soil + pan Init. __________ g Ht. __________ cm

Mass soil + pan Final __________ g Vol. __________ cm^3

Mass of Sample __________ g Density, ρ __________ g/cm^3

Constant Head

$h =$ __________ cm

Test data / *Test data used*

Test No.	t, s	Q, cm^3	T, °C	Test No.	t, s	Q, cm^3	T, °C
1							
2							
3							
4							
				Average[a]			

$k_T = \frac{QL}{Aht} =$ __________

$=$ __________ cm/s

$\alpha = \eta_t/\eta_{20} =$ __________

$k_{20} = \alpha k_T =$ __________ cm/s

Falling Head

Standpipe = [burette, other (specify)] __________ Area standpipe, $a =$ __________ cm^2

Test data[b] / *Test data used*

Test no.	h_1, cm	h_2, cm	t, s	Q_{in}, cm^3	Q_{out}, cm^3	T, °C	Test no.	h_1, cm	h_2, cm	t, s	T, °C
1											
2											
3											
4											
							Average				

$\alpha = \eta_T/\eta_{20} =$ __________

$k_T = \frac{aL}{At}\ln\frac{h_1}{h_2} =$ __________ $=$ __________ cm/s

$k_{20} = \alpha k_T =$ __________ $=$ __________ cm/s

[a]Use averaged values only if there is a small difference in test temperature, say, 1–2°C.
[b]Simplify by using the same h_1 and h_2 each time, so you can average k.

COEFFICIENT OF PERMEABILITY (Constant Head, Falling Head) Data Sheet 11, 12

Project ______________________ Job No. ______________________

Location of Project ______________________

Description of Soil ______________________

Tested by ______________________ Date of Testing ______________________

Sample Dimensions: Diam. ____________ cm; Area ____________ cm^2

Mass soil + pan Init. ____________ g Ht. ____________ cm

Mass soil + pan Final ____________ g Vol. ____________ cm^3

Mass of Sample ____________ g Density, ρ ____________ g/cm^3

Constant Head

$h =$ ____________ cm

Test data / *Test data used*

Test No.	t, s	Q, cm^3	T, °C	Test No.	t, s	Q, cm^3	T, °C
1							
2							
3							
4							
				Average[a]			

$k_T = \frac{QL}{Aht} =$ ____________ $\alpha = \eta_t/\eta_{20} =$ ____________

$=$ ____________ cm/s $k_{20} = \alpha k_T =$ ____________ cm/s

Falling Head

Standpipe = [burette, other (specify)] ____________ Area standpipe, $a =$ ____________ cm^2

Test data[b] / *Test data used*

Test no.	h_1, cm	h_2, cm	t, s	Q_{in}, cm^3	Q_{out}, cm^3	T, °C	Test no.	h_1, cm	h_2, cm	t, s	T, °C
1											
2											
3											
4											
							Average				

$\alpha = \eta_T/\eta_{20} =$ ____________

$k_T = \frac{aL}{At} \ln \frac{h_1}{h_2} =$ ____________ $=$ ____________ cm/s

$k_{20} = \alpha k_T =$ ____________ $=$ ____________ cm/s

[a]Use averaged values only if there is a small difference in test temperature, say, 1–2°C.
[b]Simplify by using the same h_1 and h_2 each time, so you can average k.

CONSOLIDATION TEST (Sample data) Data Sheet 13*a*

Project __________ Job No. __________

Location __________ Boring No. __________ Sample No. __________

Description of Soil __________ Depth of Sample __________

Tested By __________ Date of Testing __________

Ring: Diam. = __________ mm; Area, A_r = __________ cm^2; Ht. = __________ mm

Sample Ht., H_i = __________ mm Soil, G_s = __________

Equations for H_s

$$H_s = H_i - \Delta H - H_{vf}$$

also

$$H_s = M_s/(G_s \rho_w A_r)$$

Mass of ring + wet sample at start of test = __________ g

Mass of ring = __________ g

Mass of wet soil = __________ g

Initial water content[1], W_i = __________ %

Mass of dry soil computed from init. water content M_s = __________ g

Final Water Content Determination — End of Test

Mass dish + ring + wet soil cake = __________ g

Mass dish + ring + dry soil cake = __________ g

Mass dish + ring = __________ g

Mass of oven dry soil cake M_s = __________ g

Mass water in soil cake M_{wf} = __________ g

Final water content $W_f = M_{wf}/M_s$ = __________ %

Compute: Ht. of soil solids H_s = __________ cm

Initial ht. of voids, $H_{vi} = H_i - H_s$ = __________ cm

Initial void ratio $e_o = H_{vi}/H_s$ = __________ cm

Final Test Data

Dial readings: Initial D_1 = __________ Final D_2 = __________

Change in sample ht., $\Delta H = D_2 - D_1 - \Delta H_e$ = __________

Final ht. voids, $H_{vf} = M_{wf}/A_r$ = __________ cm

Final void ratio, $e_f = H_{vf}/H_s$ = __________

____

____

[1]Use water content sheet from data sheet set (Data Sheet 1).

CONSOLIDATION TEST (Sample data) Data Sheet 13*a*

Project ______________________ Job No. ______________________

Location ______________________ Boring No. ____________ Sample No. ____________

Description of Soil ______________________ Depth of Sample ____________

Tested By ______________________ Date of Testing ______________________

Ring: Diam. = ____________ mm; Area, A_r = ____________ cm^2; Ht. = ____________ mm

Sample Ht., H_i = ____________ mm Soil, G_s = ____________

Mass of ring + wet sample at start of test = ____________ g

Mass of ring = ____________ g

Mass of wet soil = ____________ g

Initial water content[1], W_i = ____________ %

Mass of dry soil computed from init. water content M_s = ____________ g

Equations for H_s

$$H_s = H_i - \Delta H - H_{vf}$$

also

$$H_s = M_s/(G_s \rho_w A_r)$$

Final Water Content Determination — End of Test

Mass dish + ring + wet soil cake = ____________ g

Mass dish + ring + dry soil cake = ____________ g

Mass dish + ring = ____________ g

Mass of oven dry soil cake M_s = ____________ g

Mass water in soil cake M_{wf} = ____________ g

Final water content $W_f = M_{wf}/M_s$ = ____________ %

Compute: Ht. of soil solids H_s = ______________________ cm

Initial ht. of voids, $H_{vi} = H_i - H_s$ = ______________________ cm

Initial void ratio $e_o = H_{vi}/H_s$ = ______________________ cm

Final Test Data

Dial readings: Initial D_1 = ____________ Final D_2 = ____________

Change in sample ht., $\Delta H = D_2 - D_1 - \Delta H_e$ = ______________________

Final ht. voids, $H_{vf} = M_{wf}/A_r$ = ______________________ cm

Final void ratio, $e_f = H_{vf}/H_s$ = ______________________

____

____

[1]Use water content sheet from data sheet set (Data Sheet 1).

CONSOLIDATION TEST (Sample data) Data Sheet 13*a*

Project ______________________ Job No. ______________________

Location ______________________ Boring No. ____________ Sample No. ____________

Description of Soil ______________________ Depth of Sample ____________

Tested By ______________________ Date of Testing ______________________

Ring: Diam. = ____________ mm; Area, A_r = ____________ cm^2; Ht. = ____________ mm

Sample Ht., H_i = ____________ mm Soil, G_s = ____________

Mass of ring + wet sample at start of test = ____________ g

Mass of ring = ____________ g

Mass of wet soil = ____________ g

Initial water content[1], W_i = ____________ %

Mass of dry soil computed from init. water content M_s = ____________ g

Equations for H_s

$$H_s = H_i - \Delta H - H_{vf}$$

also

$$H_s = M_s/(G_s \rho_w A_r)$$

Final Water Content Determination — End of Test

Mass dish + ring + wet soil cake = ____________ g

Mass dish + ring + dry soil cake = ____________ g

Mass dish + ring = ____________ g

Mass of oven dry soil cake M_s = ____________ g

Mass water in soil cake M_{wf} = ____________ g

Final water content $W_f = M_{wf}/M_s$ = ____________ %

Compute: Ht. of soil solids H_s = ______________________ cm

Initial ht. of voids, $H_{vi} = H_i - H_s$ = ______________________ cm

Initial void ratio $e_o = H_{vi}/H_s$ = ______________________ cm

Final Test Data

Dial readings: Initial D_1 = ____________ Final D_2 = ____________

Change in sample ht., $\Delta H = D_2 - D_1 - \Delta H_e$ = ______________________

Final ht. voids, $H_{vf} = M_{wf}/A_r$ = ______________________ cm

Final void ratio, $e_f = H_{vf}/H_s$ = ______________________

____

____

[1]Use water content sheet from data sheet set (Data Sheet 1).

CONSOLIDATION TEST (Time-compression data) Data Sheet 13*b*

Project ______________________ Job No. ______________________

Location of Project ______________________ Boring No. __________ Sample No. __________

Tested By ______________________ Date of Testing ______________________

Loading Test Data

Load __________ kPa, ksf

Date applied __________

Applied by __________

Clock time and date	Elapsed time, min	*Dial readings × _____	
		Original	Adjusted**
	0		
	0.1		
	0.25		
	0.5		
	1		
	2		
	4		
	8		
	15		
	30		
	60		

Load __________ kPa, ksf

Date applied __________

Applied by __________

Clock time and date	Elapsed time, min	*Dial readings × _____	
		Original	Adjusted**
	0		
	0.1		
	0.25		
	0.5		
	1		
	2		
	4		
	8		
	15		
	30		
	60		

*Insert gauge subdivisions 0.01 mm/div, etc.; if LVDT insert deformation, in or mm.

**Adjust if reset dial gauge for large compression.

CONSOLIDATION TEST (Time-compression data) Data Sheet 13*b*

Project ______________________ Job No. ______________________

Location of Project ______________________ Boring No. __________ Sample No. __________

Tested By ______________________ Date of Testing ______________________

Loading Test Data

Load __________ kPa, ksf

Date applied __________

Applied by __________

Clock time and date	Elapsed time, min	*Dial readings × _____	
		Original	Adjusted**
	0		
	0.1		
	0.25		
	0.5		
	1		
	2		
	4		
	8		
	15		
	30		
	60		

Load __________ kPa, ksf

Date applied __________

Applied by __________

Clock time and date	Elapsed time, min	*Dial readings × _____	
		Original	Adjusted**
	0		
	0.1		
	0.25		
	0.5		
	1		
	2		
	4		
	8		
	15		
	30		
	60		

*Insert gauge subdivisions 0.01 mm/div, etc.; if LVDT insert deformation, in or mm.
**Adjust if reset dial gauge for large compression.

CONSOLIDATION TEST (Time-compression data) Data Sheet 13*b*

Project ______________________ Job No. ______________________

Location of Project ______________________ Boring No. ____________ Sample No. ____________

Tested By ______________________ Date of Testing ______________________

Loading Test Data

Load ____________ kPa, ksf

Date applied ____________

Applied by ____________

Clock time and date	Elapsed time, min	*Dial readings × _____	
		Original	Adjusted**
	0		
	0.1		
	0.25		
	0.5		
	1		
	2		
	4		
	8		
	15		
	30		
	60		

Load ____________ kPa, ksf

Date applied ____________

Applied by ____________

Clock time and date	Elapsed time, min	*Dial readings × _____	
		Original	Adjusted**
	0		
	0.1		
	0.25		
	0.5		
	1		
	2		
	4		
	8		
	15		
	30		
	60		

*Insert gauge subdivisions 0.01 mm/div, etc.; if LVDT insert deformation, in or mm.
**Adjust if reset dial gauge for large compression.

CONSOLIDATION TEST (Time-compression data) Data Sheet 13*b*

Project ______________________ Job No. ______________________

Location of Project ______________________ Boring No. __________ Sample No. __________

Tested By ______________________ Date of Testing ______________________

Loading Test Data

Load __________ kPa, ksf

Date applied __________

Applied by __________

Clock time and date	Elapsed time, min	*Dial readings × _____	
		Original	Adjusted**
	0		
	0.1		
	0.25		
	0.5		
	1		
	2		
	4		
	8		
	15		
	30		
	60		

Load __________ kPa, ksf

Date applied __________

Applied by __________

Clock time and date	Elapsed time, min	*Dial readings × _____	
		Original	Adjusted**
	0		
	0.1		
	0.25		
	0.5		
	1		
	2		
	4		
	8		
	15		
	30		
	60		

*Insert gauge subdivisions 0.01 mm/div, etc.; if LVDT insert deformation, in or mm.
**Adjust if reset dial gauge for large compression.

CONSOLIDATION TEST (Time-compression data) Data Sheet 13*b*

Project ______________________ Job No. ______________________

Location of Project ______________________ Boring No. __________ Sample No. __________

Tested By ______________________ Date of Testing ______________________

Loading Test Data

Load __________ kPa, ksf

Date applied __________

Applied by __________

Clock time and date	Elapsed time, min	*Dial readings × _____	
		Original	Adjusted**
	0		
	0.1		
	0.25		
	0.5		
	1		
	2		
	4		
	8		
	15		
	30		
	60		

Load __________ kPa, ksf

Date applied __________

Applied by __________

Clock time and date	Elapsed time, min	*Dial readings × _____	
		Original	Adjusted**
	0		
	0.1		
	0.25		
	0.5		
	1		
	2		
	4		
	8		
	15		
	30		
	60		

*Insert gauge subdivisions 0.01 mm/div, etc.; if LVDT insert deformation, in or mm.
**Adjust if reset dial gauge for large compression.

CONSOLIDATION TEST (Time-compression data)

Data Sheet 13*b*

Project ________________ Job No. ________________

Location of Project ________________ Boring No. ________ Sample No. ________

Tested By ________________ Date of Testing ________________

Loading Test Data

Load ________ kPa, ksf	Load ________ kPa, ksf
Date applied ________	Date applied ________
Applied by ________	Applied by ________

Clock time and date	Elapsed time, min	*Dial readings × ____		Clock time and date	Elapsed time, min	*Dial readings × ____	
		Original	Adjusted**			Original	Adjusted**
	0				0		
	0.1				0.1		
	0.25				0.25		
	0.5				0.5		
	1				1		
	2				2		
	4				4		
	8				8		
	15				15		
	30				30		
	60				60		

*Insert gauge subdivisions 0.01 mm/div, etc.; if LVDT insert deformation, in or mm.
**Adjust if reset dial gauge for large compression.

CONSOLIDATION TEST (Time-compression data) Data Sheet 13*b*

Project ______________________ Job No. ______________________

Location of Project ______________________ Boring No. __________ Sample No. __________

Tested By ______________________ Date of Testing ______________________

Loading Test Data

Load __________ kPa, ksf

Date applied __________

Applied by __________

Clock time and date	Elapsed time, min	*Dial readings × _____	
		Original	Adjusted**
	0		
	0.1		
	0.25		
	0.5		
	1		
	2		
	4		
	8		
	15		
	30		
	60		

Load __________ kPa, ksf

Date applied __________

Applied by __________

Clock time and date	Elapsed time, min	*Dial readings × _____	
		Original	Adjusted**
	0		
	0.1		
	0.25		
	0.5		
	1		
	2		
	4		
	8		
	15		
	30		
	60		

*Insert gauge subdivisions 0.01 mm/div, etc.; if LVDT insert deformation, in or mm.
**Adjust if reset dial gauge for large compression.

CONSOLIDATION TEST (Time-compression data) Data Sheet 13*b*

Project ______________________ Job No. ______________________

Location of Project ______________________ Boring No. __________ Sample No. __________

Tested By ______________________ Date of Testing ______________________

Loading Test Data

Load __________ kPa, ksf

Date applied __________

Applied by __________

Clock time and date	Elapsed time, min	*Dial readings × ____	
		Original	Adjusted**
	0		
	0.1		
	0.25		
	0.5		
	1		
	2		
	4		
	8		
	15		
	30		
	60		

Load __________ kPa, ksf

Date applied __________

Applied by __________

Clock time and date	Elapsed time, min	*Dial readings × ____	
		Original	Adjusted**
	0		
	0.1		
	0.25		
	0.5		
	1		
	2		
	4		
	8		
	15		
	30		
	60		

*Insert gauge subdivisions 0.01 mm/div, etc.; if LVDT insert deformation, in or mm.
**Adjust if reset dial gauge for large compression.

CONSOLIDATION TEST (Time-compression data) Data Sheet 13*b*

Project ____________________ Job No. ____________________

Location of Project ____________________ Boring No. __________ Sample No. __________

Tested By ____________________ Date of Testing ____________________

Loading Test Data

Load __________ kPa, ksf

Date applied __________

Applied by __________

Clock time and date	Elapsed time, min	*Dial readings × _____	
		Original	Adjusted**
	0		
	0.1		
	0.25		
	0.5		
	1		
	2		
	4		
	8		
	15		
	30		
	60		

Load __________ kPa, ksf

Date applied __________

Applied by __________

Clock time and date	Elapsed time, min	*Dial readings × _____	
		Original	Adjusted**
	0		
	0.1		
	0.25		
	0.5		
	1		
	2		
	4		
	8		
	15		
	30		
	60		

*Insert gauge subdivisions 0.01 mm/div, etc.; if LVDT insert deformation, in or mm.
**Adjust if reset dial gauge for large compression.

CONSOLIDATION TEST (Time-compression data) Data Sheet 13*b*

Project ______________________ Job No. ______________________

Location of Project ______________________ Boring No. __________ Sample No. __________

Tested By ______________________ Date of Testing ______________________

Loading Test Data

Load __________ kPa, ksf Load __________ kPa, ksf

Date applied __________ Date applied __________

Applied by __________ Applied by __________

Clock time and date	Elapsed time, min	*Dial readings × _____		Clock time and date	Elapsed time, min	*Dial readings × _____	
		Original	Adjusted**			Original	Adjusted**
	0				0		
	0.1				0.1		
	0.25				0.25		
	0.5				0.5		
	1				1		
	2				2		
	4				4		
	8				8		
	15				15		
	30				30		
	60				60		

*Insert gauge subdivisions 0.01 mm/div, etc.; if LVDT insert deformation, in or mm.
**Adjust if reset dial gauge for large compression.

CONSOLIDATION TEST (Time-compression data) Data Sheet 13*b*

Project ____________________ Job No. ____________________

Location of Project ____________________ Boring No. __________ Sample No. __________

Tested By ____________________ Date of Testing ____________________

Loading Test Data

Load __________ kPa, ksf Load __________ kPa, ksf

Date applied __________ Date applied __________

Applied by __________ Applied by __________

Clock time and date	Elapsed time, min	*Dial readings × _____		Clock time and date	Elapsed time, min	*Dial readings × _____	
		Original	Adjusted**			Original	Adjusted**
	0				0		
	0.1				0.1		
	0.25				0.25		
	0.5				0.5		
	1				1		
	2				2		
	4				4		
	8				8		
	15				15		
	30				30		
	60				60		

*Insert gauge subdivisions 0.01 mm/div, etc.; if LVDT insert deformation, in or mm.
**Adjust if reset dial gauge for large compression.

CONSOLIDATION TEST (Time-compression data)

Data Sheet 13*b*

Project ______________________ Job No. ______________________

Location of Project ______________________ Boring No. __________ Sample No. __________

Tested By ______________________ Date of Testing ______________________

Loading Test Data

Load ________ kPa, ksf	Load ________ kPa, ksf
Date applied ________	Date applied ________
Applied by ________	Applied by ________

Clock time and date	Elapsed time, min	*Dial readings × ____		Clock time and date	Elapsed time, min	*Dial readings × ____	
		Original	Adjusted**			Original	Adjusted**
	0				0		
	0.1				0.1		
	0.25				0.25		
	0.5				0.5		
	1				1		
	2				2		
	4				4		
	8				8		
	15				15		
	30				30		
	60				60		

*Insert gauge subdivisions 0.01 mm/div, etc.; if LVDT insert deformation, in or mm.

**Adjust if reset dial gauge for large compression.

CONSOLIDATION TEST (Time-compression data) Data Sheet 13*b*

Project ____________________ Job No. ____________________

Location of Project ____________________ Boring No. ____________ Sample No. ____________

Tested By ____________________ Date of Testing ____________________

Loading Test Data

Load ____________ kPa, ksf Load ____________ kPa, ksf

Date applied ____________ Date applied ____________

Applied by ____________ Applied by ____________

Clock time and date	Elapsed time, min	*Dial readings × _____		Clock time and date	Elapsed time, min	*Dial readings × _____	
		Original	Adjusted**			Original	Adjusted**
	0				0		
	0.1				0.1		
	0.25				0.25		
	0.5				0.5		
	1				1		
	2				2		
	4				4		
	8				8		
	15				15		
	30				30		
	60				60		

*Insert gauge subdivisions 0.01 mm/div, etc.; if LVDT insert deformation, in or mm.
**Adjust if reset dial gauge for large compression.

CONSOLIDATION TEST (Time-compression data) Data Sheet 13*b*

Project ____________________ Job No. ____________________

Location of Project ____________________ Boring No. __________ Sample No. __________

Tested By ____________________ Date of Testing ____________________

Loading Test Data

Load __________ kPa, ksf

Date applied __________

Applied by __________

Clock time and date	Elapsed time, min	*Dial readings × _____	
		Original	Adjusted**
	0		
	0.1		
	0.25		
	0.5		
	1		
	2		
	4		
	8		
	15		
	30		
	60		

Load __________ kPa, ksf

Date applied __________

Applied by __________

Clock time and date	Elapsed time, min	*Dial readings × _____	
		Original	Adjusted**
	0		
	0.1		
	0.25		
	0.5		
	1		
	2		
	4		
	8		
	15		
	30		
	60		

*Insert gauge subdivisions 0.01 mm/div, etc.; if LVDT insert deformation, in or mm.

**Adjust if reset dial gauge for large compression.

CONSOLIDATION TEST (Time-compression data) Data Sheet 13*b*

Project ____________ Job No. ____________

Location of Project ____________ Boring No. ________ Sample No. ________

Tested By ____________ Date of Testing ____________

Loading Test Data

Load ________ kPa, ksf

Date applied ________

Applied by ________

Clock time and date	Elapsed time, min	*Dial readings × ____	
		Original	Adjusted**
	0		
	0.1		
	0.25		
	0.5		
	1		
	2		
	4		
	8		
	15		
	30		
	60		

Load ________ kPa, ksf

Date applied ________

Applied by ________

Clock time and date	Elapsed time, min	*Dial readings × ____	
		Original	Adjusted**
	0		
	0.1		
	0.25		
	0.5		
	1		
	2		
	4		
	8		
	15		
	30		
	60		

*Insert gauge subdivisions 0.01 mm/div, etc.; if LVDT insert deformation, in or mm.
**Adjust if reset dial gauge for large compression.

CONSOLIDATION TEST (Time-compression data)

Data Sheet 13*b*

Project ______________________ Job No. ______________________

Location of Project ______________________ Boring No. __________ Sample No. __________

Tested By ______________________ Date of Testing ______________________

Loading Test Data

Load ____________ kPa, ksf

Date applied ____________

Applied by ____________

Clock time and date	Elapsed time, min	*Dial readings × _____	
		Original	Adjusted**
	0		
	0.1		
	0.25		
	0.5		
	1		
	2		
	4		
	8		
	15		
	30		
	60		

Load ____________ kPa, ksf

Date applied ____________

Applied by ____________

Clock time and date	Elapsed time, min	*Dial readings × _____	
		Original	Adjusted**
	0		
	0.1		
	0.25		
	0.5		
	1		
	2		
	4		
	8		
	15		
	30		
	60		

*Insert gauge subdivisions 0.01 mm/div, etc.; if LVDT insert deformation, in or mm.

**Adjust if reset dial gauge for large compression.

CONSOLIDATION TEST (Computation sheet for e, $\in$ and c_v) Data Sheet 13c

Project ______________________ Job No. ______________________

Computed by ______________________ Date __________ Test Date ______________________

Sample Data (fixed ring, floating-ring) Ring diam. = ______________ mm

Sample heights: H_i = __________ mm H_v = __________ mm H_s = __________ mm

Sample void ratios: e_o = __________ e_f = __________

Method for D_{100}: (use D_{100}) __________ (D at end of load) __________ (D at specified time) __________ (check one)

Load increment ____ (1)	Def dial reading at end of load ____ /div (2)	D_{50}/D_{100} ____ (3)	Equipment deform ΔH_e ____ (4)	VR = e Strain = $\in$ VR/Strain (5)	Average sample ht H ____ (6)	H used for c_v ____ (7)	t_{50} from DR vs time curves, min (8)	c_v ____ /min (9)

Note: Insert units in column headings as necessary.

[a]Average height = $H_i - D_{50} + \Delta H_e$

$e = e_o - \Delta H/H_s$; $\in = \Delta H/H_i$; $\Delta H = D_{100} - \Delta H_e$

CONSOLIDATION TEST (Computation sheet for e, ϵ and c_v) Data Sheet 13*c*

Project ______________________ Job No. ______________________

Computed by ______________ Date __________ Test Date ______________

Sample Data (fixed ring, floating-ring) Ring diam. = ______________ mm

Sample heights: H_i = __________ mm H_v = __________ mm H_s = __________ mm

Sample void ratios: e_o = __________ e_f = __________

Method for D_{100}: (use D_{100}) ________ (D at end of load) ________ (D at specified time) ________ (check one)

Load increment ____ (1)	Def dial reading at end of load ____/div (2)	D_{50}/D_{100} ____ (3)	Equipment deform ΔH_e ____ (4)	VR = e Strain = ϵ VR/Strain (5)	Average sample ht H ____ (6)	H used for c_v ____ (7)	t_{50} from DR vs time curves, min (8)	c_v ____/min (9)

Note: Insert units in column headings as necessary.
[a]Average height = $H_i - D_{50} + \Delta H_e$
$e = e_o - \Delta H/H_s$; $\epsilon = \Delta H/H_i$; $\Delta H = D_{100} - \Delta H_e$

CONSOLIDATION TEST (Computation sheet for e, ϵ and c_v) Data Sheet 13c

Project ______________________ Job No. ______________________

Computed by ______________ Date __________ Test Date ______________

Sample Data (fixed ring, floating-ring) Ring diam. = ______________ mm

Sample heights: H_i = __________ mm H_v = __________ mm H_s = __________ mm

Sample void ratios: e_o = __________ e_f = __________

Method for D_{100}: (use D_{100}) ________ (D at end of load) ________ (D at specified time) ________ (check one)

Load increment ____ (1)	Def dial reading at end of load ____ /div (2)	D_{50}/D_{100} ____ (3)	Equipment deform ΔH_e ____ (4)	VR = e Strain = ϵ VR/Strain (5)	Average sample ht H ____ (6)	H used for c_v ____ (7)	t_{50} from DR vs time curves, min (8)	c_v ____ /min (9)

Note: Insert units in column headings as necessary.

[a]Average height = $H_i - D_{50} + \Delta H_e$

$e = e_o - \Delta H/H_s$; $\epsilon = \Delta H/H_i$; $\Delta H = D_{100} - \Delta H_e$

UNCONFINED COMPRESSION TEST Type of Sample (Undisturbed, Remolded) Data Sheet 14

Project ______________________ Job No. ______________

Location of Project ______________________

Description of Soil ______________________

Tested By ______________________ Date of Testing ______________

Sample Data

Strain rate ______________

Diam. ____________ Area A_o ____________ Ht., L_o ____________

Vol. ____________ Mass ____________ g Wet density ____________

Water content, w% ____________ Dry density, ρ_d ____________ LRC ____________

Deformation dial[1] reading ( )	Load dial[1] (units)	Sample deformation ΔL, ( )	Unit strain, ϵ $\Delta L/L_o$	Area CF $1-\epsilon$	Corrected area A'. ( )	Total load on sample[2] (col. 2 × LRC)	Sample stress, σ_1 kPa
1	2	3	4	5	6	7	8

Note: Insert units in column headings as necessary.
[1]omit if using LVDT or load cell
[2]insert direct reading of load if using load cell.

Unconfined compressive strength q_u = ____________ Cohesion $s_u = q_u/2$ = ____________

UNCONFINED COMPRESSION TEST Type of Sample (Undisturbed, Remolded) Data Sheet 14

Project ______________________ Job No. ______________________

Location of Project ______________________

Description of Soil ______________________

Tested By ______________________ Date of Testing ______________________

Sample Data

Strain rate ______________________

Diam. ____________ Area A_o ____________ Ht., L_o ____________

Vol. ____________ Mass ____________ g Wet density ____________

Water content, w% ____________ Dry density, ρ_d ____________ LRC ____________

Deformation dial[1] reading ( )	Load dial[1] (units)	Sample deformation ΔL, ( )	Unit strain, ϵ $\Delta L/L_o$	Area CF $1-\epsilon$	Corrected area A'. ( )	Total load on sample[2] (col. 2 × LRC)	Sample stress, σ_1 kPa
1	2	3	4	5	6	7	8

Note: Insert units in column headings as necessary.
[1]omit if using LVDT or load cell
[2]insert direct reading of load if using load cell.

Unconfined compressive strength q_u = ____________ Cohesion $s_u = q_u/2$ = ____________

UNCONFINED COMPRESSION TEST Type of Sample (Undisturbed, Remolded) Data Sheet 14

Project __________ Job No. __________

Location of Project __________

Description of Soil __________

Tested By __________ Date of Testing __________

Sample Data

Strain rate __________

Diam. __________ Area A_o __________ Ht., L_o __________

Vol. __________ Mass __________ g Wet density __________

Water content, w% __________ Dry density, ρ_d __________ LRC __________

Deformation dial[1] reading ( )	Load dial[1] (units)	Sample deformation ΔL, ( )	Unit strain, ϵ $\Delta L/L_o$	Area CF $1-\epsilon$	Corrected area A'. ( )	Total load on sample[2] (col. 2 × LRC)	Sample stress, σ_1 kPa
1	2	3	4	5	6	7	8

Note: Insert units in column headings as necessary.
[1]omit if using LVDT or load cell
[2]insert direct reading of load if using load cell.

Unconfined compressive strength q_u = __________ Cohesion $s_u = q_u/2$ = __________

UNCONFINED COMPRESSION TEST Type of Sample (Undisturbed, Remolded) Data Sheet 14

Project ______________________________ Job No. ______________

Location of Project ______________________________

Description of Soil ______________________________

Tested By ______________________________ Date of Testing ______________

Sample Data

Strain rate ______________

Diam. ______________ Area A_o ______________ Ht., L_o ______________

Vol. ______________ Mass ______________ g Wet density ______________

Water content, w% ______________ Dry density, ρ_d ______________ LRC ______________

Deformation dial[1] reading ( )	Load dial[1] (units)	Sample deformation ΔL, ( )	Unit strain, ϵ $\Delta L/L_o$	Area CF $1-\epsilon$	Corrected area A'. ( )	Total load on sample[2] (col. 2 × LRC)	Sample stress, σ_1 kPa
1	2	3	4	5	6	7	8

Note: Insert units in column headings as necessary.
[1]omit if using LVDT or load cell
[2]insert direct reading of load if using load cell.

Unconfined compressive strength q_u = ______________ Cohesion $s_u = q_u/2$ = ______________

TRIAXIAL COMPRESSION TEST(Cohesive, Cohesionless)

Data Sheet 15*a*

Project ______________________ Job No. ______________

Location of Project ______________________ Boring No. __________ Sample No. __________

Description of Soil ______________________ Depth of Sample __________

Tested By ______________________ Date of Testing ______________

Fill in the blanks with data and appropriate units.

Sample Data

Sample dimensions: Diam. D_0 = ____________ mm Area A_0 = ____________ cm^2 L_0 = ____________ mm

Vol. V_0 = ____________ cm^3 Water content w = ____________ % Degree of saturation S = ____________ %

Mass M_0 = ____________ g ______________________

For Cohesionless soils

Initial mass of dish + sand = ____________

Final mass of dish + sand = ____________

Mass of sand used in specimen, M_o = ____________

Specific gravity of sand G_s = ____________

Vol. of soil solids in test specimen V_s = ____________

Vol. of voids in test specimen (initial) V_v = ____________

Initial void ratio of test specimen e_i = ____________

Void ratio of sand at minimum density e_{max} = ____________

Void ratio of sand at maximum density e_{min} = ____________

Relative density of test specimen D_r = ____________

Density of test specimen (cohesive, cohesionless) $\rho = M_0/V_0$ = ____________

Machine Data

Rate of loading ________.___/min (insert mm or in)

The following data may not be applicable if machine can be adjusted to tare these effects out of the load readings.

Cross section area of loading piston, A_p = ____________ cm^2

Upward load on piston = $A_p \sigma_3$ = ____________

Mass of loading piston = ____________

Computed value of initial sample load = ____________

TRIAXIAL COMPRESSION TEST(Cohesive, Cohesionless) Data Sheet 15*a*

Project ______________________ Job No. ______________________

Location of Project ______________________ Boring No. __________ Sample No. __________

Description of Soil ______________________ Depth of Sample __________

Tested By ______________________ Date of Testing ______________________

Fill in the blanks with data and appropriate units.

Sample Data

Sample dimensions: Diam. D_0 = __________ mm Area A_0 = __________ cm^2 L_0 = __________ mm

Vol. V_0 = __________ cm^3 Water content w = __________ % Degree of saturation S = __________ %

Mass M_0 = __________ g ______________________

For Cohesionless soils

Initial mass of dish + sand = __________

Final mass of dish + sand = __________

Mass of sand used in specimen, M_o = __________

Specific gravity of sand G_s = __________

Vol. of soil solids in test specimen V_s = __________

Vol. of voids in test specimen (initial) V_v = __________

Initial void ratio of test specimen e_i = __________

Void ratio of sand at minimum density e_{max} = __________

Void ratio of sand at maximum density e_{min} = __________

Relative density of test specimen D_r = __________

Density of test specimen (cohesive, cohesionless) $\rho = M_0/V_0$ = __________

Machine Data

Rate of loading __________.____/min (insert mm or in)

The following data may not be applicable if machine can be adjusted to tare these effects out of the load readings.

Cross section area of loading piston, A_p = __________ cm^2

Upward load on piston = $A_p \sigma_3$ = __________

Mass of loading piston = __________

Computed value of initial sample load = __________

TRIAXIAL COMPRESSION TEST(Cohesive, Cohesionless) Data Sheet 15*a*

Project ______________________________ Job No. ______________________

Location of Project ______________________ Boring No. __________ Sample No. __________

Description of Soil ______________________________ Depth of Sample __________

Tested By ______________________________ Date of Testing ______________________

Fill in the blanks with data and appropriate units.

Sample Data

Sample dimensions: Diam. D_0 = ______________ mm Area A_0 = ______________ cm^2 L_0 = ______________ mm

Vol. V_0 = ______________ cm^3 Water content w = ______________ % Degree of saturation S = ______________ %

Mass M_0 = ______________ g ______________________________________

For Cohesionless soils

Initial mass of dish + sand = ______________

Final mass of dish + sand = ______________

Mass of sand used in specimen, M_o = ______________

Specific gravity of sand G_s = ______________

Vol. of soil solids in test specimen V_s = ______________

Vol. of voids in test specimen (initial) V_v = ______________

Initial void ratio of test specimen e_i = ______________

Void ratio of sand at minimum density e_{max} = ______________

Void ratio of sand at maximum density e_{min} = ______________

Relative density of test specimen D_r = ______________

Density of test specimen (cohesive, cohesionless) $\rho = M_0/V_0$ = ______________

Machine Data

Rate of loading ________.___/min (insert mm or in)

The following data may not be applicable if machine can be adjusted to tare these effects out of the load readings.

Cross section area of loading piston, A_p = ______________ cm^2

Upward load on piston = $A_p \sigma_3$ = ______________

Mass of loading piston = ______________

Computed value of initial sample load = ______________

TRIAXIAL COMPRESSION TEST(Cohesive, Cohesionless) Data Sheet 15*a*

Project ______________________ Job No. ______________________

Location of Project ______________________ Boring No. ________ Sample No. ________

Description of Soil ______________________ Depth of Sample ________

Tested By ______________________ Date of Testing ______________________

Fill in the blanks with data and appropriate units.

Sample Data

Sample dimensions: Diam. D_0 = ____________ mm Area A_0 = ____________ cm^2 L_0 = ____________ mm

Vol. V_0 = ____________ cm^3 Water content w = ____________ % Degree of saturation S = ____________ %

Mass M_0 = ____________ g ______________________

For Cohesionless soils

Initial mass of dish + sand = ____________

Final mass of dish + sand = ____________

Mass of sand used in specimen, M_o = ____________

Specific gravity of sand G_s = ____________

Vol. of soil solids in test specimen V_s = ____________

Vol. of voids in test specimen (initial) V_v = ____________

Initial void ratio of test specimen e_i = ____________

Void ratio of sand at minimum density e_{max} = ____________

Void ratio of sand at maximum density e_{min} = ____________

Relative density of test specimen D_r = ____________

Density of test specimen (cohesive, cohesionless) $\rho = M_0/V_0$ = ____________

Machine Data

Rate of loading ________.___/min (insert mm or in)

The following data may not be applicable if machine can be adjusted to tare these effects out of the load readings.

Cross section area of loading piston, A_p = ____________ cm^2

Upward load on piston = $A_p \sigma_3$ = ____________

Mass of loading piston = ____________

Computed value of initial sample load = ____________

TRIAXIAL COMPRESSION TEST(Cohesive, Cohesionless) Data Sheet 15*a*

Project ______________________________ Job No. ______________________________

Location of Project ______________________________ Boring No. ____________ Sample No. ____________

Description of Soil ______________________________ Depth of Sample ____________

Tested By ______________________________ Date of Testing ______________________________

Fill in the blanks with data and appropriate units.

Sample Data

Sample dimensions: Diam. D_0 = ______________ mm Area A_0 = ______________ cm^2 L_0 = ______________ mm

Vol. V_0 = ______________ cm^3 Water content w = ______________ % Degree of saturation S = ______________ %

Mass M_0 = ______________ g ______________________________

For Cohesionless soils

Initial mass of dish + sand = ______________

Final mass of dish + sand = ______________

Mass of sand used in specimen, M_o = ______________

Specific gravity of sand G_s = ______________

Vol. of soil solids in test specimen V_s = ______________

Vol. of voids in test specimen (initial) V_v = ______________

Initial void ratio of test specimen e_i = ______________

Void ratio of sand at minimum density e_{max} = ______________

Void ratio of sand at maximum density e_{min} = ______________

Relative density of test specimen D_r = ______________

Density of test specimen (cohesive, cohesionless) $\rho = M_0/V_0$ = ______________

Machine Data

Rate of loading ________.___/min (insert mm or in)

The following data may not be applicable if machine can be adjusted to tare these effects out of the load readings.

Cross section area of loading piston, A_p = ______________ cm^2

Upward load on piston = $A_p \sigma_3$ = ______________

Mass of loading piston = ______________

Computed value of initial sample load = ______________

TRIAXIAL COMPRESSION TEST(Cohesive, Cohesionless) Data Sheet 15*a*

Project ________________________________ Job No. ____________________

Location of Project ________________________ Boring No. __________ Sample No. __________

Description of Soil ______________________________ Depth of Sample __________

Tested By ________________________________ Date of Testing ____________________

Fill in the blanks with data and appropriate units.

Sample Data

Sample dimensions: Diam. D_0 = ______________ mm Area A_0 = ______________ cm^2 L_0 = ______________ mm

Vol. V_0 = ______________ cm^3 Water content w = ______________ % Degree of saturation S = ______________ %

Mass M_0 = ______________ g ________________________________

For Cohesionless soils

Initial mass of dish + sand = ______________

Final mass of dish + sand = ______________

Mass of sand used in specimen, M_o = ______________

Specific gravity of sand G_s = ______________

Vol. of soil solids in test specimen V_s = ______________

Vol. of voids in test specimen (initial) V_v = ______________

Initial void ratio of test specimen e_i = ______________

Void ratio of sand at minimum density e_{max} = ______________

Void ratio of sand at maximum density e_{min} = ______________

Relative density of test specimen D_r = ______________

Density of test specimen (cohesive, cohesionless) $\rho = M_0/V_0$ = ______________

Machine Data

Rate of loading ________.___/min (insert mm or in)

The following data may not be applicable if machine can be adjusted to tare these effects out of the load readings.

Cross section area of loading piston, A_p = ______________ cm^2

Upward load on piston = $A_p \sigma_3$ = ______________

Mass of loading piston = ______________

Computed value of initial sample load = ______________

TRIAXIAL COMPRESSION TEST Data Sheet 15*b*

Project ______________________________ Job No. ______________________

Tested By ______________________________ Date of Testing ______________________

Sample Data: Area A_o = ____________ cm^2 Length L_o = __________ mm

Machine Data: Load rate = ____________ /min. Load ring constant LRC = __________ /div.

______________________ Cell pressure σ_3 ____________

	Deform. dial (units)	ΔL	Load dial units	Sample load P	ϵ $\Delta L/L_o$	$1 - \epsilon$	Corr Area A^1	Deviator stress $\Delta\sigma_1 = P/A^1$	Normal stress $\Delta\sigma_1/\sigma_3$
1	2	3	4	5	6	7	8	9	10

Note: Insert units in column headings as necessary.

[a]The Deviator stress computation shown is based on taring the loading system so that the load reading is the deviator load value.

Computed Data

Maximum deviator stress (from stress-strain curve) $\Delta\sigma_1$ = ____________

Maximum value of vertical stress $\sigma_1 = \sigma_3 + \Delta\sigma_1$ = ____________

TRIAXIAL COMPRESSION TEST Data Sheet 15*b*

Project ______________________ Job No. ______________

Tested By ______________________ Date of Testing ______________

Sample Data: Area A_o = ____________ cm^2 Length L_o = ________ mm

Machine Data: Load rate = ____________ /min. Load ring constant LRC = ________ /div.

____________________ Cell pressure σ_3 ____________

	Deform. dial (units)	ΔL ____	Load dial units	Sample load P ____	ϵ $\Delta L/L_o$	$1 - \epsilon$	Corr Area A^1 ____	Deviator stress $\Delta\sigma_1 = P/A^1$ ____	Normal stress $\Delta\sigma_1/\sigma_3$
1	2	3	4	5	6	7	8	9	10

Note: Insert units in column headings as necessary.

[a]The Deviator stress computation shown is based on taring the loading system so that the load reading is the deviator load value.

Computed Data

Maximum deviator stress (from stress-strain curve) $\Delta\sigma_1$ = ____________

Maximum value of vertical stress $\sigma_1 = \sigma_3 + \Delta\sigma_1$ = ____________

TRIAXIAL COMPRESSION TEST Data Sheet 15*b*

Project ________________________________ Job No. ________________

Tested By ________________________________ Date of Testing ________________

Sample Data: Area A_o = ________________ cm^2 Length L_o = ____________ mm

Machine Data: Load rate = ________________ /min. Load ring constant LRC = ____________ /div.

________________________ Cell pressure σ_3 ________________

	Deform. dial (units)	ΔL ____	Load dial units	Sample load P ____	ϵ $\Delta L/L_o$	$1-\epsilon$	Corr Area A^1 ____	Deviator stress $\Delta\sigma_1 = P/A^1$ ____	Normal stress $\Delta\sigma_1/\sigma_3$
1	2	3	4	5	6	7	8	9	10

Note: Insert units in column headings as necessary.

[a]The Deviator stress computation shown is based on taring the loading system so that the load reading is the deviator load value.

Computed Data

Maximum deviator stress (from stress-strain curve) $\Delta\sigma_1$ = ________________

Maximum value of vertical stress $\sigma_1 = \sigma_3 + \Delta\sigma_1$ = ________________

TRIAXIAL COMPRESSION TEST Data Sheet 15*b*

Project ______________________ Job No. ______________________

Tested By ______________________ Date of Testing ______________________

Sample Data: Area A_o = ______________ cm^2 Length L_o = ______________ mm

Machine Data: Load rate = ______________ /min. Load ring constant LRC = ______________ /div.

______________________ Cell pressure σ_3 ______________

	Deform. dial (units)	ΔL ____	Load dial units	Sample load P ____	ϵ $\Delta L/L_o$	$1 - \epsilon$	Corr Area A^1 ____	Deviator stress $\Delta\sigma_1 = P/A^1$ ____	Normal stress $\Delta\sigma_1/\sigma_3$
1	2	3	4	5	6	7	8	9	10

Note: Insert units in column headings as necessary.

[a]The Deviator stress computation shown is based on taring the loading system so that the load reading is the deviator load value.

Computed Data

Maximum deviator stress (from stress-strain curve) $\Delta\sigma_1$ = ______________

Maximum value of vertical stress $\sigma_1 = \sigma_3 + \Delta\sigma_1$ = ______________

TRIAXIAL COMPRESSION TEST Data Sheet 15*b*

Project ____________________ Job No. ____________

Tested By ____________________ Date of Testing ____________

Sample Data: Area A_o = ____________ cm^2 Length L_o = ________ mm

Machine Data: Load rate = ____________ /min. Load ring constant LRC = ________ /div.

____________________ Cell pressure σ_3 ____________

	Deform. dial (units)	ΔL ____	Load dial units	Sample load P ____	$\in$ $\Delta L/L_o$	$1 - \in$	Corr Area A^1 ____	Deviator stress $\Delta\sigma_1 = P/A^1$ ____	Normal stress $\Delta\sigma_1/\sigma_3$
1	2	3	4	5	6	7	8	9	10

Note: Insert units in column headings as necessary.

[a]The Deviator stress computation shown is based on taring the loading system so that the load reading is the deviator load value.

Computed Data

Maximum deviator stress (from stress-strain curve) $\Delta\sigma_1$ = ____________

Maximum value of vertical stress $\sigma_1 = \sigma_3 + \Delta\sigma_1$ = ____________

TRIAXIAL COMPRESSION TEST Data Sheet 15*b*

Project ______________________ Job No. ______________

Tested By ______________________ Date of Testing ______________

Sample Data: Area A_o = ____________ cm^2 Length L_o = __________ mm

Machine Data: Load rate = ____________ /min. Load ring constant LRC = __________ /div.

______________________ Cell pressure σ_3 ______________

	Deform. dial (units)	ΔL	Load dial units	Sample load P	ϵ $\Delta L/L_o$	$1-\epsilon$	Corr Area A^1	Deviator stress $\Delta\sigma_1 = P/A^1$	Normal stress $\Delta\sigma_1/\sigma_3$
1	2	3	4	5	6	7	8	9	10

Note: Insert units in column headings as necessary.

[a]The Deviator stress computation shown is based on taring the loading system so that the load reading is the deviator load value.

Computed Data

Maximum deviator stress (from stress-strain curve) $\Delta\sigma_1$ = ____________

Maximum value of vertical stress $\sigma_1 = \sigma_3 + \Delta\sigma_1$ = ____________

Project ______________________ Job No. __________ Location of Project ______________________

Boring No. __________ Sample No. __________ Description of Soil ______________________

Depth of Sample ______________________ Tested By __________ Date of Testing ______________________

Area of specimen A_o = ________ Length of specimen L_o = ________ Confining Pressure σ_3 = ________ Rate of loading = ____/min Load ring constant LRC = ____

Initial burette reading = __________ cm^3 Initial sample void ratio, e_0 = ______________ Vol. of solids, V_s = ______________

Elapsed time, min.	Deform. dial[a] read. (______)	Load dial[b] read.	ΔL, (col. 2 × ____)	ϵ Unit strain $\Delta L/L_o$	Area CF $1-\epsilon$	Corr. area A'	Burette read., cm^3	ΔV, cm^3	Δe, $\Delta V/V_s$	Inst. void ratio e	Pore press. (______)	Deviator[c] load (col. 3 × LRC)		
1	2	3	4	5	6	7	8	9	10	11	12	13	14	15

Note: Insert units in column headings as necessary.
[a]omit if using LVDT and enter ΔL in col. 4.
[b]omit if using load cell and enter load in col. 13
[c]based on taring system so load ring reading is the deviator load value.

Data for Mohr's Circle:

Maximum deviator stress, $\Delta\sigma_1$ = ______________ Principal stress, σ_1 = ______________ Effective principal stress, σ'_1 = ______________

Project ______________ Job No. ______ Location of Project ______________

Boring No. ______ Sample No. ______ Description of Soil ______________

Depth of Sample ______________ Tested By ______ Date of Testing ______________

Area of specimen A_o = ______ Length of specimen L_o = ______ Confining Pressure σ_3 = ______ Rate of loading = ______/min Load ring constant LRC = ______

Initial burette reading = ______ cm^3 Initial sample void ratio, e_0 = ______ Vol. of solids, V_s = ______

Elapsed time, min.	Deform. dial[a] read. (______)	Load dial[b] read.	ΔL, (col. 2 × ______)	ϵ Unit strain $\Delta L/L_o$	Area CF $1 - \epsilon$	Corr. area A'	Burette read., cm^3	ΔV, cm^3	Δe, $\Delta V/V_s$	Inst. void ratio e	Pore press. (______)	Deviator[c] load (col. 3 × LRC)		
1	2	3	4	5	6	7	8	9	10	11	12	13	14	15

Note: Insert units in column headings as necessary.
[a]omit if using LVDT and enter ΔL in col. 4.
[b]omit if using load cell and enter load in col. 13
[c]based on taring system so load ring reading is the deviator load value.

Data for Mohr's Circle:

Maximum deviator stress, $\Delta\sigma_1$ = ______ Principal stress, σ_1 = ______ Effective principal stress, σ_1' = ______

Project ______________________ Job No. __________ Location of Project ______________________

Boring No. __________ Sample No. __________ Description of Soil ______________________

Depth of Sample ______________________ Tested By __________ Date of Testing ______________________

Area of specimen A_o = ______ Length of specimen L_o = ______ Confining Pressure σ_3 = ______ Rate of loading = ____/min Load ring constant LRC = ____

Initial burette reading = ________ cm^3 Initial sample void ratio, e_0 = ______________ Vol. of solids, V_s = ______________

Elapsed time, min.	Deform. dial[a] read. (______)	Load dial[b] read.	ΔL, (col. 2 × ____)	ϵ Unit strain $\Delta L/L_o$	Area CF 1 - ϵ	Corr. area A'	Burette read., cm^3	ΔV, cm^3	Δe, $\Delta V/V_s$	Inst. void ratio e	Pore press. (______)	Deviator[c] load (col. 3 × LRC)		
1	2	3	4	5	6	7	8	9	10	11	12	13	14	15

Note: Insert units in column headings as necessary.
[a]omit if using LVDT and enter ΔL in col. 4.
[b]omit if using load cell and enter load in col. 13
[c]based on taring system so load ring reading is the deviator load value.

Data for Mohr's Circle:

Maximum deviator stress, $\Delta\sigma_1$ = ______________ Principal stress, σ_1 = ______________ Effective principal stress, σ'_1 = ______________

Project ______________________ Job No. __________ Location of Project ______________________

Boring No. __________ Sample No. __________ Description of Soil ______________________

Depth of Sample ______________________ Tested By __________ Date of Testing ______________________

Area of specimen A_o = ________ Length of specimen L_o = ________ Confining Pressure σ_3 = ________ Rate of loading = ______/min Load ring constant LRC = ______

Initial burette reading = __________ cm^3 Initial sample void ratio, e_0 = ______________ Vol. of solids, V_s = ______________

Elapsed time, min.	Deform. dial[a] read. (______)	Load dial[b] read.	ΔL, (col. 2 ×______)	ϵ Unit strain $\Delta L/L_o$	Area CF $1 - \epsilon$	Corr. area A'	Burette read., cm^3	ΔV, cm^3	Δe, $\Delta V/V_s$	Inst. void ratio e	Pore press. (______)	Deviator[c] load (col. 3 × LRC)		
1	2	3	4	5	6	7	8	9	10	11	12	13	14	15

Note: Insert units in column headings as necessary.
[a]omit if using LVDT and enter ΔL in col. 4.
[b]omit if using load cell and enter load in col. 13
[c]based on taring system so load ring reading is the deviator load value.

Data for Mohr's Circle:

Maximum deviator stress, $\Delta\sigma_1$ = ______________ Principal stress, σ_1 = ______________ Effective principal stress, σ_1' = ______________

DIRECT SHEAR TEST (Cohesive Soil, Cohesionless Soil) Data Sheet 17

Project ______________________ Job No. ______________________

Location of Project ______________________ Boring No. ____________ Sample No. ____________

Description of Soil ______________________ Depth of Sample ____________

Tested By ______________________ Date of Testing ______________________

Sample Data: Soil state (wet, dry) Soil sample (disturbed, undisturbed)

Data to Obtain Sample Density if not an Undisturbed Sample

Initial mass container + soil = ______________

Final mass container + soil = ______________

Mass of soil used = ______________

Shear specimen data

Sample Dimensions:

Dia. or side = ______________

Ht. = ______________

Area = ______________

Vol. = ______________

Density: ρ_{wet} = ______________ g/cm^3 w = ______________ % ρ_{dry} = ______________ g/cm^3

Normal load = ______________ Normal stress σ_n = ______________

Loading rate = ______________ /min Load ring constant LRC = ______________ /div

	Vert. dial[a] reading (______)	Vert. displace. ΔV, (______)	Horiz. dial[a] reading (______)	Horiz. displace. ΔH, (______)	Corr.[b] area A'	Load dial[c] reading (______)	Horiz. shear force, (______)	Shear stress s, kPa

Note: Insert units in column headings as necessary; not all columns used with LVDT's or load cell.

[a]omit it using LVDT and enter displacements ΔV and ΔH.

[b]for square samples, may use corrected specimen area at failure as $A' = A_o - \Delta H$ to compute σ_n and s.

[c]omit if using a load cell and enter "Horiz. shear force."

DIRECT SHEAR TEST (Cohesive Soil, Cohesionless Soil) Data Sheet 17

Project ______________________ Job No. ______________________

Location of Project ______________________ Boring No. __________ Sample No. __________

Description of Soil ______________________ Depth of Sample __________

Tested By ______________________ Date of Testing ______________________

Sample Data: Soil state (wet, dry) Soil sample (disturbed, undisturbed)

Data to Obtain Sample Density if not an Undisturbed Sample

Initial mass container + soil = ______________

Final mass container + soil = ______________

Mass of soil used = ______________

Shear specimen data

Sample Dimensions:

Dia. or side = ______________

Ht. = ______________

Area = ______________

Vol. = ______________

Density: ρ_{wet} = ______________ g/cm^3 w = ______________ % ρ_{dry} = ______________ g/cm^3

Normal load = ______________ Normal stress σ_n = ______________

Loading rate = ______________ /min Load ring constant LRC = __________ ______________ /div

	Vert. dial[a] reading (____)	Vert. displace. ΔV, (____)	Horiz. dial[a] reading (____)	Horiz. displace. ΔH, (____)	Corr.[b] area A'	Load dial[c] reading (____)	Horiz. shear force, (____)	Shear stress s, kPa

Note: Insert units in column headings as necessary; not all columns used with LVDT's or load cell.

[a]omit it using LVDT and enter displacements ΔV and ΔH.

[b]for square samples, may use corrected specimen area at failure as $A' = A_o - \Delta H$ to compute σ_n and s.

[c]omit if using a load cell and enter "Horiz. shear force."

DIRECT SHEAR TEST (Cohesive Soil, Cohesionless Soil) Data Sheet 17

Project ______________________ Job No. ______________________

Location of Project ______________________ Boring No. __________ Sample No. __________

Description of Soil ______________________ Depth of Sample __________

Tested By ______________________ Date of Testing ______________________

Sample Data: Soil state (wet, dry) Soil sample (disturbed, undisturbed)

Data to Obtain Sample Density if not an Undisturbed Sample

Initial mass container + soil = __________

Final mass container + soil = __________

Mass of soil used = __________

Shear specimen data

Sample Dimensions:

Dia. or side = __________

Ht. = __________

Area = __________

Vol. = __________

Density: ρ_{wet} = __________ g/cm³ w = __________ % ρ_{dry} = __________ g/cm³

Normal load = __________ Normal stress σ_n = __________

Loading rate = __________ /min Load ring constant LRC = __________ /div

	Vert. dial[a] reading (____)	Vert. displace. ΔV, (____)	Horiz. dial[a] reading (____)	Horiz. displace. ΔH, (____)	Corr.[b] area A'	Load dial[c] reading (____)	Horiz. shear force, (____)	Shear stress s, kPa

Note: Insert units in column headings as necessary; not all columns used with LVDT's or load cell.

[a]omit it using LVDT and enter displacements ΔV and ΔH.

[b]for square samples, may use corrected specimen area at failure as $A' = A_o - \Delta H$ to compute σ_n and s.

[c]omit if using a load cell and enter "Horiz. shear force."

DIRECT SHEAR TEST (Cohesive Soil, Cohesionless Soil) Data Sheet 17

Project ______________________ Job No. ______________________

Location of Project ______________________ Boring No. __________ Sample No. __________

Description of Soil ______________________ Depth of Sample __________

Tested By ______________________ Date of Testing ______________________

Sample Data: Soil state (wet, dry) Soil sample (disturbed, undisturbed)

Data to Obtain Sample Density if not an Undisturbed Sample

Initial mass container + soil = ______________

Final mass container + soil = ______________

Mass of soil used = ______________

Shear specimen data

Sample Dimensions:

Dia. or side = ______________

Ht. = ______________

Area = ______________

Vol. = ______________

Density: ρ_{wet} = ______________ g/cm^3 w = ______________ % ρ_{dry} = ______________ g/cm^3

Normal load = ______________ Normal stress σ_n = ______________

Loading rate = ______________ /min Load ring constant LRC = ______________ /div

	Vert. dial[a] reading (______)	Vert. displace. ΔV, (______)	Horiz. dial[a] reading (______)	Horiz. displace. ΔH, (______)	Corr.[b] area A'	Load dial[c] reading (______)	Horiz. shear force, (______)	Shear stress s, kPa

Note: Insert units in column headings as necessary; not all columns used with LVDT's or load cell.

[a]omit it using LVDT and enter displacements ΔV and ΔH.

[b]for square samples, may use corrected specimen area at failure as $A' = A_o - \Delta H$ to compute σ_n and s.

[c]omit if using a load cell and enter "Horiz. shear force."

DIRECT SHEAR TEST (Cohesive Soil, Cohesionless Soil) Data Sheet 17

Project ______________________ Job No. ______________________

Location of Project ______________________ Boring No. __________ Sample No. __________

Description of Soil ______________________ Depth of Sample __________

Tested By ______________________ Date of Testing ______________________

Sample Data: Soil state (wet, dry) Soil sample (disturbed, undisturbed)

Data to Obtain Sample Density if not an Undisturbed Sample

Initial mass container + soil = __________

Final mass container + soil = __________

Mass of soil used = __________

Shear specimen data

Sample Dimensions:

Dia. or side = __________

Ht. = __________

Area = __________

Vol. = __________

Density: ρ_{wet} = __________ g/cm³ w = __________ % ρ_{dry} = __________ g/cm³

Normal load = __________ Normal stress σ_n = __________

Loading rate = __________ /min Load ring constant LRC = __________ /div

	Vert. dial[a] reading (____)	Vert. displace. ΔV, (____)	Horiz. dial[a] reading (____)	Horiz. displace. ΔH, (____)	Corr.[b] area A'	Load dial[c] reading (____)	Horiz. shear force, (____)	Shear stress s, kPa

Note: Insert units in column headings as necessary; not all columns used with LVDT's or load cell.

[a]omit it using LVDT and enter displacements ΔV and ΔH.

[b]for square samples, may use corrected specimen area at failure as $A' = A_o - \Delta H$ to compute σ_n and s.

[c]omit if using a load cell and enter "Horiz. shear force."

DIRECT SHEAR TEST (Cohesive Soil, Cohesionless Soil) Data Sheet 17

Project ______________________ Job No. ______________________

Location of Project ______________________ Boring No. ____________ Sample No. ____________

Description of Soil ______________________ Depth of Sample ____________

Tested By ______________________ Date of Testing ______________________

Sample Data: Soil state (wet, dry) Soil sample (disturbed, undisturbed)

Data to Obtain Sample Density if not an Undisturbed Sample

Initial mass container + soil = ______________

Final mass container + soil = ______________

Mass of soil used = ______________

Shear specimen data

Sample Dimensions:

Dia. or side = ______________

Ht. = ______________

Area = ______________

Vol. = ______________

Density: ρ_{wet} = ______________ g/cm^3 w = ______________ % ρ_{dry} = ______________ g/cm^3

Normal load = ______________ Normal stress σ_n = ______________

Loading rate = ______________ /min Load ring constant LRC = ______________ /div

	Vert. dial[a] reading (______)	Vert. displace. ΔV, (______)	Horiz. dial[a] reading (______)	Horiz. displace. ΔH, (______)	Corr.[b] area A'	Load dial[c] reading (______)	Horiz. shear force, (______)	Shear stress s, kPa

Note: Insert units in column headings as necessary; not all columns used with LVDT's or load cell.
[a]omit it using LVDT and enter displacements ΔV and ΔH.
[b]for square samples, may use corrected specimen area at failure as $A' = A_o - \Delta H$ to compute σ_n and s.
[c]omit if using a load cell and enter "Horiz. shear force."

RELATIVE DENSITY Data Sheet 18

Project ______________________ Job No. ______________________

Location of Project ______________________ Boring No. ____________ Sample No. ____________

Description of Soil ______________________ Depth of Sample ____________

Tested by ______________________ Date of Testing ______________________

To Use: Plot maximum γ or ρ on right and minimum on left; connect with a straight line. Plot field values on both sides and connect. At intersection of two lines drop perpendicular and read D_r.

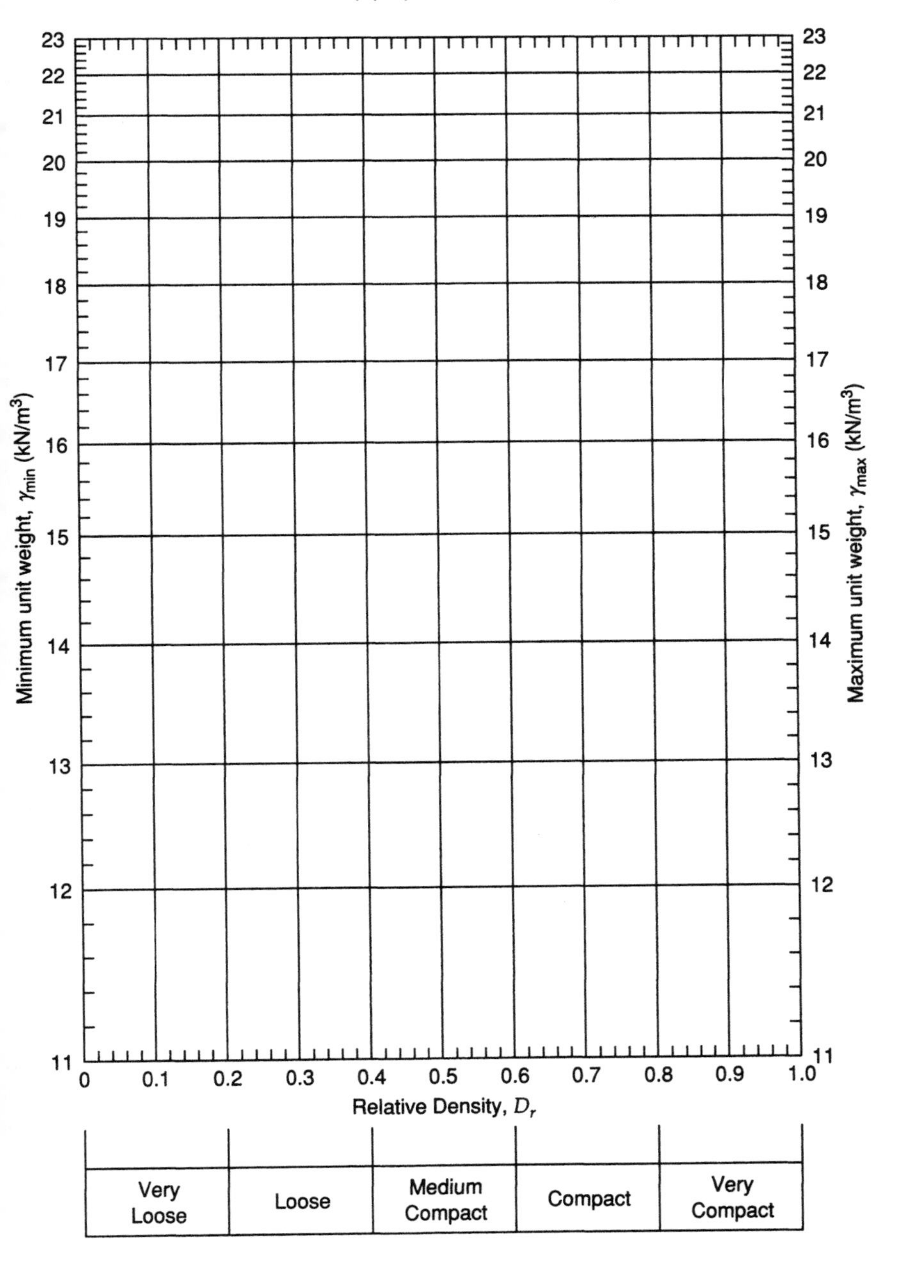

Very Loose	Loose	Medium Compact	Compact	Very Compact

RELATIVE DENSITY Data Sheet 18

Project ____________________ Job No. ____________________

Location of Project ____________________ Boring No. ____________ Sample No. ____________

Description of Soil ____________________ Depth of Sample ____________

Tested by ____________________ Date of Testing ____________________

To Use: Plot maximum γ or ρ on right and minimum on left; connect with a straight line. Plot field values on both sides and connect. At intersection of two lines drop perpendicular and read D_r.

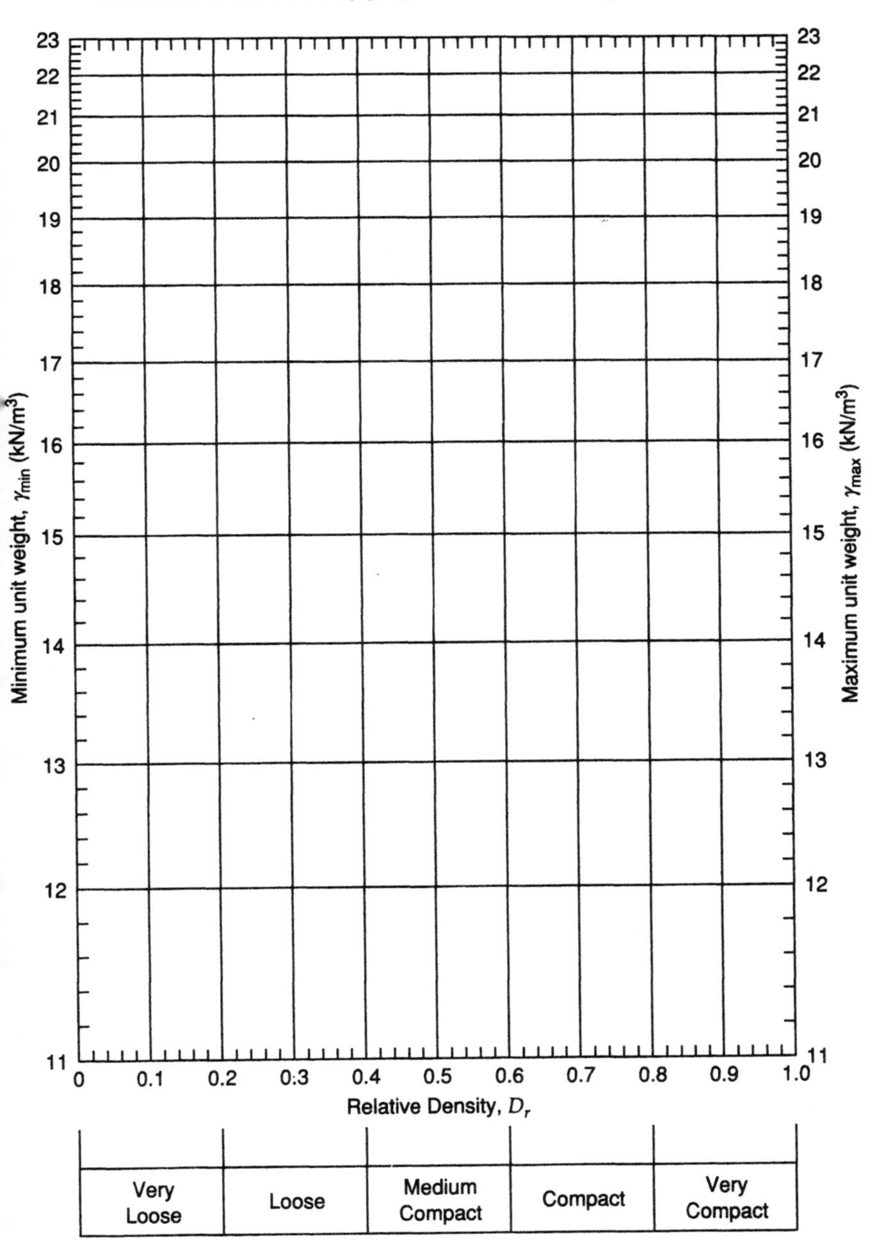

Very Loose	Loose	Medium Compact	Compact	Very Compact

RELATIVE DENSITY Data Sheet 18

Project ______________________ Job No. ______________________

Location of Project ______________________ Boring No. __________ Sample No. __________

Description of Soil ______________________ Depth of Sample __________

Tested by ______________________ Date of Testing ______________________

To Use: Plot maximum γ or ρ on right and minimum on left; connect with a straight line. Plot field values on both sides and connect. At intersection of two lines drop perpendicular and read D_r.

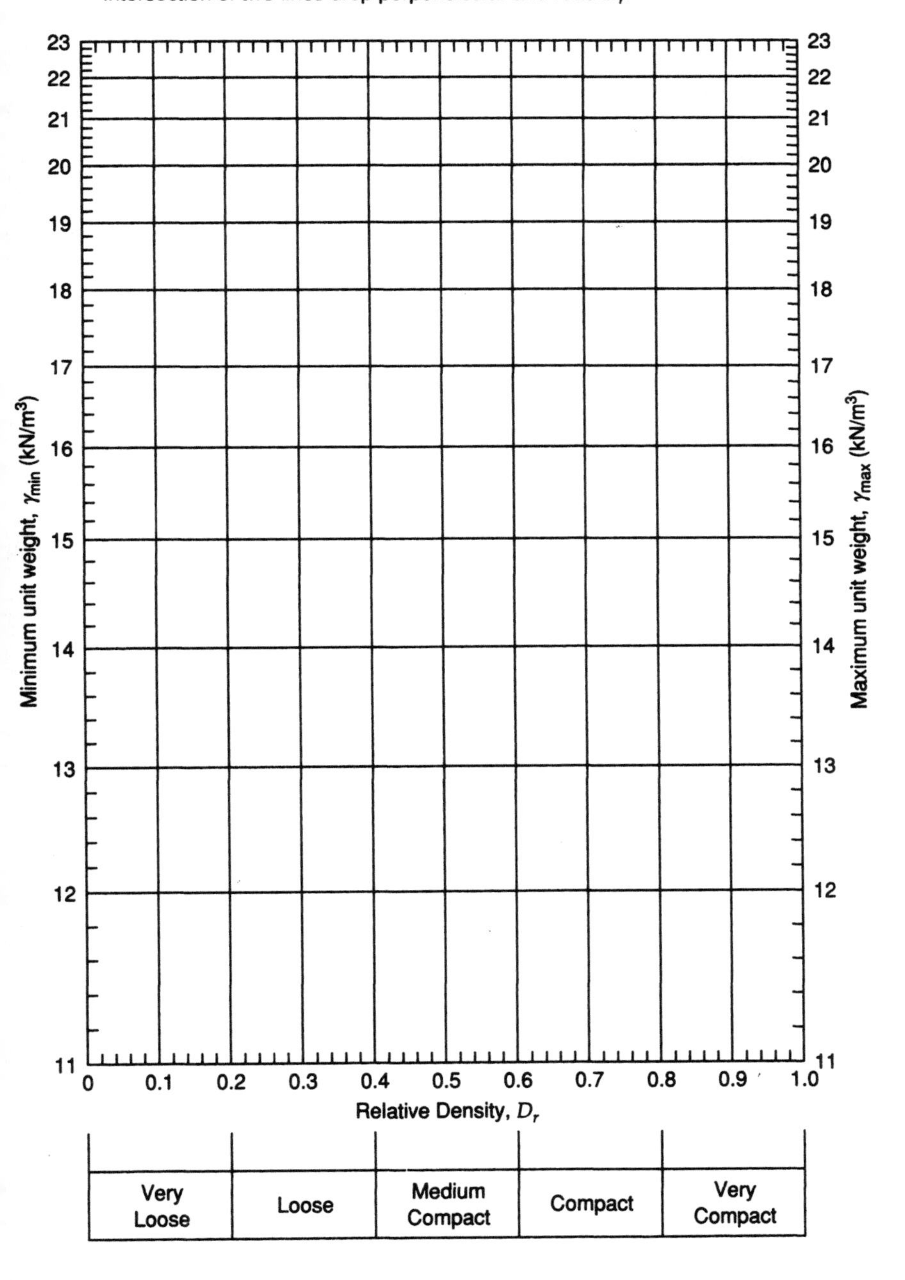

Very Loose	Loose	Medium Compact	Compact	Very Compact

BEARING RATIO TEST Data Sheet 19*a*

Project ______________________ Job No. ______________________

Location of Project ______________________

Description of Soil ______________________

Tested By ______________________ Date of Testing ______________________

Compaction Energy: Rammer __________ N No. of layers __________ Blows/layer __________
w at compaction __________ % Mold diam. __________ cm Ht. of soil __________ cm
Vol. __________ cm^3 __________

Swell Data

Starting time and date	Elapsed time	Mold no. ______ Surcharge ______ N		Mold no. ______ Surcharge ______ N		Mold no. ______ Surcharge ______ N	
		Dial reading[a] (× ______)	$\% = \frac{S}{H}(100)$	Dial reading (× ______)	$\% = \frac{S}{H}(100)$	Dial reading (× ______)	$\% = \frac{S}{H}(100)$
	0h						
	1h						
	2h						
	4h						

[a]load dial units or direct load cell readings.

After Soaking

Mold No.			
Surcharge, N			
Initial mass of wet soil + mold + base plate			
Final mass of wet soil + mold + base plate			
Mass of mold + base plate			
Initial mass of wet soil, M_i			
Initial wet density, ρ_i			
Mass of water absorbed, M_w			
% water absorbed = $M_w/M_s \times 100$			

Note: Insert units in column headings as necessary.

BEARING RATIO TEST Data Sheet 19*a*

Project ______________________ Job No. ______________________

Location of Project ______________________

Description of Soil ______________________

Tested By ______________________ Date of Testing ______________________

Compaction Energy: Rammer __________ N No. of layers __________ Blows/layer __________

w at compaction __________ % Mold diam. __________ cm Ht. of soil __________ cm

Vol. __________ cm^3 ______________________

Swell Data

Starting time and date	Elapsed time	Mold no. ______ Surcharge ______ N		Mold no. ______ Surcharge ______ N		Mold no. ______ Surcharge ______ N	
		Dial reading[a] (× ______)	$\% = \frac{S}{H}(100)$	Dial reading (× ______)	$\% = \frac{S}{H}(100)$	Dial reading (× ______)	$\% = \frac{S}{H}(100)$
	0h						
	1h						
	2h						
	4h						

[a]load dial units or direct load cell readings.

After Soaking

Mold No.			
Surcharge, N			
Initial mass of wet soil + mold + base plate			
Final mass of wet soil + mold + base plate			
Mass of mold + base plate			
Initial mass of wet soil, M_i			
Initial wet density, ρ_i			
Mass of water absorbed, M_w			
% water absorbed = $M_w/M_s \times 100$			

Note: Insert units in column headings as necessary.

BEARING RATIO TEST Data Sheet 19*b*

Project ______________________ Job No. ______________________

Test Information ______________________

Tested By ______________________ Date of Testing ______________________

CBR Test Load Data

Penetration, mm	(soaked, unsoaked) Mold no. ____ Surcharge ______		(soaked, unsoaked) Mold no. ____ Surcharge ______ N		(soaked, unsoaked) Mold no. ____ Surcharge ______ N	
	Piston load dial reading (______)	Stress, kPa	Piston load dial reading (______)	Stress, kPa	Piston load dial reading (______)	Stress, kPa
0.000						
0.5						
1.0						
1.5						
2.0						
2.5						
3.0						
4.0						
5.0						
Actual water content, w% (*use water content sheets*)	Top $\frac{1}{3}$					
	Middle $\frac{1}{3}$					
	Bottom $\frac{1}{3}$					
	Average					

Wet density, ρ_{wet} = ______________ ______________ ______________

Dry density, ρ_d = ______________ ______________ ______________

Note: Insert units in column headings as necessary.

BEARING RATIO TEST Data Sheet 19*b*

Project ______________________ Job No. ______________________

Test Information ______________________

Tested By ______________________ Date of Testing ______________________

CBR Test Load Data

	(soaked, unsoaked) Mold no. ____ Surcharge ______		(soaked, unsoaked) Mold no. ____ Surcharge ______ N		(soaked, unsoaked) Mold no. ____ Surcharge ______ N	
Penetration, mm	Piston load dial reading (______)	Stress, kPa	Piston load dial reading (______)	Stress, kPa	Piston load dial reading (______)	Stress, kPa
0.000						
0.5						
1.0						
1.5						
2.0						
2.5						
3.0						
4.0						
5.0						
Actual water content, w% (*use water content sheets*)	Top $\frac{1}{3}$					
	Middle $\frac{1}{3}$					
	Bottom $\frac{1}{3}$					
	Average					

Wet density, ρ_{wet} = ______________ ______________ ______________

Dry density, ρ_d = ______________ ______________ ______________

Note: Insert units in column headings as necessary.

VOLUMETRIC GRAVIMETRIC RELATIONSHIPS **Data Sheet 21**

Project ______________________________

Location of Project ______________________________

Description of Soil ______________________________

Tested by ______________________________ Date ______________________________

Volume of container, V_c = ______________ cm^3

Mass of dry soil + container = ______________ g

Mass of soil + water + container = ______________ g

Mass of container = ______________ g

Mass of dry soil, M_s = ______________ g

Mass of saturated soil, M_t = ______________ g

Volume of water to saturate soil, V_v = ______________ mL

Compute:

$$\rho = \frac{M}{V} \qquad e = \frac{V_v}{V_s} \qquad G_s = \frac{M_s}{V_s}$$

$V_s = V_c - V_v$ = ______________ − ______________ = ______________ cm^3

ρ_d = ______________ = ______________ g/cm^3

ρ_{sat} = ______________ = ______________ g/cm^3

$\gamma_d = \rho \times 9.807$ = ______________ kN/m^3

$\gamma_{sat} = \rho_{sat} \times 9.807$ = ______________ kN/m^3

Void ratio, e = ______________ = ______________

Specific gravity, G_s = ______________ = ______________

Project ______________________________

Location of Project ______________________________

Description of Soil ______________________________

Tested by ______________________ Date ______________________

Volume of container, V_c = ______________ cm^3

Mass of dry soil + container = ______________ g

Mass of soil + water + container = ______________ g

Mass of container = ______________ g

Mass of dry soil, M_s = ______________ g

Mass of saturated soil, M_t = ______________ g

Volume of water to saturate soil, V_v = ______________ mL

Compute:

$$\rho = \frac{M}{V} \qquad e = \frac{V_v}{V_s} \qquad G_s = \frac{M_s}{V_s}$$

$V_s = V_c - V_v$ = ______________ − ______________ = ______________ cm^3

ρ_d = ______________ = ______________ g/cm^3

ρ_{sat} = ______________ = ______________ g/cm^3

$\gamma_d = \rho \times 9.807$ = ______________ kN/m^3

$\gamma_{sat} = \rho_{sat} \times 9.807$ = ______________ kN/m^3

Void ratio, e = ______________ = ______________

Specific gravity, G_s = ______________ = ______________

UNIT WEIGHT OF A COHESIVE SOIL Data Sheet 22

Project ______________________ Job No. ______________________

Location of Project ______________________

Description of Soil ______________________

Tested by ______________________ Date ______________________

Boring No. ______________________ Boring No. ______________________

Depth ______________________ Depth ______________________

Mass of sample, M_t	= ________ g		Mass of sample, M_t	= ________ g
Vol. of container, V_c	= ________ cm^3		Vol. of container, V_c	= ________ cm^3
Initial grad. read.	= ________ ml		Initial grad. read.	= ________ ml
Final grad. read.	= ________ ml		Final grad. read.	= ________ ml
Vol. of water, V_w	= ________ ml		Vol. of water, V_w	= ________ ml
Vol. of soil, $V_s = V_c - V_w$	= ________ cm^3		Vol. of soil, $V_s = V_c - V_w$	= ________ cm^3

Wet density: Wet density:

$\rho_{wet} = M_t/V_s$	= ________ g/cm^3		$\rho_{wet} = M_t/V_s$	= ________ g/cm^3
$\gamma_{wet} = \rho \times 9.807$	= ________ kN/m^3		$\gamma_{wet} = \rho \times 9.807$	= ________ kN/m^3

Computation of Dry Unit Weight

Water content, w%	= ________		Water content, w%	= ________
$\gamma_{dry} = \dfrac{\gamma_{wet}}{1 + w}$	= ________ kN/m^3		$\gamma_{dry} = \dfrac{\gamma_{wet}}{1 + w}$	= ________ kN/m^3
$\gamma_{dry} = W_s/V_s$	= ________ kN/m^3		$\gamma_{dry} = W_s/V_s$	= ________ kN/m^3

UNIT WEIGHT OF A COHESIVE SOIL Data Sheet 22

Project ______________________ Job No. ______________________

Location of Project ______________________

Description of Soil ______________________

Tested by ______________________ Date ______________________

Boring No. ______________________ Boring No. ______________________

Depth ______________________ Depth ______________________

Mass of sample, M_t	= ________ g	Mass of sample, M_t	= ________ g
Vol. of container, V_c	= ________ cm^3	Vol. of container, V_c	= ________ cm^3
Initial grad. read.	= ________ ml	Initial grad. read.	= ________ ml
Final grad. read.	= ________ ml	Final grad. read.	= ________ ml
Vol. of water, V_w	= ________ ml	Vol. of water, V_w	= ________ ml
Vol. of soil, $V_s = V_c - V_w$	= ________ cm^3	Vol. of soil, $V_s = V_c - V_w$	= ________ cm^3
Wet density:		Wet density:	
$\rho_{wet} = M_t/V_s$	= ________ g/cm^3	$\rho_{wet} = M_t/V_s$	= ________ g/cm^3
$\gamma_{wet} = \rho \times 9.807$	= ________ kN/m^3	$\gamma_{wet} = \rho \times 9.807$	= ________ kN/m^3

Computation of Dry Unit Weight

Water content, $w\%$ = ________		Water content, $w\%$ = ________	
$\gamma_{dry} = \dfrac{\gamma_{wet}}{1 + w}$	= ________ kN/m^3	$\gamma_{dry} = \dfrac{\gamma_{wet}}{1 + w}$	= ________ kN/m^3
$\gamma_{dry} = W_s/V_s$	= ________ kN/m^3	$\gamma_{dry} = W_s/V_s$	= ________ kN/m^3

1
9
8
7
6
5
4
3
2
1
9
8
7
6
5
4
3
2
1
9
8
7
6
5
4
3
2
1

1
9
8
7
6
5
4
3
2
1
9
8
7
6
5
4
3
2
1
9
8
7
6
5
4
3
2
1

1
9
8
7
6
5
4
3
2
1
9
8
7
6
5
4
3
2
1
9
8
7
6
5
4
3
2
1

1 9 8 7 6 5 4 3 2
1 9 8 7 6 5 4 3 2
1 9 8 7 6 5 4 3 2
1 9 8 7 6 5 4 3 2
1 9 8 7 6 5 4 3 2
1

1
9
8
7
6
5
4
3
2
1
9
8
7
6
5
4
3
2
1
9
8
7
6
5
4
3
2
1
9
8
7
6
5
4
3
2
1
9
8
7
6
5
4
3
2
1

Eng. Prop. of Soils and Their Measurement

1
9
8
7
6
5
4
3
2
1
9
8
7
6
5
4
3
2
1
9
8
7
6
5
4
3
2
1
9
8
7
6
5
4
3
2
1
9
8
7
6
5
4
3
2
1

 5 Cycles × 2 mm
Eng. Prop. of Soils and Their Measurement

1
9
8
7
6
5
4
3
2
1
9
8
7
6
5
4
3
2
1
9
8
7
6
5
4
3
2
1
9
8
7
6
5
4
3
2
1
9
8
7
6
5
4
3
2
1

1
9
8
7
6
5
4
3
2
1
9
8
7
6
5
4
3
2
1
9
8
7
6
5
4
3
2
1
9
8
7
6
5
4
3
2
1
9
8
7
6
5
4
3
2
1

1
9
8
7
6
5
4
3
2
1
9
8
7
6
5
4
3
2
1
9
8
7
6
5
4
3
2
1
9
8
7
6
5
4
3
2
1
9
8
7
6
5
4
3
2
1